D1200657

Natural Science Books in English 1600–1900
David M. Knight

David M. Knight

NATURAL SCIENCE BOOKS IN ENGLISH 1600–1900

Let it be remembered . . . by controversialists on all subjects, that every speculative error which boasts a multitude of advocates, has its golden *as well as its dark side; that there is always some truth connected with it, the exclusive attention to which has misled the Understanding, some moral beauty which has given its charms for the heart.*

How great a thing the possession of any one simple truth is, and how mean a thing a mere fact is, except as seen in the light of some comprehensive truth.

S. T. Coleridge

PORTMAN BOOKS

First published 1972

This edition first published 1989

ISBN 0 7134 0728 x

Printed in Great Britain by
Courier International Ltd, Tiptree, Essex
for the Publisher
Portman Books
an imprint of
B.T. Batsford Ltd
4 Fitzhardinge Street
London W1H 0AH

Contents

Introduction to 1989 edition.

This is essentially an exact reprint of the first edition, with misprints and errors corrected. Because a good deal has happened in history and bibliography of science since 1972, lists of *Additional Publications* have been added at appropriate places at the ends of chapters; they will help the reader looking for new interpretations. In particular, the social history of science – what one might call the history of scientists – has been the focus of a great deal of attention in the last 15 or 20 years. This has led to increasing interest in relatively minor scientists, who went into science for a variety of reasons and made a career for themselves in which science was more or less prominent; we try to recover the world of the foot-soldiers as well as the generals in the war on ignorance. While therefore it is, alas, no longer true (p. 4) that one can pick up Darwin first editions quite cheaply, there are still many attractive works of more popular sciences available to the collector of modest means; and much can be learned from them.

Nobody would write quite the same book 17 years on, but on reading it through again I would not disagree too much with my younger and wiser self. On p. 57, Whewell's work on natural theology is underestimated: he believed that the existence of God cannot be proved, but that if it is accepted then the study of the world will tell us more about its Creator. He extends his idea to science generally, believing that the crucial thing is to get the right general idea, which can then be tested and refined by reasoning and experiment. Similarly, on p. 140 I did not do justice to Laurent's *Chemical Method*, where this method of hypothesis and deduction was applied in chemistry. Hitherto attempts to determine formulae and structures from experimental data had led to chaos. Gerhardt believed that the answer was to agree upon conventions, because knowledge was unavailable; but Laurent argued that the way forward was to postulate structures and see what consequences followed from them. This was the method used with triumphant success in organic chemistry, and his was therefore one of the most important of chemical books. Jane Marcet's *Conversations*, p. 203, might also have been given more prominence; especially in view of the way we are learning to see women previously invisible in the scientific world of the eighteenth and nineteenth centuries.

No book of this kind could be complete, and this one should be taken as an invitation to enter the fascinating world of scientific books and journals; for that, it will I hope continue to provide an interesting guide as far as it goes, and help the reader to place works mentioned or not mentioned here in their context, and to enjoy them the more.

David M. Knight
January 1989

Illustrations

ACKNOWLEDGEMENTS

I would like to thank the following for their help: the staffs of the Bodleian Library; British Museum; British Museum (Natural History); London Library; Radcliffe Science Library, Oxford; Royal Institution, London; Royal Society; University of Durham Library (particularly Dr A. I. Doyle) and the University of Newcastle Library.

I am very grateful to Mrs Conchie, Mrs Kilner, Mrs Walker and Mrs Williams, who typed out the manuscripts; and to the numerous friends and colleagues who suggested items for inclusion.

Introduction

This is a book about books. It does not therefore set out to be a general history of science; indeed a study limited to books in English could hardly provide a balanced account of the development of modern science. But this limitation brings with it certain advantages. There is a rich scientific literature in English – some of the most important items being translations – and if works in other languages are excluded then there is more opportunity to discuss English books not of the first rank as science but important in their day, or interesting, nevertheless, to us. The historian or bibliophile who does not feel that he must cover the world in his choice of texts, and therefore scurry from peak to peak, can find great pleasure loitering in the lusher intervening valleys of the history of science; and he will thus become much more likely to judge scientific works on their merits, and not on their supposed resemblance to the productions of the present day.

While the temptations towards the whig interpretation of history are particularly seductive to the historian of science, he can also all too easily come to believe in the autonomy of his subject matter. We shall, therefore, in this volume endeavour to mention not only works of science, or books intended to be such, but also those writings which exerted a strong influence upon men of science. We should remember that the term 'scientist' is relatively new; it only came into the language in the 1830s, and was coined by analogy with 'artist'. Before that time, scientific men called themselves 'natural philosophers', or 'natural historians'; and indeed the word 'science' meant any organized body of knowledge, down to about the same date. The natural philosophers were often reflective about what they were doing, and prepared to put their insights down on paper. To later generations, and especially to those who are not professional scientists, what great scientists wrote is often at least as interesting as what they did; and the productions of their leisure hours often cast light upon their more weighty publications.

We shall, therefore, be looking at original scientific books, whether important, interesting, or curious; at philosophical writings which have interested scientists; and at books written by scientists which illuminate their scientific work. This last class could include the *Lives and Letters*, usually in two or more volumes, of eminent Victorian scientists. There are also some books by professional popularizers which deserve not to be forgotten. Most of these are, like the modern scientific textbook, of relatively recent origin. Professionalization of science came late to the British Isles. During Sir Joseph Banks' Presidency of the Royal Society, which lasted from 1778 to 1820, there was, as there had been from its foundation, a majority of dilettantes in the society, and even on its policy-making council. The number of active men of science in the country was small, and the number who were professionals in the strict sense of being paid as scientists was very small indeed. Their books, and to some extent their papers too, had therefore to be written with an inexpert readership in mind. In the second half of the nineteenth century, the number of scientists increased, and scientific books and articles were, as a rule, directed at professionals. The popularizer, and the textbook-author, interpreters of the work of others, became increasingly necessary, and original scientific writings ceased in the main to be distinguished as works of literature. It is safe to say that few textbooks of this kind make particularly interesting reading; but Fontenelle, Voltaire, and the English Newtonians count among the popularizers of science new in their day, and other men of science and of letters since have written agreeable works in this genre.

This professionalization provides some justification for stopping about 1900, although handsome and eloquent scientific books have appeared since then. The reason for starting at 1600 is less arbitrary, for it was about then that books in English on science began to flow from the presses in some profusion. Even so, it would seem to be true, broadly speaking, that it was not until after the Restoration that works indubitably of the first rank – such as those of Boyle and Hooke – began to appear for the first time in English. Before that time there had been translations of Latin books, which had sometimes of course been written by Englishmen. Scientific books first published in English in the first half of the seventeenth century were as a rule attempts to spread the New Philosophy: the doctrines of Bacon, Galileo, Descartes, and Gassendi. They belong more to the argument over whether modern times could hope to compete with antiquity, than to an account of the progress of science. But in these books we see the experimental and mechanical philosophy gradually prevailing both against the Aristotelian learning of the Schools, and against its more dangerous and equally modern adversary, natural magic and Paracelsian chemistry. Involved with these, we find the perennial argument between utilitarians and devotees of pure science.

When, after the Restoration, the experimentalists had triumphed, their polemic turned against the scoffers who found them comic, and against critics who saw their activities as leading to atheism. This latter development distinguished the scientific attitude in Britain from that on the Continent. From the late seventeenth century until the 1850s, numerous scientists of some distinction published works of natural theology, in which the existence of

God, and his attributes, were demonstrated using evidence accumulated in the various sciences. This tradition reached its greatest flowering – if that is the right word – after Hume and Kant had shown its unsoundness; it survived the alarming discoveries of early nineteenth-century geology, and only collapsed noisily on the publication of *The Origin of Species* in 1859.

From the early seventeenth century, and indeed even earlier, numerous scientific books have been translated into English, from Latin and from modern languages. Some, such as Philemon Holland's translation of Pliny, are famous as literature; perhaps the majority are somewhat pedestrian. But, since the British have never been noted for their command of foreign tongues, if we know what was available in translation then we know, within limits, what was widely read, or was felt to be important by somebody. It would not be true to say that it was always the most intrinsically interesting books which were translated.

Of books by Englishmen, it is very surprising that Gilbert's treatise on the magnet, published in Latin in 1600, did not find a translator for nearly 300 years, when two translations appeared in rapid succession, although it was frequently referred to. Harvey, on the other hand, was translated into English; but Newton's *Principia*, perhaps the last really important treatise on physics to be written by an Englishman in Latin, was not Englished for forty years. Naturally, if a book was too difficult to be read by any but advanced mathematicians, there was little need for an English edition. Latin continued rather longer to be employed in the biological sciences, and in philosophy; and in the nineteenth century, some authors preferred to put some footnotes into the relative obscurity of the learned tongue. In the eighteenth century several handsome works of natural history appeared with English and French texts. Among the works of foreign scientists, one finds some anomalies too; thus although Huygens is perhaps nowadays remembered chiefly for his wave theory of light, the book in which this appeared was not translated until the twentieth century. His treatise on the inhabitants of the celestial worlds, on the other hand, was translated at once. Kepler found no translator; Galileo's writings were translated fairly fully by Thomas Salusbury. Many translations are anonymous; where there is a preface by the translator or publisher, it is often entertaining reading.

In 1665 the *Philosophical Transactions* of the Royal Society began to appear; though it was not the first scientific journal, it was the first in English and the first to publish original scientific papers of the modern type rather than review articles or collaborative works by a whole society. T. H. Huxley, in a moment of enthusiasm, wrote: 'If all the books in the world, except the *Philosophical Transactions*, were destroyed, it is safe to say that the foundations of physical science would remain unshaken, and that the vast intellectual progress of the last two centuries would be largely, though incompletely, recorded.'

This is perhaps going a little far; but it does indicate the importance of scientific journals in the dissemination of science. Particularly since the beginning of the nineteenth century, when the *Philosophical Transactions* was joined by a rapidly increasing number of other learned journals, has it been true to say that almost every important advance has been brought before the world in an article rather than a book. One should remark, though, that the theory of evolution by natural selection aroused no great stir when Darwin and Wallace proposed it in papers in 1858, and only made its mark when it appeared in *The Origin of Species* in the following year. In this study we shall not examine journals in any detail, trying only to indicate what kind of papers any particular periodical carried. The papers of eminent scientists were frequently collected into one or more volumes, usually long after their impact had been made; but this does make them much more accessible to us.

It is to be hoped that from a study of this kind it will prove possible to catch something of the flavour of science in Britain, and to perceive something of its idiosyncratic character. For though science is in a sense no respecter of national or linguistic barriers, there are styles in science, and metaphysical assumptions which are associated with different countries at different periods; and both scientific thought and institutions in any nation have a certain autonomy. A study of scientific books can also indicate which sciences were most popular – in the sense of having most workers, or a greater lay following – at a given time. It is perhaps too ambitious at this stage to hope to give an account of the Englishness of English science, but this is a task worth bearing in mind.

This discussion has been concerned with science in Britain, and yet it is true that scientific books in English also appeared in America. It would be worth having a full-scale bibliographical study of science in America, but the general picture seems to be clear: that down to the latter part of the nineteenth century American science can best be seen as a provincial part of British science, and American books usually reprints, authorized or pirated, of those originating in this country. Isolated scientists of international importance lacked, before this date, an American scientific tradition of sufficient strength to support them by itself.

Finally, the 'natural science' of the title will be taken to include the physical and biological sciences, but not in general medicine nor social sciences. And one may add that while most seventeenth- and eighteenth-century scientific books command high and rising prices, works of natural theology or those by scientists whose names are less familiar need not be very expensive. And books of nineteenth- or early twentieth-century science are not yet prohibitive in price, provided that the really great names are avoided; even first editions of Darwin need not be expensive, provided that one avoids *The Origin of Species*, and Bewick's *Birds* and *Quadrupeds* are not too dear. Luckily for the historian, many rare scientific books are now appearing in facsimile; such editions can form the nucleus of an excellent working library.

General note

The arrangement of the chapters and bibliographies is bound to be somewhat arbitrary; some authors appear in two or more chapters because they wrote on a variety of topics, or at various levels; and the divisions between subjects, and between periods, are blurred. No doubt there are errors of interpretation and of omission; these may be due to prejudice, but are more likely to be the result of ignorance.

In the bibliographies, which follow each chapter, only first editions are as a rule mentioned. Where a later edition is also referred to, this is because it is greatly expanded, contains more plates, or is in some way of unusual interest. Some bibliophiles prefer the last edition prepared in the author's lifetime; but for books of science, the marginally greater modernity or accuracy thus achieved is usually offset by a loss of the original warmth and coherence. It is the first edition of any work of originality which makes the greatest impact, and is therefore the most interesting; though the devotee may well wish to collect other editions of favourite works also. This applies particularly, for example, to *The Origin of Species*, where the incorporation of new material, and the attempt to deal with new critics, blurred the clarity of the argument in later editions; with Newton's *Opticks*, on the other hand, and with John Snow's splendid inductive study of cholera, the later editions are valuable because of the new material they contain.

After each chapter a list first of original sources, and then of more recent publications, appears; the latter items have been chosen primarily for their bibliographical usefulness, or because they reprint or translate primary materials, and therefore do not include all the modern works to which I am indebted. Where no place of publication is given in the bibliographies, this means that the book was published in London.

Among the many valuable journals covering the history of science, the annual *History of Science* is devoted to sources and problems; and the quarterly *Isis* publishes each year a critical bibliography of books and articles on the subject. The history of science is still a field in which there is enormous scope, and need, for detailed monographs and bibliographies, as well as for works of wider compass relating developments in the sciences to those in other regions of intellectual history; this is particularly true for the nineteenth century.

In addition to the works listed in the bibliographies, the catalogues of the British Museum Library and of the Library of Congress are invaluable.

GENERAL WORKS FOR REFERENCE

Biographia Britannica, 6 vols., 1747–66.

Bolton, H. C., *A Catalogue of Scientific and Technical Periodicals, 1665–1882*, Washington, 1885.

A Select Bibliography of Chemistry, 4 pts, Washington, 1893–1904.

Burland, C. A., *The arts of the alchemists*, 1967.

Caillet, A. L., *Manuel Bibliographique des Sciences Physiques ou Occultes*, 3 vols., Paris, 1912.

Dictionary of National Biography, 16 vols., 1970–80.
Dictionary of Scientific Biography, New York, 1970.
Duveen, D. I., *Bibliotheca Alchemica et Chemica*, 1949.
Ferguson, E. S., *A Bibliography of the History of Technology*, Cambridge, Mass., 1968.
Ferguson, J., *Bibliotheca Chemica*, 2 vols., Glasgow, 1906.
Fussell, G. E., *The Old English Farming Books . . ., 1523–1730*, 1947.
More Old English Farming Books . . ., 1731–93, 1950.
Gardner, F. L., *A Catalogue Raisonné of Works on the Occult Sciences*, 2 vols., 1903–11.
Gunther, R. W. T., *Early Science in Oxford*, 14 vols., Oxford, 1923–54.
Early Science in Cambridge, Oxford, 1937.
Hurlbut, C. S. jr, *Minerals and Man*, 1969.
Macphail, I. (ed.), *Alchemy and the Occult, A Catalogue of Books and Manuscripts from the Paul and Mary Mellon Collection*, 2 vols., New Haven, 1969.
Mead, G. R. S., *Thrice Great Hermes*, 3 vols., 1906.
Michel, H., *Scientific Instruments in Art and History*, tr. R. E. W. and F. R. Maddison, 1967.
Mottelay, P. F., *Bibliographical History of Electricity and Magnetism*, 1922.
Multhauf, R. P., *The Origins of Chemistry*, 1966.
Partington, G. R., *A History of Chemistry*, vols. 2–4, 1961–4.
Pledge, H. T., *Science since 1500*, 1939.
Taylor, E. G. R., *The Mathematical Practitioners of Tudor and Stuart England*, Cambridge, 1954.
The Mathematical Practitioners of Hanoverian England, Cambridge, 1966.
Thorndike, L., *A History of Magic and Experimental Science*, 8 vols., New York, 1929–58.
Thornton, J. L., and Tully, R. I. J., *Scientific books, libraries, and collectors . . .*, 2nd ed., 1962.
Tooley, R. V., *English Books with Coloured Plates, 1790–1860*, 1954 (excludes botany, mineralogy, and ornithology).
and Bricker, C., *A History of Cartography*, 1969.
Waite, A. E., *Lives of Alchymistical Philosophers*, 1888.
Watt, R., *Bibliotheca Britannica*, 4 vols., Edinburgh, 1824.
Willey, B., *The Seventeenth Century Background*, 1934.
The Eighteenth Century Background, 1940.
Nineteenth Century Studies, 1949.
Williams, T. I. (ed.), *A Biographical Dictionary of Scientists*, 1969.
Wing, D., *Short-Title Catalogue of Books . . . 1641–1700*, New York, 1945–51.
Wolf, A., *A History of Science, Technology, and Philosophy in the 16th and 17th Centuries*, 2nd ed., 1950.
A History of Science, Technology, and Philosophy in the eighteenth century, 2nd ed., 1952.

Additional Reading

Catalog of the Sidney M. Edelstein Collection, Jerusalem, 1981.
Cole, W. A., *Chemical Literature 1700–1860: A Bibliograpy*, 1988.
Knight, D. M., *A Companion to the Physical Sciences*, 1989.
Porter, R. (ed.), *Man Masters Nature*, 1987.

Those interested in the History and Bibliography of Science will want to belong to the History of Science Society, or to the British Society for the History of Science, the Society for the History of Natural History, the Society for the History of Alchemy and Chemistry; and to read their Journals, *Isis*, *British Journal for the History of Science*, *Annals of Natural History*, and *Ambix*.

1 The Scientific Movement

The Renaissance had been, as far as the sciences are concerned, in some ways a conservative movement. The recovery of the original texts of the ancients was overwhelming for those nourished only on various encyclopedic compilations; and reverence for the ancient texts reached a peak. It seems that at the universities in England in the first half of the seventeenth century, such science as was taught was, in general, old-fashioned and taught simply as book-learning. There seems to have been little appreciation among scholars – denounced as 'pedants' by their opponents – of the rôle which science could play in education or in society. The sixteenth century had seen the beginnings of the Scientific Revolution with the work of such men as Vesalius and Copernicus; but, ironically, in England the Copernican system seems to have become better known and accepted among practical men than among scholars. The mathematical approach to nature, made possible by the availability of the works of Plato and Archimedes, began in England in the second half of the seventeenth century, when both the new knowledge and adequate mathematics appeared in university courses. In the Renaissance, not only had the works of the major scientists and philosophers been translated, but the occult Hermetic Corpus too; and the separation of science and magic was one of the problems to be tackled in the first half of the century. It would seem to be a feature of this period, as far as England is concerned, that science was more talked about than advanced.

By the end of the century, the foundations of modern science had been laid. Those qualified to form an opinion were all Copernicans; the Empiricist philosophy had replaced the Aristotelian; the atomic – or 'corpuscularian' – view of nature was almost universally held; and Newton's work had revolutionized optics, mechanics, and astronomy. The Royal Society had been in existence for nearly forty years, and a knowledge of recent scientific discoveries had become essential for anybody who wished to keep up with the times. The plain language demanded for scientific papers had begun to influence the style of literary English, and the witty conceits of the seventeenth century gave way to the elegance of the eighteenth.

Two great discoveries made about the beginning of the seventeenth century – the telescope and the microscope – had revealed how circumscribed men's knowledge had hitherto been. In the first half of the century these instruments were used essentially in a qualitative way, and the virtuosi marvelled at the new worlds revealed to their sight. By the end of the century, astronomical observations were being made through telescopes – notably at Greenwich – and Antony van Leeuwenhoek had begun to publish in the *Philosophical Transactions* his papers describing bacteria and protozoa. Marjorie Nicholson has demonstrated the enormous effect exerted by the telescope and microscope on the imaginations of literary men in the seventeenth and early eighteenth centuries. And with Newton's researches in optics, theory had at last caught up with practice. With the telescope, the height of the lunar mountains had been estimated, Saturn's rings observed, and the velocity of light determined; and the microscope had provided conclusive evidence for the circulation of the blood, and cast light on the reproductive process in animals.

Certainly the experiments on atmospheric pressure and on air-pumps, which began with Evangelista Torricelli, were of, at least, equal importance in practical results. Whereas at the beginning of the seventeenth century it was generally supposed that nature abhorred a vacuum, by the end of the century the nature of atmospheric pressure was clearly understood, and barometers were in use. The habit of keeping systematic records of the weather was encouraged by the Royal Society. Boyle's experiments with the air-pump were often demonstrated before distinguished visitors to the Royal Society. The pressure-cooker, in the form of Denis Papin's 'digester', had been invented; the principle behind it being that liquids under pressure boil at a higher temperature. By 1700 the first steps had been taken towards the steam engine, which became a practical proposition with Thomas Newcomen shortly after the turn of the century. Robert Boyle's experiments, and the arguments of the atomists, had produced in Britain general agreement among scientists of the real existence of vacuum; on the Continent the arguments for a plenum advanced by René Descartes continued to appear forceful for another generation.

In natural history, the seventeenth century saw the transition from the world of bestiaries and herbals to that of John Ray, with his accurate descriptions and his attempt to achieve a natural classification of organisms. The microscopists demonstrated the life-cycle of insects such as the gnat; and mineralogy and geology may be said to have begun to take their modern form, although the speculative geological works of Thomas Burnet and William Whiston seem to have made more impact than the rather plainer discussions of Nicholas Steno and Robert Hooke.

But as far as science in Britain was concerned, most of these great developments came in the second half of the century, and the first half was important for the wide dissemination of interest in science and for the creation of an atmosphere in which original scientific work could be carried on. Christopher Hill stresses the large number of scientific books in English which appeared in the last part of the sixteenth century and the first part of the seventeenth. Copernicus' theory was announced in English shortly after its first appearance, and books by John Dee, Thomas Digges, and Robert Recorde made available, in the vernacular, the discoveries of modern science, rebutting the charge that they were thus casting pearls before swine. Digges suggested the infinity of the universe – thus going beyond Copernicus – in one of his publications, and also seems to have used a telescope to scan the heavens some years before Galileo's far more important observations were made. The great voyages of discovery described in Hakluyt had revealed how small a fraction of the globe had been known in antiquity, had established that the Earth was a sphere – as Aristotle and Ptolemy had of course held – and had exploded the theory that the torrid zone about the Equator was impassable. The discovery of the magnetic compass was another respect in which the moderns were superior to the ancients; and when Gresham College was founded in 1598 it is not surprising that the scientific courses available dealt with astronomy and navigation.

It appears to have been generally felt, in the discussions as to whether modern learning could hope to excel ancient, that in literature there was little hope of measuring up to antiquity, whereas in technical accomplishments men had already clearly surpassed all predecessors. George Hakewill's

1 Gresham College, London (Ward, J., *Lives of the Professors of Gresham College*, 1740, pl facing p 33)

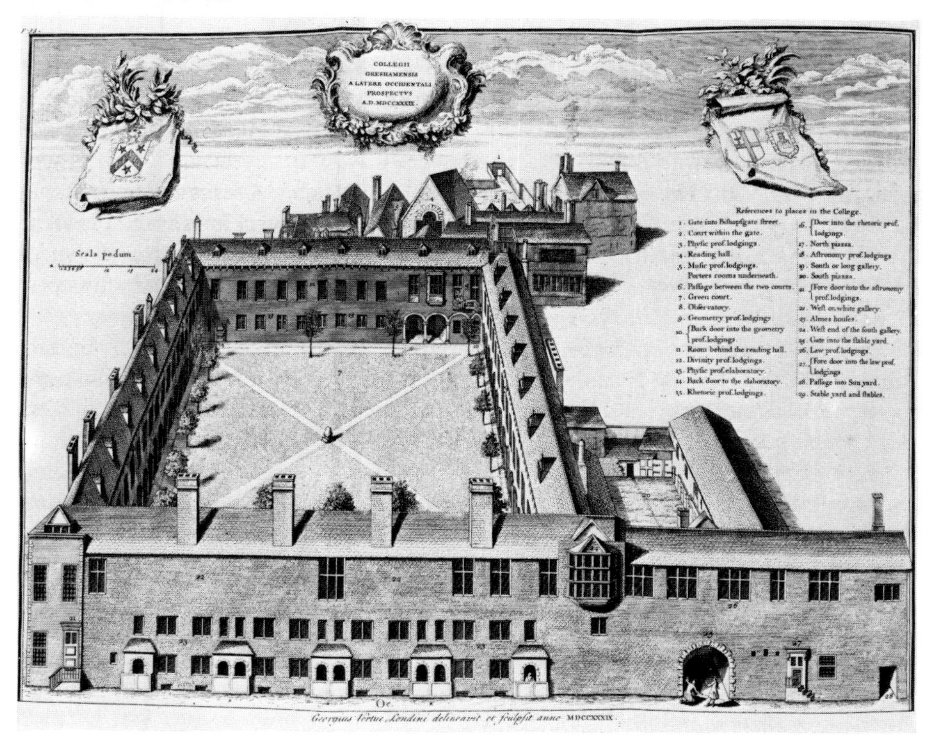

Apologie, 1627, initiated this debate, which proceeded through the whole seventeenth century, and clearly contributed to the rise of interest in science and technology, and particularly to the emphasis on utility which is so characteristic of those who write about science, though not usually of original scientists. Among propagandists for the new science, we find this emphasis; with contempt for the supposedly *a priori* and heathenish science of antiquity went a dislike of pure science and pure mathematics in favour of the applied branches. Elegant devices that saved the appearances were unsatisfactory to those who wanted facts. The relationship between these attitudes and the rise of Puritanism and Capitalism has been explored by a number of historians, and is still a matter of controversy. But it does seem to be the case that in England there was a vague correlation between moderate Puritanism and the rise of science; some of those who interested themselves in the new science inclined to some form of Puritanism, but by no means all did so. The 'new sect of latitude-men' were much more important by the Restoration. Fortunately, in discussing books, we shall not need to explore this question in general terms.

Although in the first forty years of the century, it seems that Gilbert was the major influence on men of science, with his experimental method and relatively cautious drawing of conclusions, by the mid-century Gilbert's theories of electricity and magnetism had been absorbed into the main body of science, and Bacon began to be recognized as the greatest philosopher of science of the period. This reputation he has continued to enjoy, despite attacks. Although Gilbert was so influential, his *de Magnete* was not translated into English and can therefore only have been studied by those who knew Latin; and his name was probably more familiar than his book. Bacon was unfairly critical of Gilbert, and wrote on scientific method essentially as an amateur; as Harvey remarked, he discussed science like a Lord Chancellor. But his writings were mostly soon available in English, and some indeed appeared first in the vernacular; and whether or not we care for his philosophy of science it is historically true that he rang the bell that called the wits together. His was the greatest single influence in the founding of the Royal Society, the *New Atlantis* being seen as a blue-print for a scientific society: and the early Fellows tried to put into practice his precepts, without, it must be admitted, making any very great discoveries thereby. His influence did not wane with the end of the seventeenth century, for the foundation of the British Association for the Advancement of Science was also inspired by – among other influences – the *New Atlantis*, and Bacon's works on scientific method.

Part of the appeal of Bacon's writings both to the founders of the Royal Society and to those who organized the British Association was that he attached great importance to the rôle of the hard-working but not necessarily highly educated – or anyway not professionally educated – scientist. The progress of science depended upon the collection of reliable data, and from the mass of facts, 'forms' or theory and laws, could be deduced by the application of rules. The whole process, given large numbers of people enthusiastically working together, could proceed almost mechanically, for progress depended not upon the speculations of great geniuses but on steady advance by a large body of men, checking their facts and practising the division of labour. This scheme does, *mutatis mutandis*, fit quite reasonably some aspects of modern science. Although research is not split up between different workers in quite the way set out in the *New Atlantis*, nevertheless we do find a formal separation between experimentalists and theoreticians in various sciences. And while it is difficult to fit Einstein or Planck into the Baconian scheme, what Thomas Kuhn calls 'normal science' – periods of steady and fairly predictable growth, between revolutions – does fit the Baconian picture.

The *Advancement of Learning* first appeared in 1605; in 1623 it was translated into Latin, in an expanded form and with some tactless references to Roman Catholicism deleted; and this was translated back into English in 1640. The *Advancement* was intended as the first part of Bacon's *Instauratio Magna*, which was never completed; part of its object was to clear the ground from lumber, but one nevertheless finds in it Bacon's precepts, which appeared in more detail in his *Novum Organum*. Thus there is emphasis on being open-minded: 'if a man will begin with certainties, he shall end in doubts; but if he will be content to begin with doubts, he shall end in certainties.' Theology and science are to be kept apart, for a dogmatic attitude cannot but be fatal to science; which, properly pursued, will lead to piety by revealing God in Nature. Theology and science are complementary. Bacon warns of the danger of premature theorizing; of the futility of logic-chopping in the manner of the schools; and of the foolishness of setting ancient authors up as dictators in the sciences, which should be progressive bodies of knowledge. He discusses induction, stressing the need to seek diligently for counter-instances, for otherwise one is simply conjecturing. The rules of induction are not as clearly and formally set out by Bacon as by John Herschel and by John Stuart Mill in the nineteenth century; but both these authors had two centuries of modern science from which to derive examples, whereas Bacon stood at the threshold.

The problems of Bacon's methods in sciences like physics and chemistry were to some extent revealed in the work of the less talented members of the Royal Society in the Restoration period, but, in general, it seems fair to say that in a time when little exact information was available his stress on the collection of facts and the inductive method was very valuable. In his judgements of contemporaries, Bacon is notorious for his strictures on Gilbert, and also for his equivocal attitude – to say the least – towards the Copernican hypothesis. Both these mistaken evaluations seem to stem from Bacon's reluctance to allow theorizing on the grand scale. Gilbert he accused of 'making a philosophy' out of a few observations on the loadstone. Galileo and Kepler recognized Gilbert's achievement, and the magnetic theory of gravity was a step in the groping towards Newtonian mechanics. Bacon's apologists in later generations had to admit also that he had failed to realize the extremely important rôle played by mathematics – a pure and abstract discipline not depending upon experiment – in the physical sciences.

The *New Atlantis*, written in the style of Utopian literature, was an appeal for collective work in the sciences, and for some kind of research institute in which large-scale experiments could be carried on, with the public good always in mind: 'The end of our foundation is the knowledge of causes, and secret motions of things; and the enlarging of the bounds of human empire, to the effecting of all things possible.' Because they so hated lies, the fellows of Saloman's House were severely forbidden to 'show any natural work or thing adorned or swelling, but only pure as it is, and without all affectation or strangeness'; and it duly became part of the programme of the Royal Society to disseminate scientific knowledge in plain ordinary language, that of artisans and merchants rather than of wits and scholars. In fact, not surprisingly, the language of Bacon himself, and of Wilkins, Sprat, Hooke, and Boyle, is hardly that of artisans; but it is far less elaborate than that of, say, Sir Thomas Browne. Bacon's own writings in particular are sparkling and avoid the matter-of-factness which one associates with the Baconian programme.

Translations or paraphrases of Descartes and Gassendi appeared during the Commonwealth, but in general it seems fair to say that down to the founding of the Royal Society Bacon was the most important and influential philosopher of science for those whose first language was English; though the Continental philosophers gave great prestige to mechanical explanations of phenomena.

We can now therefore leave philosophy of science for science itself. Astronomy was popularized in numerous treatises, naturally often dealing with navigation; but the most entertaining for the modern reader are those which discuss the plurality of worlds. Armchair astronomy has its attractions – after all, John Aubrey tells how Lawrence Rooke caught his death of cold observing the heavens – and speculation about the inhabitants of other heavenly bodies has always been particularly popular. Whereas in physics, and particularly chemistry, the advantage has always lain with the sooty empirics in their laboratories, in this region of astronomy – the borderline territory between science and science fiction – only *a priori* science was possible until the present day. The information brough back by space probes about the other solar planets is discouraging, but speculators appear to be undaunted and hope to contact dwellers in the depths of space by radio. They are in a tradition going back at least to the seventeenth century; and among the bold spirits who intruded into this region English authors were well represented.

Their works fall into various classes. Some, like Cyrano de Bergerac's, are to be placed in the tradition of Utopias, set in the Moon, the Sun, or a planet rather than in some island. These do not set out to teach astronomy any more than *Gulliver's Travels* tell us about geography. Others were written by scientists of various degrees of originality and eminence, from Kepler and Huygens to William Herschel and Humphry Davy, to propagate scientific or metaphysical doctrines. The telescope of Galileo confirmed what Plutarch had conjectured, that the Moon was a body resembling the Earth, with mountains and valleys and even perhaps seas. This seemed to be evidence for the Copernican system, in which there was no distinction between the Earth and other heavenly bodies; and that the telescope revealed moons encircling other solar planets was another point in the Copernicans' favour. Kepler's *Somnus* was written to show that moon-dwellers would, as we do, suppose themselves to be at rest, while the Earth circled around them and rotated before their eyes; just as the Sun appears to move in orbit around us.

Bishop Godwin's was among the early English works on this subject; it appeared in 1638 under the pseudonym of Domingo Gonsales. Marooned in St Helena, attended only by an indigenous blackamoor, Diego, the hero trained swans to carry burdens, and eventually twenty-five of them carried him through the air. He was rescued; and when the ship was attacked by the English, he escaped into the air with his swans. They flew towards the top of the Pico Teneriffe, and then on towards the Moon; gravity, being a magnetic phenomenon, soon ceased to operate, and they were weightless. After eleven or twelve days' travel they reached the Moon; by this time the Earth looked to them like the Moon does to us, providing the Ptolemaic system 'a very absurd conceit'. The Moon's gravitational field was so much less than the Earth's that the hero could jump clear of it, and fly. The Moon was a Paradise; and Domingo was sorry to leave, but was overcome by homesickness.

This production is an entertaining work of science fiction, in which the propagation of the Copernican system looms less large than the telling of a good story. In the slightly later works of John Wilkins the scientific theory is the most important feature. Wilkins was at this time a divine of Puritan tendencies; under the Commonwealth he became Warden of Wadham College, Oxford, and was responsible for attracting to Oxford in the 1650s a group of important scientists including John Wallis, Seth Ward, Christopher Wren, William Petty, and ultimately Robert Boyle. Wilkins married Oliver Cromwell's sister, but nevertheless – being 'a trimmer' – he became Bishop of Chester a few years after the Restoration. He was the most important single figure in the setting-up of the Royal Society; and became its Secretary. His major works are popularizations of science, the *Essay towards a Real Character* – which was an attempt to achieve a universal written language – and a number of sermons, which need not concern us.

Wilkins, arguing in perhaps a surprisingly teleological way, suggested that it would be most economical to use the Moon both as a luminary and as an inhabited globe; and that whereas the blotches on the Moon appear pointless blemishes, if it be a world then they represent land and sea, and therefore have a purpose. The analogies between the Earth and other planets made it sufficiently probable that they were worlds, too. He expected the inhabitants of other heavenly bodies to resemble us; indeed it seems to have been in treatises on the plurality of worlds that the idea of the uniformity of nature was first applied to the heavens. Wilkins, like Godwin, held the magnetic theory of gravitation, and believed that twenty miles up the force of gravity

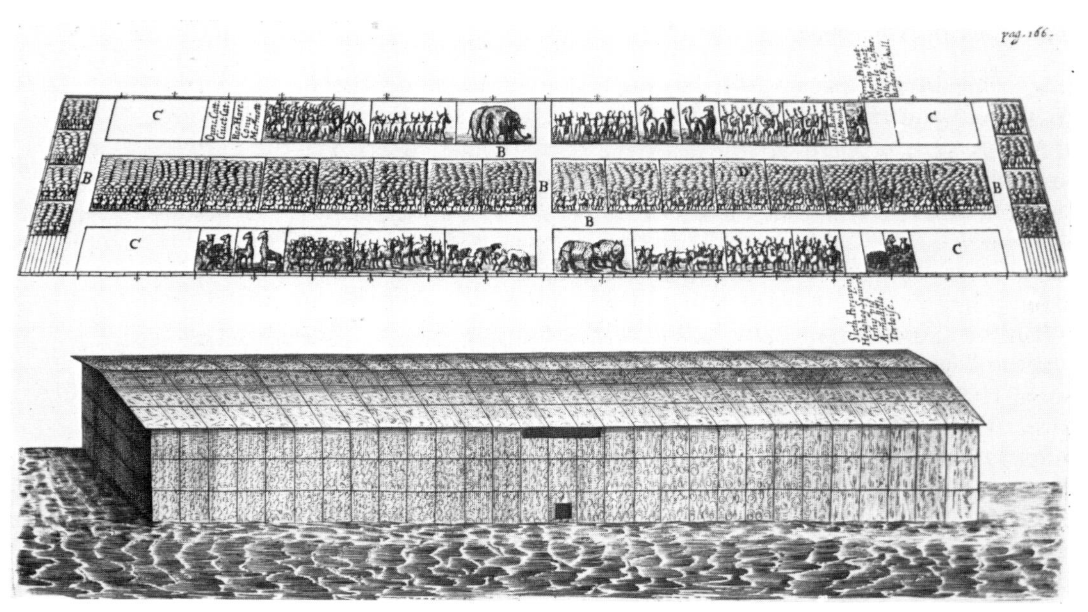

2 Noah's Ark, showing the supply of spare sheep for the carnivores (Wilkins, J., *Essay towards a Real Character*, 1668, pl facing p 166)

ceased completely to act. Kepler had transported his hero to the Moon by means of spirits, and Godwin his with harnessed birds; but Wilkins considered that the most hopeful project would be a flying carriage. Birds, which are heavier than air, fly effortlessly; and small, weak birds cross the North Sea. If we could find out the trick, we could do it too. It would probably be best to start from a mountain top; and once the carriage had climbed laboriously up a few miles it would be easily and effortlessly made to go in any desired direction, in a windless and temperate region.

Wilkins' *Discoverie of a World in the moone* and *Discourse concerning a New Planet* were chiefly intended to establish the Copernican system: the 'new planet' of the second work is in fact the Earth, recognized as a planet rather than the centre of the universe by Copernicus. Wilkins was ruthless with theologians who attempted to lay down the law in sciences; this would be expected from a Baconian, but it is surprising in a future bishop. He argued lengthily, following Augustine as Galileo had done, that too literal exegesis of passages in the Bible and the Fathers led to absurdities in physics. This kind of assault, separating the spheres of science and dogma, helped to bring about in England a situation where, right down to the middle of the nineteenth century, it was generally held that religion and science were complementary; that the study of Nature could not but lead to the worship of the God revealed in the Scriptures. Wilkins' new astronomy was attacked with spirit by Rosse.

Wilkins' other works concerned with natural philosophy were a book on mechanics, called *Mathematicall Magick* – disappointingly, it has no magic in it – and another on codes and ciphers. During his time at Wadham College, an attack on the universities for their failure to teach the new philosophy was made by John Webster, a Puritan divine. It was answered anonymously; in fact by Seth Ward, a friend and colleague of Wilkins who also eventually became a bishop, and who proposed Newton for a Fellowship of the Royal Society. Webster's complaints about university courses were by no means new, nor were they entirely up to date; but his book, and Ward's reply do give us a useful insight into the rôle played by the sciences in the higher education of the time.

Baconians like Webster demanded a utilitarian emphasis in science, and in education generally. Linguistic studies were useless and should not be pursued. Similarly, applied mathematics should be studied to the exclusion of pure mathematics. The arguments have a familiar ring today, and it is perhaps worth remarking that the really fundamental work in astronomy, for example, was all done by men with a thorough training in pure mathematics; and that then as now, the scope for the self-taught, or practical, man in the developed sciences was small. Again, the sciences in the seventeenth century were not particularly fruitful in useful applications. Technology in most cases draws upon the science of previous generations; and innovation depends far more upon economic circumstances than upon the current level of knowledge in pure science. Well within a century, the discovery of atmospheric pressure by Torricelli led to the first practicable steam engine; but the discovery of the circulation of the blood led to no radical innovations in treatment, and the Copernican system did not lead directly to any great improvements in navigation. The 'experiments of light' of the seventeenth century did not, broadly speaking, generate 'experiments of fruit' until the eighteenth century.

It was not even generally accepted how science could be expected to produce practical results. About 1600 Renaissance ideas of natural magic were current, and those bogus technologies, astrology and alchemy, continued to flourish. Frances Yates, in her book on Bruno, has shown how the diagrams of Copernicus could be interpreted as magical, giving one some kind of power over the influences of the stars and planets. Alchemy or magic could only be practised by one who had subjected himself to various disciplines; and the methods employed were concealed from all but a chosen few. The Rosicrucians began at this period; Robert Fludd was perhaps the most notable Englishman in this tradition, but his books were published abroad and not in English until 1659. To some contemporaries, Fludd seemed to be the outstanding scientist of the day.

Two English works which do illustrate this tradition – which also showed itself strongly in Paracelsian medical works – are della Porta's *Natural Magick*, translated into English in 1658; and Ashmole's *Theatrum Chemicum Britannicum*, a handsomely illustrated edition of English alchemical poetry which appeared in 1652. The latter was clearly an antiquarian work rather than a contribution to living science; and although isolated alchemists are to be found in the eighteenth century – the most distinguished in England being Peter Woulfe, F.R.S. – the corpuscular chemistry of Boyle was to triumph.

Nevertheless, certain of the attitudes and emphases of the alchemists did remain active, and Newton and Davy both show signs of them. Della Porta's book, which had appeared in Italian in 1558, is a strange and delightful compilation. It contains one of the first scientific descriptions of lenses, which had hitherto been entirely a craft preserve; and also the first description of a plane-surface kite in a European language, though this is thoroughly garbled in the English translation, where it appears on the last page. Besides these it contains an account of the loadstone, of invisible writing, and of numerous other more curious topics, such as the beautifying of women. The magician, according to Porta, must be genuinely religious, and must survey the whole creation, and understand the doctrines of sympathy and of signatures as well as mathematics and natural philosophy. The *Natural Magick* was in fact the product of a society, the forerunner of the Lincei and the Cimento in Italy, and of the Royal Society; but to compare it with the productions of Boyle or Hooke, which appeared only a few years after the English translation, is like setting a bestiary beside the works of John Ray. In the first part of the

3 (*right*) Apparatus for distillation (Glauber, J. R., *The Works*, 1689, pl facing p 189)

4 (*below*) Alchemists at work (Ashmole, E., *Theatrum Chemicum Britannicum*, 1652, pl facing p 103)

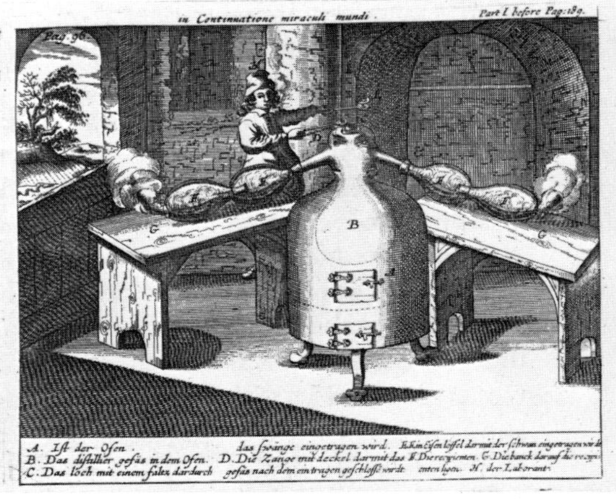

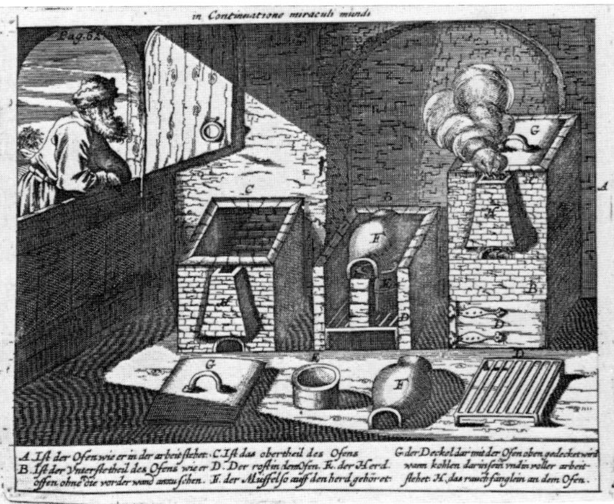

seventeenth century a frontier zone was crossed, and della Porta's *otiosi* with their astrological and magical interests, and their sheer breadth of subject matter, are on the far side of it.

All this has taken us some way from John Webster, but not as far as might appear; for Webster, in deploring that more science was not taught in universities, had in mind as much the science of Paracelsus, Fludd, and the magicians as that of Galileo and Kepler. In referring to the learning of the ancients as heathen philosophy, he is squarely in the magical tradition, in which God's creation of the world was seen as a chemical process; and in which only those with a satisfactory world-view and a blameless life could hope to bring their experiments to fruition. Webster also shows the Baconian utilitarian spirit, and suspicion of 'systems' and of the hypothetico-deductive method. Ward was able to pick, in his reply, upon Webster's lack of mathematics, and his sympathy to natural magic and astrology; and to contrast this with the true science that was then in fact being taught at universities. For Oxford, in the period when it was the Royalist capital, had begun to be a scientific centre, and in the Commonwealth this beginning came to fruition when a number of Gresham College scientists took up residence in Oxford. Ward thought it a misfortune that Bacon lacked mathematics; and he hoped for a deductive rather than a purely experimental physics. He was therefore happy to have Ptolemy taught, because his book was an excellent example of applied mathematics. He believed that students should have both the Ptolemaic and Copernican systems presented to them, so that they could choose which to believe; whereas Webster thought Copernicus' only should be taught, as the true system. Webster and Ward agreed that Descartes should be read, though they disagreed over what would be found in his writings, since both considered that Descartes would support them.

Neither author emphasized the best argument for the Copernican system, that provided by Kepler who showed that all the machinery of epicycles – which Copernicus and Ptolemy alike had employed – was unnecessary if one supposed that the planets moved in elliptical orbits around the Sun. The elliptical hypothesis is mentioned in the Inaugural Lecture which Christopher Wren delivered when, at the age of twenty-five, he was appointed Professor of Astronomy at Gresham College in 1657. This lecture is reprinted in *Parentalia*, the memoir of the Wren family edited by the architect's grandson and published in 1750. It, and the remarks about Wren in Sprat's *History of the Royal Society*, show that Wren would have made himself a reputation as a scientist, even if the Great Fire had not given him opportunities as an architect. Wren's mention of Kepler is accompanied by the reflection that his work marked only a beginning, but that the perfection of his work was to be expected from living Englishmen; a prescient observation, but Wren could hardly have had Newton, then a schoolboy, in mind. The patriotism or nationalism of authors of this period writing in English is notable; with confidence that the intellectual achievements of antiquity could be surpassed went the conviction that this country could outdo the rest.

If astronomy was the most spectacularly progressive science in the early seventeenth century, an equally important development was the revival of atomism. At the beginning of the century, atomism was strongly associated with atheism, and would not have been accepted by most of the learned. It was associated, for example, with the entourage of the 'Wizard Earl' of Northumberland. At the end of the century, the researches of Boyle had given some scientific content to the atomic theory, and in the writings of Ralph Cudworth and Richard Bentley atomism was presented as a buttress to religion. The major part of this story dates from the second half of the century when Gassendi's writings became available in Charleton's paraphrase, and when the books of Descartes had shown the desirability of mechanical explanations. Charleton came to Gassendi after translating Helmont, for whose quasi-magical doctrines he began to feel an aversion; but some Helmontian elements, such as action at a distance, insinuated themselves into Gassendi's text. The most interesting part is the argument for absolute space and time, and for the definition of matter in terms of impenetrability, in opposition to the Cartesian doctrines that all motion is relative, and that matter is identical with extension. Descartes' *Discourse* appeared in a pleasant English translation in 1649: but none of the more specifically scientific works – on mathematics, optics, and 'meteors' – for which the *Discourse* was a preface were Englished until recently. It is possible, in discussing the scientists of this period, to separate the mechanical philosophers, who desired in the manner of the atomists or of Descartes to explain phenomena in terms of unobservable mechanisms, and the experimental philosophers, to whom hypothesis is repugnant, and who wished only for laws built up from observations by induction. In the latter part of the century, this distinction becomes easier to make in principle than in practice.

In the first half of the century, the problem was to make atomism respectable, for its scientific importance lay far in the future. Thomas Hariot was perhaps the most important English atomist in the first decade of the seventeenth century; but his work was not published, and even now his manuscipts have not been fully investigated. Hariot was imprisoned on suspicion of astrological necromancy in 1605. His atomism, which was very close to that of Democritus and Lucretius, increased the suspicions of his enemies. Some of the letters of Hariot's disciples were printed in the nineteenth century by Rigaud and Halliwell. Thomas Hobbes appears to have learned of the atomic theory through this school; the opening sentences of *Leviathan* are a fine statement of the mechanical view of the world. Bacon seems to have toyed with atomism, and expected great things of Hariot, but in his later writings he showed distaste for it as hypothetical, an example of the wrong way to proceed in the sciences. Hobbes also, after writing a work on atomism which was not published until the nineteenth century, in his *De Corpore*, translated 1650, advocated a plenum, with an all-pervading ether, somewhat in the manner of Descartes. Hobbes' embracing of mechanical accounts of phenomena led him to deny the possibility of immaterial substances and to affirm the corporeality even of God. This doctrine was thoroughly alarming, and in the second half of the century scientists were at pains to show that their views had nothing in common with this. Hobbes also believed that he had squared the circle, which led to acrimonious exchanges with John Wallis, the mathematician.

It seems not unfair to say, then, that the history of atomic theory down to 1650 is rather obscure, in that atomic speculations were little published, and believed subversive or too hypothetical. Certainly no progress was made in the atomist programme of showing how, given various arrangements of given particles or corpuscles endowed with very simple properties, all the variety of things we see in the world might be, in some detail, explained. In other words, atomism was a metaphysical doctrine of very marginal scientific usefulness. It is possible to argue that this state of affairs persisted until about 1900; but even such a sceptic would have to admit that in the writings of Boyle, Newton, and their successors the atomism was in some way relevant to the science, although perhaps only as a heuristic belief. A critic of atomism in the 1640s was Sir Kenelm Digby, in his treatise *Of Bodies.* He argued that lengths can be divided into inches in just the same way as bodies into particles, and yet we do not hold that the inches are more real. An apple, he remarked, is a whole: when it is cut up the parts come into existence, but they are themselves wholes. He added that atoms, if they occupied space, must be in principle divisible, as Euclid bisected lines; and if they are mere points, then however many there are they cannot occupy a volume. Finally, he remarked that sharp boundaries – as between an atom and void – are foreign to nature; we cannot tell, for example, precisely where the finger ends and the hand begins. Digby would therefore only allow a physics of divisible particles.

In the biological sciences, the first part of the century is notable for the publication of Harvey's work on the circulation of the blood, in Latin in 1628 and in English in 1653; for the continuing debate between Paracelsian and Galenic physicians; and for the revolution in natural history from the world of herbals and bestiaries, of Pliny and Topsell, to that of John Ray and Nehemiah Grew. The watershed here is Sir Thomas Browne's *Vulgar Errors*, which appeared in 1646; and we shall leave Ray and Grew for the next chapter. Harvey's work is so well known, and the biography and bibliography of him by Geoffrey Keynes so authoritative, that there is little that needs to be said. The Preface of the 1653 edition of the *Anatomical Exercitationes Concerning the Generation of Living Creatures* is particularly interesting, for Harvey then gives us his philosophy of science. Harvey was conservative and cited Aristotle with great frequency in his works; but he stated firmly that knowledge is to be sought in things and not in books, and the revolution which he brought about in physiology was far greater than anything achieved by the Paracelsians, although they did introduce new therapeutic measures.

The Paracelsian cosmology, in which all parts were interrelated, and chemical analogies prominent, did not, it appears, enjoy any popularity in sixteenth-century England. It represented a fusion of occult elements with practical chemistry, and accommodated such notions as the macrocosm-microcosm analogy, doctrine of signatures, and such forms of action at a distance as cures by sympathy; this last point came up in England in the weapon-salve controversy. In 1605 Thomas Tymme translated passages from a celebrated Paracelsian, Duchesne; in 1585 R. Bostocke had published an apology for Paracelsian theory, but his book went through only one edition, as did a shorter work by I.W. which appeared in the following year. In the 1650s a great number of such works, including those of Hermes, Helmont, Croll, Paracelsus, and Lemnius appeared. Although to us much of what these authors write seems hypothetical, occult and wild, their appeals to experiment did impress some contemporaries, to whom chemistry – which meant iatrochemistry or alchemy – became the type of experimental science. Paracelsians had advocated three elements, the *tria prima* – salt, sulphur, and mercury – as the basic substances of chemistry. They were not our substances of the same name, for our mercury, for example, would contain some of the ideal elements salt and sulphur as well as mercury. Their relationship to the four Aristotelian elements – earth, air, fire, and water –

Non invicta recedo. 106

To my Scholler Mr. HANNIBAL BASKERVILE.

THIS *Indian* beaſt, by Nature armed ſo,
That ſcarce the Steele can peirce his ſcalie ſide:
Aſſaulteth oft the *Elephant* his foe,
And either doth the conqueror abide,
Or by his mightie combatant is ſlaine,
For never vanquiſht, he returnes againe.

So you that muſt encounter Want, and Care,
To overcome your hard, and crabbed skill,
Take courage, and treade vnder foote diſpaire,
For better hap, attendes the vent'rous ſtill:
And ſooner leaue, your bodie in the place,
Then back returne, vnletter'd with diſgrace.

This Embleme was deviſed at firſt by Paulus Iovius.

A Rhinoceros was ſét to Rome by Emanuel king of Portingal who fought with it cōming on land thorough Provence: but by the waie, by hard fortune it was drowned neere Porto Venere: ſeeking a long time to ſaue it ſelfe amōg the Rocks. *Paulus Iovius.*

Q3. *Non*

5 Natural history as an emblem; the rhinoceros. Drawn by John Abbot. (Peacham, H., *Minerva Britanna*, p 106)

6 Swallowtail butterfly (Topsell, E., *History of four-footed beasts* . . ., 1658, p 967)

was unclear, and Paracelsian authors seem to disagree about it. Boyle's *Sceptical Chymist* was written to show that neither theory, nor yet a combination of them, could stand up to critical examination. It is by no means easy to separate the Paracelsians from the alchemists, but in fact, as Allen Debus points out, the former wished to be understood in their writings while the latter wrote only for adepts. The controversy over the weapon-salve concerned the possibility of curing wounds by applying medicaments to the weapon which had caused the wound, rather than to the wound itself, which was simply bound up with clean linen. When one looks at some of the Paracelsian remedies, or indeed the herbal or animal medicines to be found in Dioscorides or Topsell, one cannot but agree that they would be better applied to a weapon than to a patient.

This brings us to natural history proper. Aristotle and Pliny were the best sources for zoology and we should remember, if tempted to deride Pliny, that John Ray praised him. There were also bestiaries, designed to edify rather than to inform; and Topsell represents a delightful compromise between fable and biology, with splendid woodcuts of dragons, the mantichora and the hydra as well as delightful ones of such familiar creatures as the donkey and the hedgehog. Topsell intended his book for Sunday reading, or so he declared, and hence fables about the animals, and rebukes for those 'manichees' who believe that some animals are useless or harmful, are jumbled most astonishingly with practical advice on the cure of diseases of horses. Topsell, like Pliny, sometimes made it clear that he did not know whether to believe some of the information he presented; but the compilation of such works was always felt to demand erudition rather than observation. His source was Conrad Gesner, a Swiss, whose illustrations – including Dürer's rhinoceros – he reproduced and whose enormous Latin text he translated and condensed. To the second edition of his work was added Mouffet's *Theater of Insects*; some insects had been treated – following a draft of Mouffet – in Topsell's second volume, as 'serpents', but Mouffet's own treatment is much fuller and is in the same spirit as Topsell's own volumes. Many of the tall tales, as of the phoenix, the beaver castrating himself to avoid his hunters, and the salamander living in the fire, were exploded by Sir Thomas Browne in his *Vulgar Errors*; a book which, in Browne's inimitable style, cleared the way for the period of a more sober genuine natural history; where creatures just as surprising and delightful were in fact found to exist.

These proto-zoological works are a feast for the reader; Pliny contains some botany too, but the major work of antiquity on this subject, Dioscorides' *Herbal*, though it was translated by John Goodyear in 1655 was not published until R. T. Gunther discovered it in 1934, and embellished it with reproductions of illustrations from a Byzantine manuscript of the early sixth century now in Vienna. Herbals were intended as guides in the preparation of drugs rather than as works of botany for its own sake. But with the Renaissance pictures of plants drawn from nature began to appear in them, long after plants had begun to be drawn and coloured accurately by artists. With the Renaissance came the great herbals of Brunfels and Fuchs, which are described in Agnes Arber's book, and whose plates were copied in the English herbals of William Turner, the so-called father of British botany, and John Gerard. Turner's *Herball* was published between 1551 and 1568,

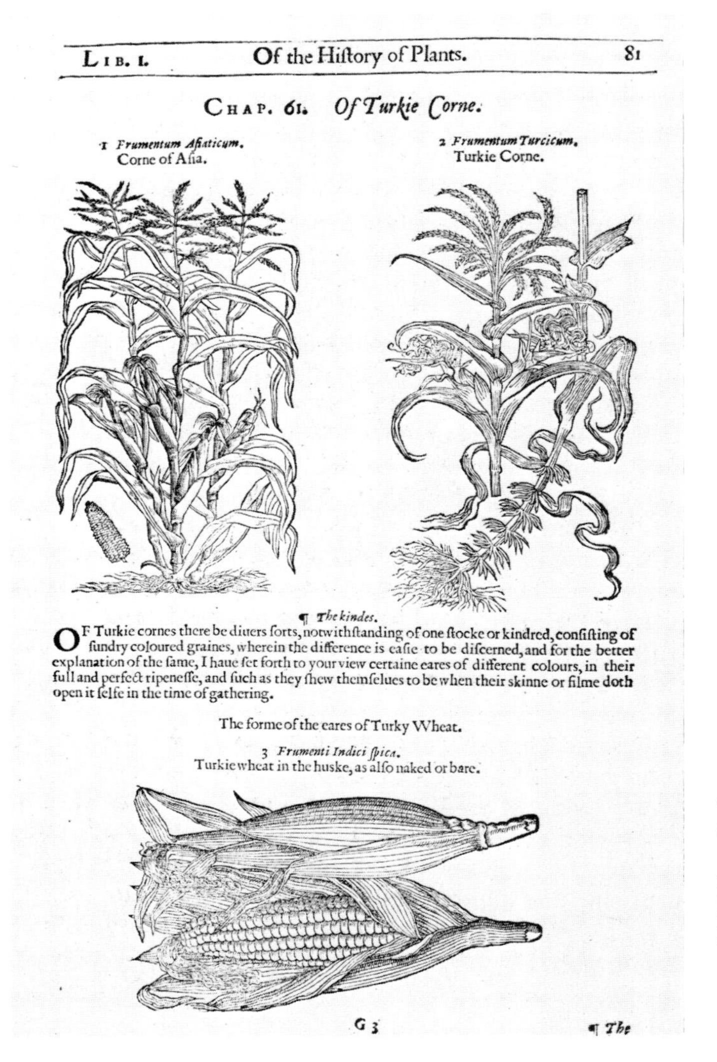
LIB. I. Of the Hiſtory of Plants. 81

CHAP. 61. *Of Turkie Corne.*

1 *Frumentum Aſiaticum.*
Corne of Aſia.

2 *Frumentum Turcicum.*
Turkie Corne.

¶ *The kindes.*

OF Turkie cornes there be diuers ſorts, notwithſtanding of one ſtocke or kindred, conſiſting of ſundry coloured graines, wherein the difference is eaſie to be diſcerned, and for the better explanation of the ſame, I haue ſet forth to your view certaine eares of different colours, in their full and perfect ripeneſſe, and ſuch as they ſhew themſelues to be when their skinne or filme doth open it ſelfe in the time of gathering.

The forme of the eares of Turky Wheat.

3 *Frumenti Indici ſpica.*
Turkie wheat in the huske, as alſo naked or bare.

G 3 ¶ *The*

7 Maize, or Turkey Corn (Gerard, J., *Herball*, 1633, p 81)

8 (*opposite*) Adam and Eve in Paradise, with the Vegetable lamb (Parkinson, J., *Paradisi in Sole*, 1629, frontispiece)

PARADISI IN SOLE
Paradisus Terrestris.
or
A Garden of all sorts of pleasant flowers which our
English ayre will permitt to be noursed vp:
with
A Kitchen garden of all manner of herbes, rootes, & fruites,
for meate or sause vsed with vs,
and
An Orchard of all sorte of fruitbearing Trees
and shrubbes fit for our Land
together
With the right orderinge planting & preseruing
of them and their vses & vertues
Collected by John Parkinson
Apothecary of London
1629
Qui veut parangonner l'artifice a Nature
Et nos parcs a l'Eden indiscret il mesure.
Le pas de l'elephant par le pas du ciron,
Et de l'Aigle le vol par cil du mouscheron,

9 Oaks (Parkinson, J., *Theatrum Botanicum*, 1640, p 1386)

with illustrations from Fuchs' octavo work of 1545; though no herbals are to be described as works of science *tout court*, Turner is generally praised for his scientific attitude. Gerard, on the other hand, is censored for his ignorance and for publishing as his own one Dr Priest's translation of Dodoens' herbal; but the result is a very handsome and delightful book for anyone who is not too concerned about botanical accuracy or morality. Gerard's *Herball* was printed in Roman type, and has a more modern look than Turner's; and it contained the first illustrations of the potato. The second edition was prepared by Thomas Johnson and came out in 1633; it was botanically greatly superior to the first edition, and Gerard remained so popular that

John Parkinson's writing never became so well known. In 1629 Parkinson produced his *Paradisus*, a charming work on gardening whose frontispiece illustrated the 'vegetable lamb'; and in 1640 *Theatrum Botanicum*, the most complete English treatise on plants of its day. All these herbals, like the bestiaries and Topsell, are discursive and full of surprising folklore, as well as handsome plant-illustrations; and they paved the way for modern natural history when the critical spirit of the scientific revolution began to affect their readers.

1 THE SCIENTIFIC MOVEMENT

Ashmole, E., *Theatrum Chemicum Britannicum* . . . , 1652.
Bacon, F., *Twoo Bookes. . . . Of the proficiencie and advancement of Learning* . . . , 1605.
Of the Advancement and Proficiencie of Learning . . . , tr. G. Wats, Oxford, 1640.
The Wisedom of the Ancients, tr. A. Gorges, 1619.
The Naturall and Experimental History of Winds, tr. R. G., 1653.
Sylva Sylvarum . . . , ed. W. Rawley, 1629; incl. *New Atlantis*.
The History of Life and Death, 1638; *History, Naturall and Experimentall, of Life and Death*; two translations of the same work.
(Bostocke, R.), *The Difference between the Auncient Physicke . . . and the latter Physicke proceeding from Idolators* . . . , 1585.
Browne, T., *Pseudoxia Epidemica* . . . , 1646.
The Works, ed. G. Keynes, 4 vols., 1964.
Burton, R., *The Anatomy of Melancholy*, Oxford, 1621.
Camden, W. *Britain* . . . , tr. P. Holland, 1610; new tr., ed. E. Gibson, 1695.
Charleton, W., *Physiologia Epicuro-Gassendo-Charletoniana* . . . , 1654.
Coles, W., *The Art of Simpling*, 1656.
Adam in Eden, 1657.
Comenius, J. A., *Naturall Philosophie reformed by divine light*, 1651.
Crollius, O., in: *Philosophy Reformed and Improved in Four Profound Tractates*, tr. and ed. H. Pinnell, 1657.
Bazilica Chemica, & Praxis Chymiatricae . . . , 1670.
Culpeper, N., *The English Physitian enlarged*, 1653.
Descartes, R., *A Discourse of a Method* . . . , 1649.
The Passions of the Soule . . . , 1650.
Digby, K., *Two Treatises, in the one of which the nature of bodies; in the other, the nature of mans soule; is looked into* . . . , Paris, 1644.
Digges, L., *A Prognostication of right good effect* . . . , 1555.
Digges, T., *A Geometrical Practise, named Pantometria* . . . , 1571.
Dodoens, R., *A niewe herball* . . . , tr. H. Lyte, 1578.
Rams little Dodoen . . . , 1606. (W. Ram's abridgement.)
Donne, J., *Ignatius his conclave* . . . , 1634.
Duchesne, J., *The Practise of Chymicall and Hermeticall Physicke* . . . , tr. T. Thymme, 1605.
Fludd, R., *Mosaicall Philosophy* . . . , 1659.
Gerard, J., *The herball or generall history of plantes*, 1597; 2nd ed., ed. T. Johnson, 1633.
Glauber, J. R., *The Works*, tr. C. Packe, 1689.
(Godwin, F.), *The Man in the Moone: or a discourse of a voyage thither by Domingo Gonsales* . . . , 1638.
Hakewill, G., *An Apologie of the power and providence of God in the government of the world* . . . , Oxford, 1627.
Hakluyt, R., The *Principall Navigations Voiages and Discoveries of the English Nation*, 1589; 2nd ed., 3 vols., 1598–1600.

Harvey, W., *Anatomical Exercitationes Concerning the Generation of Living Creatures . . .*, 1653.

Helmont, F. M. van, *The Paradoxical Discourses . . . concerning the Macrocosm and Microcosm . . .*, ed. J. B., 1685.

Helmont, J. B. van, *A Ternary of Paradoxes . . .*, tr. W. Charleton, 1650.

Oriatricke, or Physick refined . . ., tr. J. C., 2 pts, 1662.

An Hundred and Fifty-Three Chymical Aphorisms . . ., tr. N. N., 1689 or 1690. (English and Latin text; two title pages.)

Hermes trismegistus, *The Divine Pymander . . .*, tr. Dr Everard, 1650.

Hobbes, T., *Elements of Philosophy, the first section concerning Body. . . . To which are added Six Lessons to the Professors of Mathematics . . . in the Univeristy of Oxford*, 1656.

Leviathan . . ., 1651.

Decameron Physiologium, or ten dialogues of natural philosophy, 1678.

Lemnius, C., *The Secret Miracles of Nature . . .*, 1658.

Mouffet, T., *The Theater of Insects*, 1658; see Topsell, E.

Napier, J., *A Description of the admirable Table of Logarithmes . . .*, tr. E. Wright, 1616; also tr. W. R. Macdonald, Edinburgh, 1818.

The Art of numbering by speaking rods: vulgarly termed Napier's bones, 1667.

Paracelsus (P. A. T. Bombast von Hohenheim), *Of the nature of things, nine books . . .*, tr. J. F(rench), 1650.

A Chymicall Dictionary . . ., tr. J. F(rench), 1650.

in: *Philosophy Reformed and Improved in Four Profound Tractates*, tr. and ed. H. Pinnell, 1657.

The hermetic and alchemical writings, ed. A. W. Waite, 2 vols., 1894.

Parkinson, J., *Paradisi in Sole . . .*, 1629.

Theatrum Botanicum . . ., 1640.

Peacham, H., *Minerva Britanna . . .*, 1612.

Pliny, *The Historie of the World . . .*, tr. P. Holland, 2 vols., 1601.

Porta, J. B., *Natural Magick*, 1658.

Ramus, P., *The Logicke*, tr. M. R. Macmillan, 1574.

Recorde, R., *The Grounde of Artes. . .* (1542).

The Castle of Knowledge . . ., 1556.

Rosse, A., *The New Planet no Planet: or the Earth no wandring Star; except in the wandring heads of Galileans . . .*, 1646.

Topsell, E., *The Historie of Fore-footed Beastes*, 1607.

The Historie of Serpents, 1608. A new edition of these, with Mouffet's *Insects*, appeared in 1658.

Tradescant, J., *Musaeum Tradescantium . . .*, 1656.

Turner, W., *A New herball*, 3 pts, 1551–68.

Tymme, T., *A Dialogue Philosophicall*, 1612.

Wallis, J., *Due Correction for Mr Hobbes . . .*, 1656.

Ward, J., *The Lives of the Professors of Gresham College . . .*, 1740.

(Ward, S.), *Vindicae Academiorum . . .*, Oxford, 1654.

Webster, J., *Academiarum Examen . . .*, 1654.

(Wilkins, J.), *The discoverie of a world in the moone . . .*, 1638.

A Discourse concerning a New Planet . . ., 1640.

Mercury, or the Secret and Swift Messenger . . ., by I. W., 1641.

Mathematicall Magick . . ., by I. W., 1648.

An Essay towards a real character . . ., 1668.

Mathematical and Philosophical Works . . ., 1708.

Wood, A. à, *Athenae Oxonienses . . .*, 2 vols., 1691–2.

Wren, S., *Parentalia . . .*, 1750.

Recent Publications

Arber, A., *Herbals, Their Origin and Evolution*, Cambridge, 1912.
Aubrey, J., *The Natural History of Wiltshire*, ed., J. Britton, 1847.
Brief Lives . . . , ed. A. Clark, 2 vols., Oxford, 1898.
Browne, T., *A Bibliography*, by G. Keynes, 2nd ed., 1968.
Campanella, T., in: H. Morley (ed.), *Ideal Commonwealths*, 1885.
Crombie, A. C., *Augustine to Galileo*, 1952.
Curtis, M. H., *Oxford and Cambridge in Transition*, 1558–1642, Oxford, 1959.
Debus, A. G., *The English Paracelsians*, 1965.
Science and Education in the Seventeenth Century, 1970.
Dioscorides, *The Greek Herbal*, tr. J. Goodyear, ed. R. T. Gunther, Oxford, 1934.
Evans, A. H., *Turner on Birds* . . . , Cambridge, 1903.
George, W., *Animals and Maps*, 1969.
Grigson, G., *The Englishman's Flora*, 1958.
Harvey, W., *A Bibliography*, by G. Keynes, 2nd ed., Cambridge, 1953.
Hill, C., *Intellectual Origins of the English Revolution*, Oxford, 1965.
Johnson, F. R., *Astronomical Thought in Renaissance England*, Baltimore, 1937.
Jones, R. F., *Ancients and Moderns*, 2nd ed., St Louis, 1961.
Jung, C. G., *Psychology and Alchemy*, tr. R. F. C. Hull, 1953.
Kargon, R. H., *Atomism in England* . . . , Oxford, 1966.
Kuhn, T. S., *The Copernican Revolution*, Cambridge, Mass., 1957.
Merton, R., *Science, Technology and Society in 17th Century England, Osiris*, IV (1938), 360–632; 2nd ed., New York, 1970.
Pollard, A. W. and Redgrave, G. R., *A Short-Title Catalogue of Books Printed in England and Ireland, and of English Books Printed Abroad, 1475–1640*, 1926.
Raven, C. E., *English Naturalists* . . . , Cambridge, 1947.
Rohde, E. S. R., *The Old English Herbals*, 1922.
Thomas, K., *Religion and the Decline of Magic*, 1971.
Walker, J., and Bliss, P., *Letters written by eminent persons in the 17th and 18th centuries* . . . , 2 vols., 1813.
Webster, C., *Samuel Hartlib and the Advancement of Learning*, Cambridge, 1970.
White, T. H., *The Book of Beasts*, 1954.
Yates, F. A., *Giordano Bruno and the Hermetic Tradition*, 1964.
The Art of Memory, 1966.

Additional Reading

Clarke, T. H., *The Rhinoceros from Durer to Stubbs*, 1986.
Feingold, M., *The Mathematician's Apprenticeship*,. Cambridge, 1984.
Hunter, M., *Science and Society in Restoration England*, Cambridge, 1981.
Pagel, W., *Joan Baptista van Helmont*, Cambridge, 1982.
Shapin S. & Schaffer, S., *"Leviathan" and the Air Pump*, Princeton, 1985.
Shirley, J. W., *Thomas Harriot*, Oxford, 1983.
Shumaker, W., *The Occult Sciences in the Renaissance*, Berkeley, 1972.
Webster, C., *From Paracelsus to Newton*, Cambridge, 1982.

2 *The Royal Society*

On 28 November 1660 at Gresham College a society was inaugurated 'for the promoting of Experimentall Philosophy, and a designe to founding a Colledge for the Promoting of Physico-Mathematicall, Experimentall learning'. Wilkins was chairman. This was really the beginning of the Royal Society, though the formal charters were not granted until 1662 and 1663. The founding of the Royal Society was clearly a step of enormous importance in the history of science in England, particularly as Restoration Oxford ceased to be the centre of scientific interest which it had been in the 1640s and 50s. Until the latter part of the eighteenth century, the Royal Society was the only scientific society in the country; and it provided a meeting ground for those engaged in research and those interested in the sciences. In the early years it was dominated by the dilettantes, and this remained true until the middle of the nineteenth century: not until after the death of Sir Joseph Banks in 1820 was there even a majority of active scientists on the policy-making Council of the Society, and not for many years longer were they in a majority in the Society as a whole.

For the early years of the Society, the sources are the writings of John Wallis; Sprat's *History*; Birch's *History* which reprints the minutes and register-books of the Society; and various writings of Glanvill, Boyle, and Hooke. Sorbière describes a meeting in his *Voyage to England*. Sprat's *History* was written as an apology; its appearance was delayed by the Great Fire, and when the first chapters at least were written the Society's history was still extremely short. The book seems to have been read in manuscript by a Committee including Wilkins, and represents in some sense an official view. It is divided into three parts: the first devoted to ancient philosophy, the second to the Royal Society, and the third to a defence of experimental science. The book is prefaced with an ode by Cowley which stresses the importance of Bacon, the Moses of natural philosophy who had seen, in the *New Atlantis*, the Promised Land of the Royal Society. Sprat had no doubt that the moderns were surpassing antiquity; and they were doing so because they were industri-

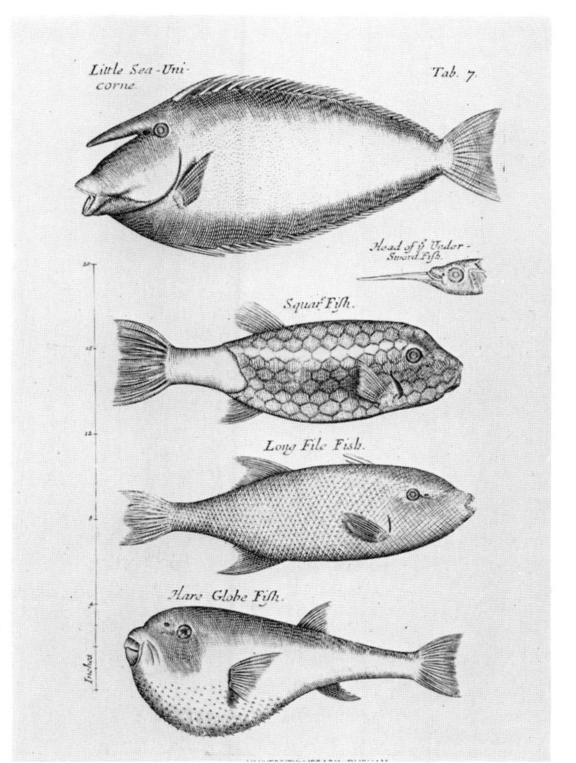

10 (*left*) Fishes from the Royal Society's collection (Grew, N., *Musaeum regalis societatis*, 1681, pl 7)

11 (*right*) Second frontispiece to Sprat's *History of the Royal Society*, 1667; only found in large-paper copies

ous and undogmatic rather than disputatious, and because they were working together.

The mode of procedure of the Royal Society was to discuss a matter, and then get somebody to prepare experiments on the questions, and report back; it was Robert Hooke who usually did the experiments, and his labours are reported in his *Posthumous Works, Cutler Lectures*, and in his diary. Sprat gives us some of the questionnaires sent out by the Society, and Hooke's recommendations on the keeping of weather reports. In the third part of his book, we find a defence of natural philosophy from charges of atheism and magic, and arguments for the value of experimental training in education; and also the celebrated resolve of the Royal Society to exact a simple prose style from their members. Sir Thomas Browne had remarked, in the Preface to his *Vulgar Errors*, that 'if elegancy still proceedeth, and English Pens maintain that stream we have of late observed, to flow from many, we shall, within few years be fain to learn Latin to understand English'; but this trend was deliberately reversed.

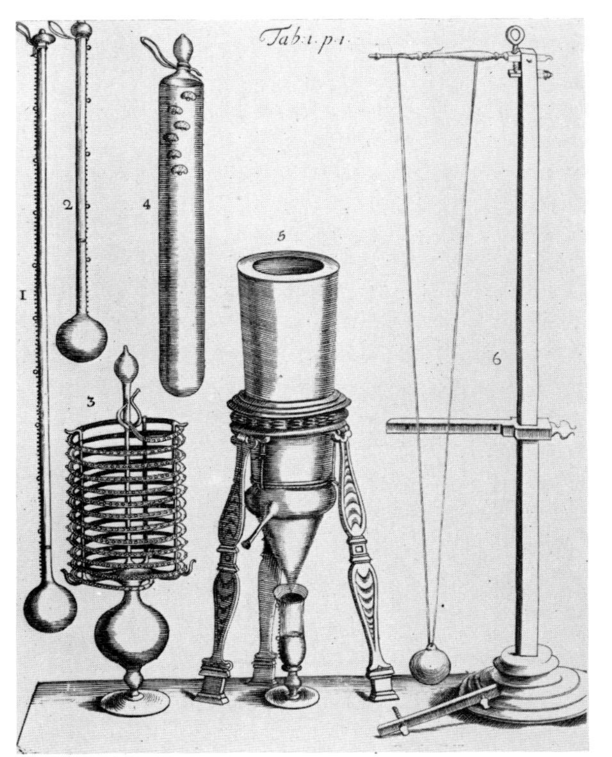

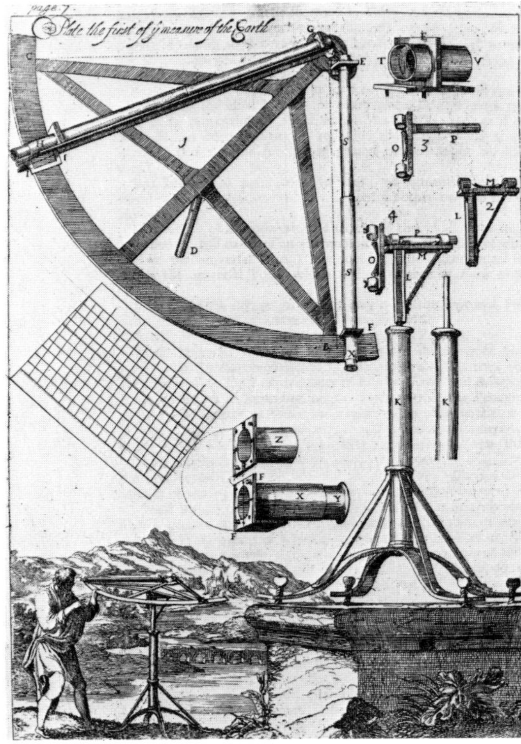

12 (*left*) Thermometers, hygroscope, and pendulum (Academie del Cimento, *Essayes . . .*, 1684, pl 1)

13 (*right*) Surveyor with his equipment (Royal Academy of Sciences, *Memoirs for a Natural History of Animals*, 1688; pl facing p 7 of the section on the Measure of the Earth)

While the Royal Society began as a cooperative enterprise, it soon became an association of individuals working separately at different problems; and when the *Philosophical Transactions* were begun by Henry Oldenberg in 1665 the articles resembled those in modern journals in that they were, for the most part, signed papers describing some discovery or observation. This journal has continued to this day; when none appeared between 1677 and 1683 Hooke edited *Philosophical Collections* in seven parts, between 1679 and 1682, on the same plan. The transactions of the Academie del Cimento, a Florentine group which met under the patronage of Leopold de Medici between 1657 and 1667, are very different, in that the experiments described are not assigned to individuals. A consequence perhaps of this is that we do not find any hypotheses, but simply experimental results; and when the work was translated into English in 1684 it was apparently felt to be rather out of date, though an admirable example of scientific method. The contributions to the early *Philosophical Transactions* ranged from the curious, such as Sir Robert Moray's description of the barnacles which turned into geese, to Newton's great paper on light and colours. It is quite difficult to keep a balanced view of the Royal Society, composed as it was of relatively few men of science of the first rank, and considerable numbers of virtuosi and noblemen whose ideas on the sciences were much less critical.

Certainly, to dip into the *Philosophical Transactions* is to capture the spirit of the science of the later seventeenth century in the same way that browsing in Topsell reveals that of an earlier period. Agricultural questionnaires, and others for sea-voyagers; experiments on blood-transfusion; problems of gunnery; how to find the longitude at sea; Leeuwenhoek's microscopic observations; and a host of other topics all appear in its pages, together with book-reviews, intelligence from other countries, and analyses of mineral-waters. *Memoirs* of the French virtuosi of the Royal Academy of Sciences were also translated into English and are not very different.

It appears from Pepys' *Diary* – and Pepys as President gave the *imprimatur* for Newton's *Principia* – that at Court the scientists were regarded as figures of fun, and in particular were laughed at for weighing air. From less-sophisticated critics they met sterner disapproval; and in particular came under virulent attack from one Robert Crosse and from Henry Stubbe, a physician of Warwick, whose assault was particularly directed at Sprat's book and also at the works of Joseph Glanvill. Glanvill emerged as one of the champions of the Royal Society and the moderns; and also of belief in witchcraft and spirits against those modern atheistic Sadducees who denied their existence. His output is bibliographically confusing because of his habit of giving new titles to what are essentially different editions of his books, sometimes even when the contents were unchanged; fortunately Professor J. I. Cope can be our guide. Glanvill's most famous work is his *Vanity of Dogmatizing*, of which the second edition had the title *Scepsis Scientifica*; and parts then also appeared in his *Essays on Several Important Subjects. The Vanity of Dogmatizing* appeared in 1661, and was an appeal, in a style reminiscent of Sir Thomas Browne's, for an empirical, open-minded approach to science. Glanvill's style changed with the times, and became more sober as he got older; and in many ways this exuberant first work is his most enjoyable to read today. His *Plus Ultra*, to which Stubbe took such exception, appeared in 1668 and was intended as a supplement to Sprat's *History*; it was quite heavily used by William Wotton in his *Reflections upon Ancient and Modern Learning* of 1694. It seems to have been felt that Sprat's *History* had been too general; that more details of new discoveries were required to refute those like Crosse who believed that Aristotle knew more than the moderns. Accordingly, Glanvill tells us much more than Sprat had of the discoveries of the Fellows of the Royal Society, and particularly those of Robert Boyle.

Stubbe was a physician of a contentious disposition who was, it appears, persuaded to launch his attack on the Royal Society partly from the belief that that body constituted a threat to the Royal College of Physicians. He was able to show that some of the papers published by the Royal Society would have been rejected by a competent referee. He suggested that the moderns had done no more than provide fresh hypotheses to explain well-known facts and cures.

He argued, directly and in a series of innuendoes, that the new philosophy was politically subversive and would lead to Popery or to a second Revolution, and at the same time that the vaunted experimental method was no more than day-labouring. He poured out a stream of pamphlets in 1670 and 1671, to which Glanvill gave answers in the same vein; then the controversy seems to have died down. Glanvill's replies included increasingly prominent invocations of the great name of Bacon. Another, but more courteous, attack on Glanvill came from the humanist Meric Casaubon, who remarked that his materialist emphasis and the importance given to facts rather than ethics, could only lead to Hobbesian atheism. This charge was one which Glanvill, and other scientists of the period, were particularly anxious to refute; we shall look at books on this question in more detail in the chapter on Natural Theology.

The labours of Sprat and Glanvill were successful in that the world was made safe for the Royal Society; though its survival and flourishing was perhaps due less to such apologists and popularizers than to the brilliant labours of a few Fellows. We should therefore turn to the works of original science of this period; beginning perhaps with the works on the microscope. Of these the most famous is Hooke's *Micrographia*: but it appeared after Henry Power's *Experimental Philosophy*, the fame of which has been somewhat unjustly eclipsed by Hooke's larger and more handsomely illustrated volume.

Power was a protégé of Sir Thomas Browne, and had studied medicine at Cambridge. His book represents the kind of thoroughly experimental treatment which the Royal Society was set up to encourage. The first book, concerning microscopic observations, is of more interest to us than the other books, on mercurial experiments – that is, those in the Torricellian vacuum at the top of a barometer tube – and magnetical ones; for these later books are less original. Dr M. B. Hall, in her introduction to the reprint of Power's book, remarks that in his experiments on air pressure Power reveals himself to be in the French tradition, for like Descartes he refused to believe that the space above the mercury was empty. A vacuum, for Cartesians, was inconceivable, for 'extension' is identical with 'matter'; if anybody were to say that there was 'nothing' in a tube, he could only mean that its sides had come together. Boyle argued on the other hand for the existence of vacuum, as part of his general programme of explaining all phenomena in terms of corpuscles, or atoms, and the void. Power, like Boyle and the Academie del Cimento, gave full details of his experiments so that other workers could repeat them; this practice was by no means universal at this period.

Power's microscopical observations include descriptions of the flea and louse which are as amusing as Hooke's; the early microscopists could not conceal their delight at the neatness and adaptation to function which they found in the parts of the little creatures they examined, and Power, marvelling at 'the mechanick power which Providence has immured within these living walls of Jet' in the flea, is no exception. Besides insects, spiders, and mites,

Power examined grains of sand, a blob of quicksilver, and seeds of various plants; and he hoped, in vain, to be able to view the 'effluxions' from aromatic, electrical, or magnetic bodies. Indeed, he believed that a microscope did and would furnish the best evidence for the atomic view of matter. All the experiments, and the digressions and conclusions, and the preface, are in an agreeable easy style which tempts the reader on.

Hooke's *Micrographia* is so well known, and the plates so often reproduced, that it is not necessary to say very much about it; except perhaps to remind those who have only seen the paperback facsimile that the original is about four times as large. We should also remember that the descriptions of insects seen through the microscope form only a relatively small part of the book; historians of science today, and presumably natural philosophers of the seventeenth century, read the work as a treatise on experimental physics as much as on natural history. Hooke even demonstrated how various crystal forms could be built up from heaps of lead shot, as part of a general argument for atomism; but this passage was completely neglected until 1813, when William Hyde Wollaston independently proposed the same view of crystal structures in the *Philosophical Transactions.* Hooke's experimental facility is apparent throughout the book, and so is the range of interests which has probably made his reputation less than it should be. Hooke's wide interests, and his facility, are further illustrated in the *Posthumous Works*, a splendid folio volume of which

14 Ammonites (Hooke, R., *Posthumous Works*, 1705, pl 1)

perhaps the highlight is the section on earthquakes, in which he proposed, in lectures before the Royal Society, a thoroughly empirical attitude to geology; but it was to be at least another century before the geology foreshadowed by Hooke and his contemporary Nicholas Steno was to be born. Further papers of Hooke appeared nearly a quarter of a century after his death, edited by William Derham; this volume includes an article on the improvement of the barometer – it is to Hooke that we owe the first barometers with a dial – and others on carriages, on ice, and on various mechanical devices. His paper on springs is to be found in the *Cutler Lectures* of 1679.

Hooke began his scientific career as Robert Boyle's assistant, and was employed by the Royal Society as curator of experiments on Boyle's recommendation. Together, Boyle and Hooke can stand as the great exponents of Baconian scientific tradition. Both display extremely broad interests, and experimental prowess of a very high order; but both also show the limitations of the non-mathematical approach to the physical sciences, and the relative neglect of the hypothetico-deductive method. Boyle was a most voluminous writer, and the standard edition of his works compiled by Thomas Birch is somewhat daunting; and it must be confessed that a number of Boyle's books – especially the *Sceptical Chymist*, which is far more often referred to than read – would have gained considerably in readability with heavy cutting. The shorter treatises, particularly those dealing with experimental science, are quite readable, although they lack the charm or organization of some contemporary works. It is a surprise to turn from the great volumes of the works to the original editions, which are quite small in format.

The *Sceptical Chymist* was – and still is – often referred to as containing the first statement of the definition of a chemical element which prevailed from the time of Lavoisier to that of Moseley and Bohr, that an 'element' is the limit of analysis. Examination of the famous definition in its context shows, however, that this is not the case. The book was written to combat the view that everything was composed of the four 'elements', earth, air, fire, and water, or the three 'principles', sulphur, salt, and mercury. All bodies were, on current theory, supposed to contain all four, or all three, of these substances. Boyle argued that some bodies clearly have more constituents than three or four, and others – such as gold – have less, as is revealed on analysis. But instead of arguing that substances like gold, which cannot be analysed, are therefore elements, Boyle advanced 'corpuscular' explanations, and urged chemists to follow him and thus link their science to mechanics. Atomists in the seventeenth century and beyond did not believe that the world was composed of atoms of gold, atoms of iron, atoms of lead, and so on, but simply of atoms of matter. These atoms might differ in size or shape, but not in the stuff of which they were composed, which had simply the 'primary qualities' of impenetrability, hardness, mobility, and inertia. To suppose otherwise would have seemed to be resurrecting the *homoiomeria* of Anaxagoras. For a corpuscularian such as Boyle – and later for Newton – there were therefore no elements; some arrangements of atoms were more stable than others, so that transmutation might be more difficult than some chemical reactions, but in principle anything could be made from anything, given the right experimental conditions.

The *Sceptical Chymist* was, as the title implied, a work intended to clear the ground by extirpating false hypotheses rather than to construct a new chemistry. Boyle published more detailed arguments in favour of atomism in *The Origine of Formes and Qualities* of 1666; in *The Mechanical Origine and Production of Qualities*, of 1675; and the *Vulgarly Received Notion of Nature* of 1685. Indeed it could be said without serious exaggeration that all Boyle's theoretical works were directed at spreading the doctrines of atomism. He is probably most famous for his work on air pressure and the air-pump, in his experiments with which he hoped to prove the existence of a vacuum or void. Boyle's Law – that the pressure of a mass of gas varies inversely as its volume – was first announced in the second edition of *New Experiments Physico-Mechanical, 1662*; the law is sometimes ascribed to Mariotte, who himself explicitly acknowledged Boyle's priority. Boyle published his results in tabular form – he was one of the first men of science to do so – and compared the observed ratio with that predicted by his hypothesis, finding good agreement. His air-pumps, and experiments on respiration, on the measurement of low pressures, and on the falling of bodies *in vacuo* made with them, are also described in the various editions of this book.

His numerous other books also repay study, covering as they do the physical science of the seventeenth century and its relations with theology. Of perhaps particular interest is his book on colours, which is thoroughly experimental and as such met with the approval of Goethe, who denounced Newton's theory of light and colours as hypothetical. Boyle demonstrated that black

15 The first pressure cooker, designed by a pupil of Boyle (Papin, D., *A New Digester*, 1687, frontispiece)

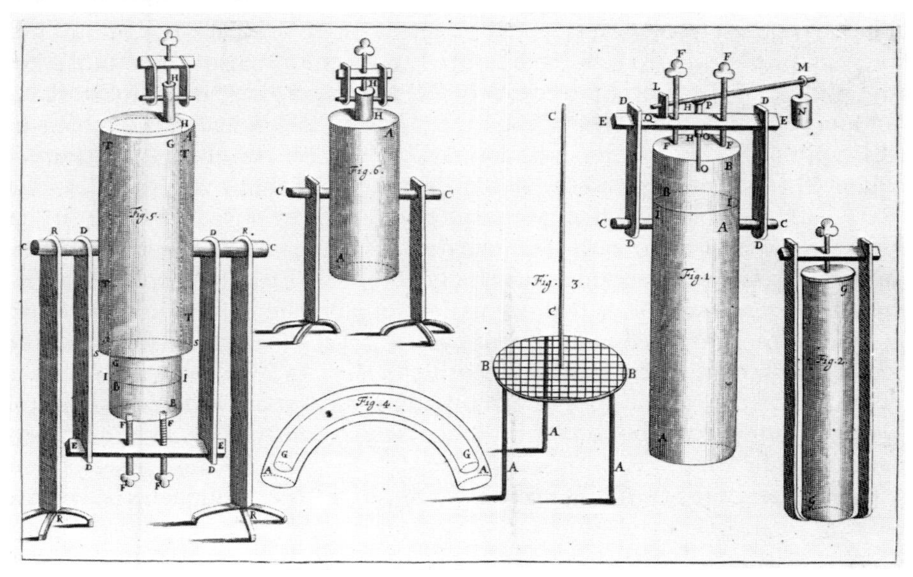

bodies exposed to radiation get hot much more quickly than white ones; he made experiments with the prism; and he discussed the colour changes which accompany certain chemical reactions. In all Boyle's scientific works there is a certain reluctance to come to a firm conclusion; the way is always left open for further experiments. One could say that Boyle kept facts and metaphysics distinct to an unusual degree; certainly it seems to be the case that his influence on chemical theory was less than it might have been. The corpuscular philosophy was stated in such general terms that anything could be compatible with it; and chemists did not, until the nineteenth century, possess a testable atomic theory.

The authority of Boyle was clearly of great importance in the general acceptance of the atomic theory which came about in the latter part of the seventeenth century. Descartes was considered an atomist of a sort, and Gassendi's writings were also known; but atomism in England took a rather different form from the mechanical philosophy which prevailed on the Continent. Thus in Charleton's paraphrase of Gassendi there appeared the sort of action at a distance which would have horrified Gassendi himself; and Boyle's atomism had a theological component which was absent from that of Continental thinkers. More important, Boyle rejected *systems* which could not be tested: and he sought to examine by experiment such tenets of the Cartesians as that void was impossible.

It has been suggested that in discussing the interaction between Chinese and Western science, it is helpful to bear in mind that there has been a point at which the Western science has overtaken the Chinese. Then, after a short period in the case of mathematics and astronomy and a long one in the biological sciences, has come the time of fusion when the Western and Chinese traditions have become integrated into cosmopolitan modern science. It may be helpful to apply a similar analysis to the competition between the mechanical, atomic science and older views. Since the seventeenth century, at least, there have been few thinkers who have seriously advocated a continuum; but there have been many who have refused to accept atomism. Not until after the middle of the nineteenth century did atomic explanations in chemistry and in physics become clearly superior to accounts in which the question of the composition of bodies was not raised. The kinetic theory of gases of Maxwell and Clausius, and the stereochemistry of Kekulé, van 't Hoff, and le Bel, mark the point at which atomism began to draw ahead; and the fusion point would have come in the first two decades of the twentieth century, when atomic explanations became universally accepted. From our point of view, what emerges from this is that discussions of atomism before the nineteenth century are concerned with metaphysics, or a programme for scientific progress, and not with a testable scientific theory. As Whewell was to declare in 1840, whether one was for or against atoms did not depend upon chemical facts, but upon one's notion of substance; which controversy, as he said, was ancient and curious.

It is, therefore, to philosophical works, or to the writings of scientists

16 The defeat of atheism (Cudworth, R., *True Intellectual System*, 1678, frontispiece)

writing as natural philosophers in the strict sense, that we must turn. Ralph Cudworth's *True Intellectual System* is largely concerned with the problem of reconciling atomism with the belief that man has an immortal soul, and that his actions are not governed by necessity. Cudworth argued for the extreme antiquity of atomism, and sought to show that Leucippus, Democritus, Epicurus, and Lucretius represent a falling-away from the true doctrine. The founder of atomism was said to be Moschus, a Phoenician; and Cudworth had no hesitation about identifying him with Moses, and hence claiming for atomism an almost divine authority. The axioms of the atomists – the distinction between primary qualities, which really belong to bodies, and secondary ones which are the product of our minds; and the belief that nothing can come of nothing or go to nothing – Cudworth enthusiastically accepted. But whereas mechanical philosophers had thought that from this basis they could construct the world given only matter and motion, for Cudworth these axioms entailed the existence of incorporeal substance; a concept that would be meaningless to a Hobbesian. Cudworth refused to accept that animals were machines, and man a machine with a ghost inside. He believed that matter was by itself inert and dead; that living organisms must therefore have a soul which directs their brute matter; and indeed that the cause of all motion was immaterial.

His book represents an attempt to synthesize Platonism and atomism, for he believed that Plato had gone too far in his reaction against the atheistic atomists, and that the philosophy of substantial forms was a mistake. He saw the axiom that nothing can be created *ex nihilo*, nor annihilated, as the basis of Plato's belief in the pre-existence and transmigration of souls; but believed that for the Christian it was not logically necessary to believe in the pre-existence of souls, but only in their immortality. He considered that the world displayed clear evidence of design rather than chance; and that it was impossible rightly to understand the doctrines of atomism and to embrace atheism. The book is a handsome and curious piece of sustained argument, which deserves to be read not only for its intrinsic merits but also as one of the works of the Cambridge Platonists, the influence of whom on Newton is being increasingly uncovered. Newton's ideas on gravitation, and particularly his famous remark about the impossibility of 'inanimate brute matter' acting at a distance, may indicate that he was thinking in a framework not far removed from Cudworth's.

Other important writers in this group were Henry More, quondam disciple of Descartes, and – a generation later – Joseph Raphson. More had already, in his first work, the Copernican poem *Democritus Platonissans*, tried some thirty years before Cudworth to achieve a similar synthesis. One of his most interesting works is his exchange of letters with Descartes, published in 1662, which Koyré has recently analysed, in which More criticized, as he had already in his poetry, the Cartesian identification of space or extension with matter. More believed in immaterial substances, rejecting Descartes' dualism, and made some moves towards identifying space and God. His later works show more influence of neo-Platonic and Hermetic doctrines; and he is famous, like Glanvill and Browne, as one of the supporters of the New Philosophy who nevertheless maintained a fervent belief in witchcraft. His writings help to cast light on Newton's strange remark about space being the *sensorium* of God; and in the writings of Raphson – which appeared after Newton's *Principia* – the parallels in the attributes of God and of Absolute Space are made explicit. Newton does not mention More in his writings, but the connections seem to be established beyond reasonable doubt; and Newton's interest in Hermetic literature is well known.

If the Cambridge Platonists illuminate one side of Newton, it is in the writings of John Locke that we find the other side, the side which appealed to the eighteenth century. It was not until the early nineteenth century that we find a resurgence of interest among major scientists and philosophers in Hermetic and neo-Platonic writings. Locke's *Essay* is so familiar that it is not necessary to dilate upon it here; in the chapter on the Newtonians we shall look in more detail at Locke and at Empiricism generally. To consider later seventeenth-century science in English without dealing properly with Newton would seem paradoxical; but Newton's *Opticks* and the English translation of the *Principia* did not appear until the eighteenth century, and can therefore be considered in the fourth chapter.

Nevertheless, there are certain problems raised by the general method of approach of Newton which cannot be any longer avoided. The mechanical philosophers Gassendi and Descartes had sought to achieve certainty by means of a physics based on clear and distinct ideas. This hypothetico-deductive method can be justified by arguments of a transcendental kind; that given unobservable particles and void, or other hypothetical mechanisms, the consequences which are then in fact observed will follow as a matter of course. To be convincing, such arguments must be self-consistent: and as Henry More and Hooke pointed out, Descartes sometimes used atomic mechanisms, and sometimes those based on a continuum. In particular in his *Dioptrique* he employed two incompatible mechanisms for light; in one part of the book it is laid down that light travels with infinite velocity, and in another that it travels faster in water than in air. He can be taken, in modern terms, to be using two models; but this cannot lead to a science in which the conclusions are certain, and it was Descartes' aim to achieve in all branches of knowledge the certainty which characterizes mathematics. It became clear, too, that Descartes had made a number of mistakes – for example, in his controversy, set out in the *Discourse*, with Harvey – and also that it might be possible to derive true conclusions from false hypotheses. The mathematical physics of Newton, and of his teacher Isaac Barrow, can be seen as an attempt to avoid these problems by returning to the examples of Kepler and, particularly, Galileo.

Galileo's law of falling bodies, which was translated into English by Thomas Salusbury, was derived when, instead of considering why bodies gravitated or how motion was possible, Galileo simply investigated how fast things fell. And here again he did not begin with measurements or experiments as a Baconian might have done. For he believed that the book of nature was written in mathematical language, and that natural laws would therefore be characterized by extreme mathematical simplicity. So he defined uniform acceleration, and calculated the laws of motion of a body being uniformly accelerated; and decided that because this was the simplest non-uniform motion, it must describe the falling of bodies. His experiments – which aroused the suspicions of contemporaries – were designed simply to test this conclusion. He realized that laws thus derived are not always exactly obeyed, corrections must be made for such things as air resistance, but the method of proceeding is first to find out the law, and then to explain deviations from it in terms of adventitious circumstances. This approach was much older than Galileo, and can be seen beautifully developed in the hydrostatics, and in the treatment of the balance and the lever, by Archimedes. The scientist who follows this method does not attempt to explain everything; he reckons simply to produce the equations to which the phenomena conform. He is therefore open to the charge of explaining nothing if the cry for an explanation is interpreted not as a request to be told a law of nature but as a demand for the description of a mechanism.

Barrow's mathematical lectures were not translated until 1734; and by that time the authority of Newton had secured for the mathematical approach a position of unrivalled authority. Newton's 'hypotheses non fingo' was remembered, whereas it was forgotten that in the first edition of the *Principia* there were a number of propositions explicitly labelled 'hypotheses'. Newton's opposition was only to vicious hypotheses, and he himself alternated, it seems, between ethereal mechanisms for gravitation, and a belief in some immaterial cause. His followers, as we shall see, went beyond him in declaring that it was futile to look for a cause when all the equations were already known; when detailed predictions were possible already, what advantage, they asked, would there be in knowing the mechanisms? The problem of maintaining a balance, so that one follows just analogies without falling into the sterilities of positivism or into the emptiness of *a priori* reasoning, is one of the most difficult which faces the theoretical scientist.

Cartesian astronomy was popularized in Fontenelle's *Plurality of Worlds* which appeared in English translation very shortly after its first appearance in French in 1686. Fontenelle had hitherto been known only as a literary man, but largely because of the success of this book he became in 1697 Perpetual Secretary of the Académie des Sciences, and his *Eloges* of Academicians, including Newton, became justly famous. In the *Plurality of Worlds* Fontenelle argued for the Copernican system, for a mechanical (Cartesian) view of the world, and for the idea that because nature is uniform the other solar planets must be inhabited by rational beings. The climate would, he believed, dictate the dispositions of these folk, so Venusians would be negroid and amorous, while Jovians would be phlegmatic. Fontenelle ingeniously threw on to any opponents the onus of defending themselves, arguing that his views were so reasonable that it was up to them to try to prove him wrong. Arguments for a plurality of worlds also appear in Christian Huygens' *Celestial Worlds Discover'd*, the only book of this great scientist to be translated into English on its appearance. It begins with an interesting discussion of probability; then comes the argument, essentially teleological, for peopling the cosmos. Stars that can only be seen through the telescope must have some function. The obvious one is to be a sun for other inhabited planets. This argument had been used by Henry More half a century earlier. And it would be unfair if certain planets had beings of a higher grade of intelligence or civilization than others; so, continues Huygens, all must be approximately on a level, and all must have telescopes, Euclidean geometry, jesting, amours, and shows. He modified somewhat the Cartesian mechanism of vortices, and his account of gravitation appears in non-mathematical form in this little book. For although he admired Newton's *Principia*, he never wholly deserted the Cartesian scheme. The result of these books, and those of Wilkins and others, was that by the end of the seventeenth century it had become as heretical to doubt that there were other inhabited globes in the heavens, as it had been early in the century to believe it.

In the biological sciences development was also dramatic: separation of

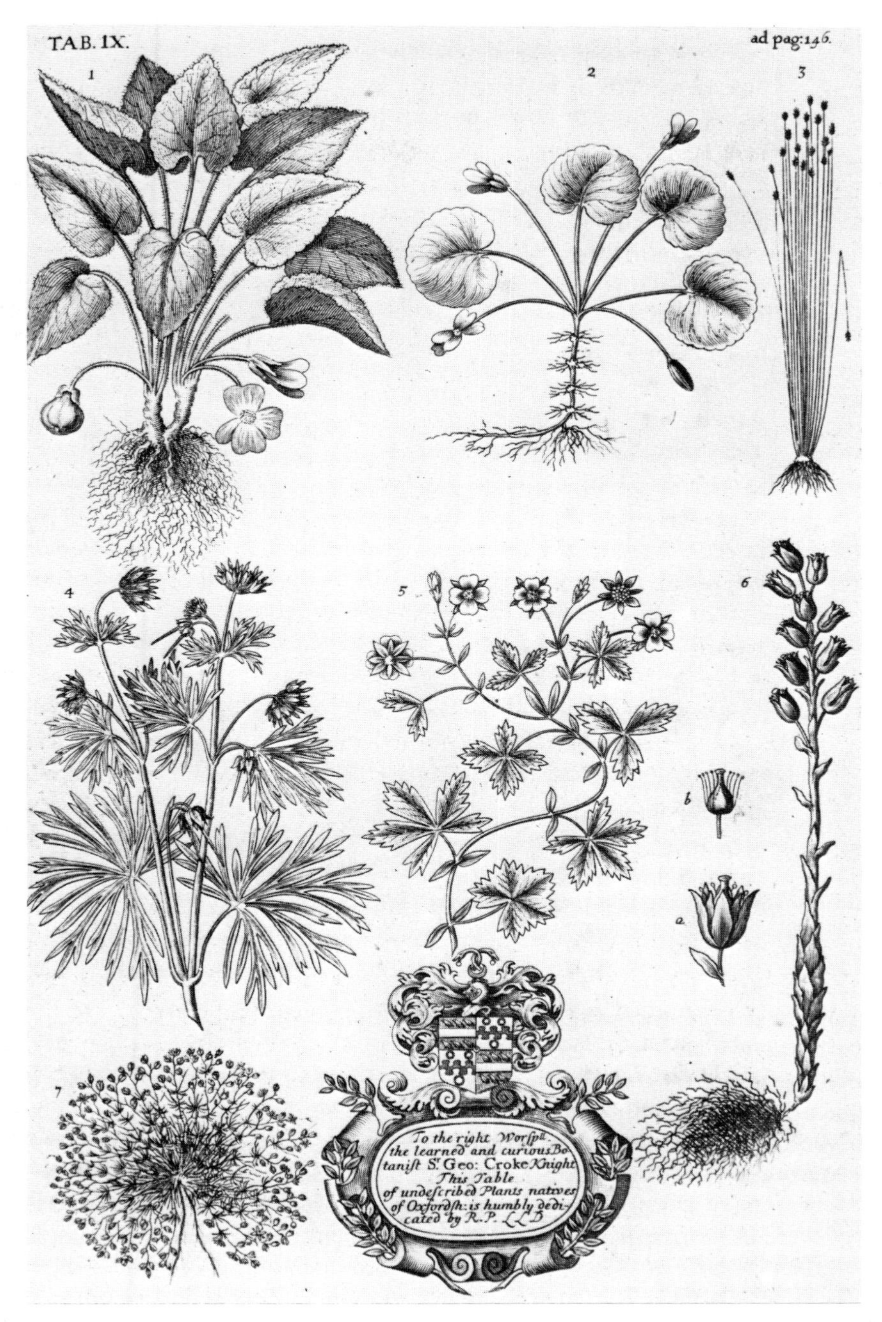

17 Plants from Oxfordshire (Plot, R., *Natural History of Oxfordshire*, 1677, pl 9)

plants into mono- and dicotyledons – those that produce one or two seed-leaves or lobes from the seed – and the discovery of sex in plants, being as

18 (*left*) Jackdaw, chough, magpie, and jay (Willughby, F., *Ornithology*, 1678, pl 9)
19 (*right*) Section of a vine branch (Grew, N., *Anatomy of Plants*, 1682, pl 36)

important in their way as that of the circulation of the blood. Descriptions of natural history can be found in Camden, Plot, and Aubrey. John Evelyn's *Sylva* is essentially a descriptive work; the writings of John Ray and Nehemiah Grew are on a quite different level. Ray, who belonged to the circle of Cambridge Platonists, was among the greatest field naturalists ever; his object was to replace the chaotic arrangements and unclear descriptions provided by his predecessors by an arrangement into natural groups. Aristotle had argued that there were in nature groups of related things, and that it was the biologist's task to sort them out. The groups shaded into one another, and it would not be possible to achieve the natural classification by following one, or only a few characteristics; a creature would be judged a member of a class if in the majority of its characteristics it displayed analogies with members of that class. Ray had helped Wilkins with his *Essay Towards a Real Character*, in which everything was grouped in threes; and this experience convinced him that this was a hopeless approach to biology. In his earlier works he cooperated with Willughby, and produced splendid *Histories* of birds and of fishes; but his own interests seem to have been chiefly directed to botany. The natural classification proved very difficult to operate, particularly since what was then required was a quick method of assigning an individual to its species, or a new species to its genus; and the artificial but convenient system of Linnaeus soon carried all before it in the latter part of the eighteenth century. By the mid-nineteenth century, there was again a realization of the essential correctness of Ray's approach to classification.

Grew's interests were very different: he employed the microscope and investigated the comparative anatomy of plants. His book on the subject is illustrated with splendid plates giving three-dimensional sections through various parts of plants. He announced the sexual nature of plant reproduction; a discovery independently made at the same time by Marcello Malpighi. Grew's book consists of a series of essays, of which the first, 'An Idea of a Philosophical History of Plants', sets out his programme, and gives an excellent idea of the state of botany at that time. The next sections describe roots, trunks, leaves, flowers, fruits, and seeds, first in each case observed with the naked eye, and then with the microscope; various hypothetical mechanical explanations also appear. Finally comes a chemical section, on mixtures, on salts found in plants, and the colours and tastes of plants.

Another important work was Edward Tyson's *Orang-Outang*, a study of an anthropoid ape. He regarded it as intermediate between a monkey and a man, and compared the various parts of its anatomy with corresponding parts in men and monkeys. In calling his ape 'homo silvestris' – though in fact he distinguished pygmies, cynocephali, and satyrs from men, believing them all to be apes or monkeys – he inaugurated the process culminating in Lord Monboddo's idea that an orang-outang might be turned into a gentleman if suitably brought up. It was this notion upon which Peacock seized in *Melincourt*, in which an amiable ape, Sir Oran Haut-ton, is bought a baronetcy and a seat in parliament. Tyson remarks that some of Galen's errors could have been avoided if he had dissected apes instead of monkeys. His book is a forerunner of T. H. Huxley's *Man's Place in Nature*; but perhaps its greatest contemporary importance was that it, in describing a link between man and monkeys, provided support for the view then coming into vogue that there was a Great Chain of Being, linking all creatures from the highest to the lowest. We shall find this idea of particular importance in eighteenth-century biological thought.

Seventeenth-century Britain also possessed, in William Petty and John Graunt, pioneers in the application of new methods in social sciences. Petty was a polymath. He was trained as a doctor, and held an academic post in Oxford during the Commonwealth. He then went to Ireland to survey the country in order that lands might be distributed among the soldiers of the Cromwellian army; and thus began his interests in the economic problems of Ireland. He believed that good government demanded accurate knowledge of vital statistics; and in such works as *Political Arithmetick* he tried to deduce such information from the Bills of Mortality of London and Dublin, and other records. The Bills of Mortality of London date from the late sixteenth century, and become continuous in the early seventeenth century. They seem to have begun as a report of the number of deaths from plague during years of epidemics; but later all deaths were included, under various headings. From these, and from the records of christenings, it was possible to calculate rather crudely what the population of cities probably was, and what immigration from the country was taking place. Graunt's treatment of these Bills – which appeared weekly, and were bound up as a book at the end of the year – is a classic for its critical handling of its dubious statistics. The causes of death were recorded by old women appointed by each parish, and Graunt remarked upon the low death rate assigned to syphilis, and that many other diseases caused a large number of deaths in plague epidemics: his suggestion was that old women could be bribed, and the dishonour of having a death from syphilis or the annoyance of being quarantined thus avoided. It has been suggested that this book was in fact written by Petty, but the external and internal evidence both fail to confirm this. Petty's interests were very wide – he was famous for his allegedly unsinkable catamaran boat – but not very deep, and Graunt's treatment of his data is more thoroughgoing than one would expect from Petty. Statistical methods needed to be made more sophisticated by mathematicians, and by astronomers concerned with averaging-out observations, and much better records were needed, before it became possible to improve greatly upon Graunt's labours. The *Observations* complement the contemporary clinical studies of Thomas Sydenham, the 'English Hippocrates', and the economic writings of Gregory King.

2 THE ROYAL SOCIETY

Academie del Cimento, *Essayes of Natural Experiments . . .* , tr. R. Waller, 1684.
Anon., *A Pleasant and Compendious History of the first Inventers and Instituters . . .* , 1685.
Barrow, I., *The Usefulness of Mathematical Learning . . .* , tr. J. Kirby, 1734.
Bergerac, C. de, *The comical history . . . of the worlds of the Moon and the Sun*, tr. A. Lovell, 1687.
Birch, T., *The History of the Royal Society . . .* , 4 vols., 1756–7.
Boyle, R., *New Experiments . . . touching the spring of air*, 1660; 2nd ed., 1662, contains Boyle's Law.
The Sceptical Chymist, 1661.
Experiments and Considerations touching Colours, 1664.
The Origine of Formes and Qualities, 1666.
Experiments, Notes, &c about the Mechanical Origine . . . of Qualities . . . , 1675.
A Free Enquiry into the Vulgarly Received Notion of Nature, 1685.
Butler, S., *Hudibras*, 3 pts, 1663–78.
Casaubon, M., *A Letter to Peter du Moulin . . . concerning natural experimental philosophy . . .* , Cambridge, 1669.

Cavendish, M., Duchess of Newcastle, *Philosophical and Physical Opinions*, 1655; revised as: *Grounds of Natural Philosophy*, 1668.
Philosophical Letters . . ., 1664.
Observations upon Experimental Philosophy, 4 pts, 1666.
(Charleton, W.), *Two Discourses. I. Concerning the Different Wits of Men: II. Of the Mysterie of Vintners*, 1669.
Natural History of the Passions, 1674.
Cudworth, R., *The True Intellectual System of the Universe . . .*, 1678.
Evelyn, J., *Sylva: or discourse of forest trees . . .*, 1664.
A philosophical discourse of earth . . ., 1676.
Fontenelle, B. le B. de, *A discourse of the plurality of worlds . . .*, tr. Sir W. D., Dublin, 1687; also tr. (J.) Glanvill, 1688; Mrs Behn, 1700; W. Gardiner, 1715.
The Lives of the French, Italian, and German Philosophers . . ., tr. J. Chamberlayne, 1717.
(Galileo), Salusbury, T., *Mathematical Collections and Translations*, 2 vols., 1661–5.
Glanvill, J., *The Vanity of Dogmatizing*, 1661; new ed., called *Scepsis Scientifica*, 1664.
Plus Ultra, 1668.
Essays on several important subjects, 1676.
Saducismus Triumphatus . . ., 1681, 3rd ed., 1689.
Graunt, J., *Natural and Political Observations upon the Bills of Mortality*, 1662.
Grew, N., *Musaeum Regalis Societatis . . .*, 1681.
The Anatomy of Plants . . ., 1682.
Hooke, R., Micrographia . . ., 1665.
Lectiones Cutlerianae . . ., 1679.
The Posthumous Works . . ., ed. R. Waller, 1705.
Philosophical Experiments and Observations, ed. W. Derham, 1726.
Huygens, C., *The Celestial Worlds Discover'd . . .*, 1698.
King, G.: in Chalmers, G., *Estimate of the Comparative Strength of Great Britain . . .*, 1804.
Locke, J., *An Essay concerning Humane Understanding*, 1690.
(Monboddo, J. Burnet, Lord), *Ancient Metaphysics . . .*, 6 vols., Edinburgh, 1779–99.
More, H., *Democritus Platonissans . . .*, Cambridge, 1646.
An Antidote against Atheisme . . ., 1653.
A collection of several philosophical writings . . ., 2nd ed., 1662; 1st ed. not seen.
Remarks upon two late ingenious discourses; the one touching the gravitation . . . of fluid bodies; the other touching the Torricellian experiment, 1676.
Moxon, J., *Mechanick Exercises . . .*, 2 vols., 1677–83.
Mathematicks made easie . . . a mathematical dictionary . . ., 1679.
North, R., *The Life of the Rt Hon. Francis North*, 1742; *The Life of the Hon. Sir Dudley North*, 1744; both ed. M. North.
Papin, D., *A New Digestor or Engine for softening Bones . . .*, 1681.
A Continuation of the new Digester . . ., 1687.
Petty, W., *Political Arithmetick*, 1690.
Economic Writings, ed. C. H. Hull, 2 vols, New York, 1899.
Plot, R., *The Natural History of Oxfordshire*, Oxford, 1677.
The Natural History of Staffordshire, Oxford, 1686.
Power, H., *Experimental Philosophy . . .*, 1664.
Raphson, J., *A Mathematical Dictionary . . .*, 1702.
The History of Fluxions, 1715.
Ray, J., *A Collection of English Proverbs*, 1670.
Observations . . . made in a journey through part of the Low-Countries, Germany, Italy and France . . ., 2 pts, 1673.
A Collection of English words not generally used . . . with catalogues of English birds and fishes . . ., 1674.

A Collection of Curious Travels and Voyages . . . , 2 vols., 1693.
Philosophical Letters . . . , ed. W. Derham, 1718.
Royal Academy of Sciences, *Memoirs for a Natural History of Animals* . . . , tr. R. Waller, 1668.
Another Collection of Philosophical Conferences of the French Virtuosi . . . , tr. G. Haven and J. Davies, 1665.
Royal Society, *Philosophical Transactions, 1* (1665)–*178* (1886); pts *A* and *B* from 1887.
Philosophical Collections, 7 pts, 1679–82.
Miscellanea Curiosa, 1705; 3rd ed., 3 vols., 1723–7.
Abstracts of the Papers Printed in the Philosophical Transactions, 1 (1832); vols. *5* and *6* have the title *Abstracts of the Papers Communicated to the Royal Society of London*; from vol. *7*, the title is *Proceedings of the Royal Society*; pts *A* and *B* from vol. *75* (1905).
Sorbière, S., *A Voyage to England* . . . , 1709.
Sprat, T., *The History of the Royal Society of London* . . . , 1667.
Stubbe, H., *Legends no Histories* . . . , 1670.
Tyson, T., *Phocaena, or the Anatomy of a Porpess* . . . , 1680.
Orang-Outang, sive Homo Sylvestris . . . , 1699.
Wallis, J., *A Defence of the Royal Society* . . . , 1678.
in: *Peter Langtoft's Chronicle*, ed. T. Hearne, 2 vols., Oxford, 1725.
(Walton, I.), *The Compleat Angler*, 1653.
Willughby, F., *The Ornithology*, tr. and ed. J. Ray, 1678.
Wotton, W., *Reflections upon Ancient and Modern Learning*, 1694.

Recent Publications

Ashmole, E., *His Autobiographical and Historical Notes* . . . , ed. C. H. Josten, 5 vols., Oxford, 1966.
Boyle, R., *A Bibliography*, by J. F. Fulton, 2nd ed., Oxford, 1961.
Bradbury, S., and Turner, G. L. E. (ed.), *Historical Aspects of Microscopy*, 1967.
Clark, G. N., *Science and Social Welfare in the Age of Newton*, 2nd ed., Oxford, 1949.
Cope, J. I., *Joseph Glanvill, Anglican Apologist*, St Louis, 1956.
Dewhurst, K., *Dr Thomas Sydenham (1624–89) : His Life and Original Writings*, 1966.
Evelyn, J., *A Study in Bibliophily with a Bibliography*, by G. Keynes, 2nd ed., Oxford, 1968.
Hooke, R. *The Diary*, ed. H. W. Robinson and W. Adams, 1935.
A Bibliography, by G. Keynes, Oxford, 1960.
Huygens, C., *Treatise on Light*, tr. S. P. Thompson, 1912.
Koyré, A., *From the Closed World to the Infinite Universe*, Baltimore, 1957.
Leyden, W. von, *Seventeenth Century Metaphysics*, 1968.
Middleton, W. E. K., *The History of the Barometer*, Baltimore, 1964.
The History of the Thermometer, 1967.
Nicolson, M., *Science and Imagination*, Ithaca, 1956.
Oldenburg, H., *The Correspondence*, ed. A. R. and M. B. Hall, Madison, Wis., 1965–.
Purver, M., *The Royal Society : Concept and Creation*, 1967.
Raven, C. E., *John Ray, Naturalist* . . . , Cambridge, 1942.
Ray, J., *A Bibliography*, by G. Keynes, 1951.
Rhys, H. H., *Seventeenth Century Science and the Arts*, Princeton, 1961.

Additional Reading

Hall, A. R., *Philosophers At War*, Cambridge, 1980.
Hunter, M., *The Royal Society and Its Fellows 1660–1700*, 1982.
Jacob, J. R., *Henry Stubbe, Radical Protestantism and the Early Enlightenment*, Cambridge, 1983.
Jacob, M. C., *The Newtonians and the English Revolution 1689–1720*, Hassocks, 1976.
Stewart, M.A. (ed.), *Selected Philosophical Papers of Robert Boyle*, Manchester, 1979.
Robinson, N. H., *The Royal Society Catalogue of Portraits*, 1980.

20 Noah's Ark (Burnet, T., *Theory of the Earth*, 1684, pl facing p 101)

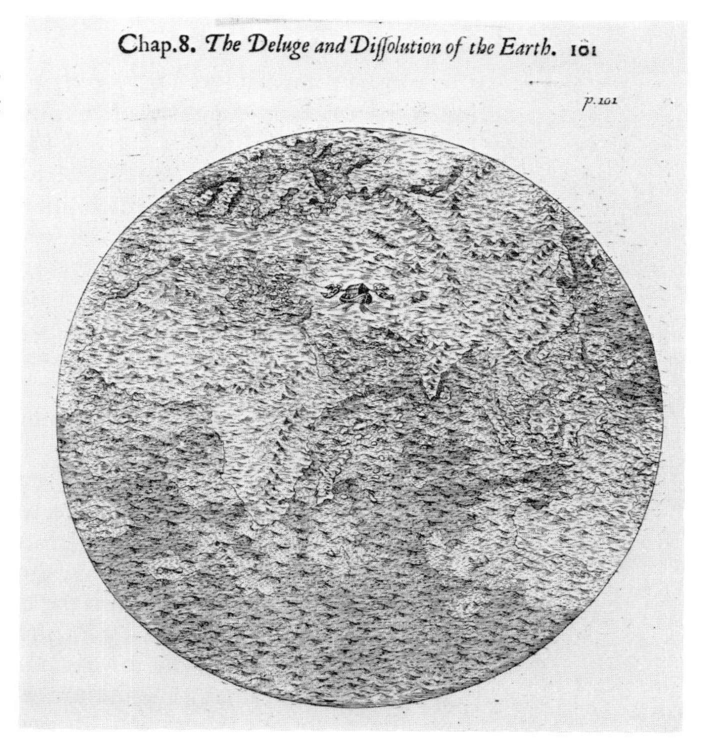

3 *The Rise and Fall of Natural Theology*

We have seen Bacon and Wilkins trying to keep theology in its place, and leave empirical investigations to be freely performed by scientists. In this objective they achieved success; and it would seem to be the case that until the latter part of the nineteenth century there was nothing that could be described as opposition by the Church to science in England. Bacon, Bishops Wilkins and Sprat, and the founders of the Royal Society generally, believed that God was revealed in the Creation as well as in Revelation; and that therefore this increase in scientific knowledge could not but lead to the fortifying of religion. Natural Theology – the demonstration of the existence of God, and of his attributes, from the study of the world – began to have an importance, both in scientific literature and in works designed for edification, which it had not had before, and has not had for the last hundred years. Perhaps the majority of scientific books written in English between 1650 and 1850 contain references to the Creator, and praise for his capacities as a Designer; however, we shall not treat as genuine Natural Theology those which confine such remarks to the Preface or concluding paragraph, but only those which contain more detailed and sustained argument.

It would be a mistake to suppose that Natural Theology, though in this period a particularly English phenomenon, arose out of the Baconian philosophy of science. Its roots are to be sought in antiquity, in the writings of Plato and the neo-Platonists, and in Aristotle and Galen. In the ever popular *Consolations* of Boethius, of which numerous English translations were made by authors beginning with Chaucer, we find arguments for the existence of God from the design of the cosmos very similar to those used by Cudworth and the Newtonian Richard Bentley over a thousand years later. And the remarks of Aristotle, on the absurd hypothesis of Empedocles that the parts of animals might have come together by chance, provided ammunition ready to hand for those who, in the seventeenth and eighteenth centuries, wished to counter the mechanistic biologies of Descartes and Hobbes. It was upon the orderly courses of the stars and planets, the convenience of our terraqueous globe and the design and contrivance manifest in organisms, that the natural theologians chose, in general, to dilate.

Usually, in accordance with the Platonic tradition, they argued that the World clearly exemplified the Divine Order and Goodness, and that Evil was unreal or only apparent because the consequence of some greater good. An exception was Thomas Burnet who, in his proto-geological *Sacred Theory of the Earth*, tried to show why it was no longer, since the Fall, a completely satisfactory globe on which to live. Burnet's book well illustrates the difficulties of doing geology at this period, for the words of *Genesis* loom larger than any empirical evidence. It is strange that, following Augustine on the exposition of the Scriptures, divines in England should have made no serious difficulties over the reception of the Copernican System, and yet continued to credit the verbal inerrancy of Biblical geology and biology. Burnet felt that an ideal Earth would not be disfigured by mountains, and would not have its axis inclined. He was also concerned at the problem of the origin of all the water of the Flood; for forty days of very hard rain would hardly cover mountains even of very moderate size. And if this water had appeared from somewhere, it was unclear where it could have gone as the Flood receded. The clue was that the fountains of the deep were said to have been opened; and following this Burnet was able to construct a self-consistent deductive system which accounted for the Flood and even allowed for the second catastrophe, by fire.

According to Burnet, as the Earth had been formed, the earthy material had gone to the centre, and the water had collected evenly on top of it. A scum of fatty material had floated to the surface of the water, and this, mixed with dust, congealed to form a thick, spherical skin, which formed the surface of the primeval Earth. The Earth's axis was vertical, and the zone around the Equator therefore torrid and impassable; the Garden of Eden was in one hemisphere, and on their expulsion from it Adam and Eve had been driven to the other. Though not so agreeable as the Garden, even this hemisphere was a pleasant enough place to live: flat, well watered, fertile, and so equable in climate – because the Earth's axis was not inclined – that the Patriarchs lived to the great ages described in Scripture. Then came the catastrophe. Softened by rain, and cracked by heat, the fatty crust collapsed, and the mountains were formed from its debris. The Earth as we see it is a ruin, and will only be restored to a state close to its paradisiacal condition – but stable for a thousand years – following its conflagration. Finally it will become a fixed star. The book is illustrated with figures, including a delightful one of the tiny Ark among gigantic waves, coming to rest upon a mountain-top. The work was assailed by Keill, Croft, and Warren; and defended by Beverley.

Burnet's whole approach ran counter to the general tradition of natural theology, the object of which was to establish that this was the best of all possible worlds; and the idea that the Earth was a ruin proved generally unacceptable. In Cudworth's *True Intellectual System* we find many of the arguments for Design in the Universe which were to prove important in later authors, as well as very interesting discussions on teleology. But it was to be the scientists themselves who were most influential; Boyle and Newton – through Bentley – for arguments for the Divine existence from the physical sciences, and John Ray for biological evidence. Cudworth wrote at the end of his book: 'To *Believe a God*, is to *Believe* the *Existence* of all *Possible Good* and *Perfection* in the Universe; It is to Believe, That *things are as they Should be*, and that the World is so well *Framed* and *Governed*, as that the Whole System thereof, could not Possibly have been Better.' There can be little doubt that this tenet guided scientists in England – who did believe in God – well into the nineteenth century. They therefore took particular pains to seek out Order and Design in the Universe; and, conversely, as they discovered fresh domains of law and regularity or new laws of greater simplicity, they eagerly pointed them out as new evidence for the goodness of God.

In Plato's *Laws*, in Boethius, in Cudworth, and in numerous other works, it was argued that the order which is manifest in the planetary motions could not possibly have arisen from the chance motions of atoms. But the progress of astronomy, particularly in the seventeenth century, revealed how much simpler were the laws governing the motions of these bodies than had previously been supposed. Newton's Laws of Motion and his inverse square Law of Universal Gravitation accounted for all terrestrial and celestial movements. That this degree of order should prevail throughout the cosmos was even clearer evidence for God's existence, wisdom, and power, as Richard Bentley perceived. He was a young classicist who was to become master of Trinity College, Cambridge, but was in 1691 chaplain to Stillingfleet, bishop of Worcester. Boyle, who died at the end of 1691, left money to found a lectureship in which the new science was to be used to establish the existence of God; and Bentley was asked to deliver the first series of eight in 1692. The fourth lecture was concerned with biology, and particularly with recent microscopical discoveries demonstrating Design among the minutest of creatures. But the final lectures were the most interesting and aroused most attention; for in them he gave one of the first popular accounts of the

Newtonian system, and argued from it to God. Since he knew little physics, he had written to Newton asking for help; and Newton's letters, which were published in 1756, are more interesting than Bentley's lectures. Dr Johnson's remark about them is well known; that they show how even the mind of Newton gained ground gradually upon the darkness. In particular, they contain Newton's famous remark that inanimate brute matter could not act at a distance; the remark which Faraday loved to quote as he worked out his ideas on lines of force. Newton and Berkeley were among the subscribers to Richard Cumberland's *Laws of Nature*, refuting Hobbes' materialism.

Newton's ideas were quite well reproduced in Bentley's lectures, which seem to have been accounted very successful; particularly by Bentley himself, who declared that atheists no longer dared to show themselves under their true colours, but called themselves deists instead. It is hard now to feel much sympathy with lectures of this kind, and easy to see how Benjamin Franklin could have been converted to deism by a Boyle lecturer's attempted refutation of it; but Bentley's lectures are an excellent example of the genre, and can be read with some pleasure. The same is true of another examination of the Newtonian system, the correspondence between Leibniz and Samuel Clarke. This raises philosophical problems, especially concerning space and time, which are still of great interest. The occasion of the correspondence was a letter from Leibniz to Princess Caroline – who became George II's queen – alleging that Newtonian physics weakened religion. Leibniz's first objection was that God, according to Newton, had to intervene to prevent the attractions between the planets from causing instability in the solar system. Laplace a century later showed that in the long run these perturbations would cancel out without Divine interference; and he therefore 'had no need of that hypothesis'. Leibniz simply argued that we expect a well-made clock to keep time without the maker having to tinker with it; and how much more would we expect the best of all possible worlds to run without needing its Maker's interference. Leibniz next objected to Newton's remark that space was, as it were, God's sensorium, in which he perceived all that was going on; and to Newton's view that space and time were absolute, Leibniz argued that space and time were simply relations between things and events, and could have no independent existence.

Clarke replied on behalf of Newton, who seems, nevertheless, to have had a hand in drafting the replies. He stuck to the view that God needed to intervene in the universe to prevent perturbations getting out of hand and to maintain constant the quantity of 'motion', which would otherwise decay; by 'motion' Newtonians meant 'momentum'. Liebniz insisted upon the importance of 'vis viva', which corresponds to what we call kinetic energy. The controversy as to which was the more important was not resolved until d'Alembert showed that in some interactions momentum is conserved, and in others energy. Clarke argued that to postulate a world in which God never intervened was a half-way stage to atheism. The correspondence ended at the death of Liebniz, in what has been described as a draw. The progress of physics in the eighteenth and nineteenth centuries made it increasingly difficult to credit Divine interventions of the kind envisaged by Newton and Clarke; and most writers on natural theology continued in the Platonic tradition to deduce God's existence and goodness from the orderliness of nature rather than from sudden exertions of Divine power. The exceptions to this were the alleged creations from time to time of new species of animals and plants: and that spectacular catastrophe, Noah's Flood. But these things were not of great importance in writings on natural theology until the later eighteenth century, when Whitehurst and Howard made great use of the Deluge, while Douglas tried to show that it could not explain all the facts.

In geology, there was the system of William Whiston which resembled that of Burnet in being thoroughly deductive. But Whiston accounted for the phenomena in terms of a near-collision between the Earth and a comet. Whiston was Newton's successor in the Lucasian Chair of Mathematics at Cambridge. He was dismissed for his over-zealous propagation of unitarian beliefs; Newton shared these heterodox tendencies, but prudently kept them to himself. Whiston's writings seem to fall better under the heading of natural theology than under science, since their empirical basis is not great, and they are full of the emphases we have already met in this chapter. In the seventeenth century there was considerable interest in fossils, and antiquaries did a certain amount of what one could call field geology; but a true science of geology was still to be born. In 1695 Woodward proposed a system in which all minerals had crystallized from solution. In natural history generally, and particularly in botany, the situation was very different. Animals and plants had been considered as emblems or signatures; and whether or not a unicorn or a basilisk or some kind of plant could actually be found was of relatively minor interest. With the seventeenth century came the return to the Aristotelian interest in the functions of the various parts of animals; and it became clear to John Ray and the microscopists that it was impossible to suppose that animals, in which every part appeared to make its contribution to the whole, could have been generated by mere chance. In his *Wisdom of God Manifested* and in his *Miscellaneous Discourses*, which appeared in 1691 and 1692 respectively, Ray argued for God's existence and attributes from the realm of biology. As time passed, arguments which were fresh and genuine in Ray became, on repetition by other authors, increasingly bland and sometimes almost perfunctory.

Examples and arguments derived from Ray and from Newtonian physics formed the basis of two of the most popular books on natural theology of the eighteenth century: the *Physico-Theology* and *Astro-Theology* of William Derham. Derham was himself a naturalist; he was the vicar of Upminster, near London, and a Fellow of the Royal Society. He edited works of Ray, Hooke and Albin, and published a number of scientific papers in the *Philosophical Transactions*. He was therefore well qualified to write on natural theology; nevertheless, although there is little point in mocking long-dead authors, it is difficult always to take his books seriously as contemporaries evidently did. This is particularly true of the *Physico-Theology*, which begins with reflections upon the Earth's shape, the happily chosen density of the atmosphere, and the admirable distribution of land and sea. Volcanoes were indeed horrid; but it seemed possible that they were safety-valves, and that their eruptions prevented worse catastrophes which would otherwise have occurred. Mountains, which were 'generally found fault with', Derham believed to be, on balance, beneficial. They excited noble thoughts; suited some constitutions, and some species of plants and animals; and made rivers flow quickly, so that they did not stagnate or stink. Providence maintained the balance of nature, and among men kept populations approximately constant, providing a surplus of boys to fight wars. Ichneumons, which lay their eggs in living caterpillars, were 'an interesting provision'; poisonous snakes were indeed disagreeable, but they were mostly found in heathen countries, and rattlesnakes at least gave a warning.

Derham is more interesting when he makes the point that nature is far more varied than it might have been. A great variety of soils and climates throughout the world support an enormous variety of plants and animals. Atheists could not explain why in similar situations the same fauna and flora is not always found; nor why the faces, voices, and handwritings of men all differ. The instincts of animals – particularly that of bees leading them to make cells of the precise form a mathematician would choose – and the way their parts worked together proved that they had been designed by a beneficent Creator. The beauty of feathers and scales under the microscope made our best efforts look crude, bungled, and botched. It is by Design that animals all eat different foods; that their mouths, limbs, and digestive organs are adjusted to their food; that they have specialized organs, like the beaks of the woodpecker and crossbill; and that living creatures are all provided with means of self-preservation. Within the framework of a work of edification, Derham succeeded in providing a considerable amount of solid natural history; and anybody who read this book would have learned quite a lot of biology. While it does not have the freshness of the writings of Gilbert White, or of Ray, it smells much less of the lamp than do the writings of most of Derham's successors.

Works dealing with natural history from the point of view of natural theology seem designed to send one to one's knees rather than to the laboratory or into the fields; but this tendency is much stronger in astronomical works. Those without a rigorous mathematical training, or genius for the making and using of telescopes, could not hope to contribute seriously to astronomy; whereas a country clergyman could advance biology by careful observations. Popular astronomy therefore concentrated on the sublime aspects of the science, and on the splendid order revealed among the celestial orbs by the researches of Newton. At most, one might be expected to contemplate the starry heavens on a clear night as evidence of God's majesty and wisdom; and one might, in the tradition of Fontenelle, speculate about the inhabitants of those distant globes. Whereas natural theology might lead to active nature study, it was unlikely to produce anything other than the passive pursuit of astronomy.

Derham's *Astro-Theology* reveals this tendency. It contains the interesting distinction between the Ptolemaic, Copernican, and 'new' systems; the first two are both 'closed' systems, bounded by a sphere of fixed stars, and differed only in that in the one the Earth and in the other the Sun was at the centre. In the new system, the universe was infinite, and stars – of which the Sun was typical – were scattered about in this infinite Euclidean space. Derham had made certain observations of the heavens himself, using Huygens' telescopes, and he believed the solar planets and even the Moon to be the abode of rational beings, for much the same reasons as Huygens had. He disagreed with Huygens in supposing that the Moon had seas on it and was therefore habitable. There is little suggestion in Derham's works, and in those of his successors, of the doctrine of the inscrutability of God; in this he differs from such precursors as Henry More. Writers on natural theology felt no qualms about reading God's mind, and giving praise where they felt it was due, for some particularly ingenious solution to a problem. Such an attitude does not lead to a very profound religion; and it was therefore criticized from about 1800 by those influenced by the Romantic Movement. It had earlier come under attack from David Hume; but many Christians today might well feel that the infidel Hume was closer to genuine theology than the bland natural theologians.

Hume's *Dialogue* represents a masterly demolition of natural theology in the manner of Derham; and yet Paley's works, which Darwin had to study at Cambridge and which he found to be useful, come after Hume's. It was the patient researches of Darwin and not Hume's cleverness which rendered nugatory the theologizing of authors like Derham. Hume argued that to pry into God's essence was next only in impiety to denying his existence. He suggested that to compare God to men was a degradation; that whereas a house indicates a builder, the analogy between a house and the universe is imperfect. Raising his familiar doubts about causation and inductive reasoning, Hume went on to remark that what happens at present in a small corner of the universe is by no means a certain guide to what happened long ago, or what goes on in distant parts. The Divine Mind, he added, could not really be called infinite if it could be argued to only by analogy with human minds. This recalls the seventeenth-century distinction between the infinity of God and the indefiniteness of the universe.

Hume went on to demonstrate that even if one accepted that it was possible to argue for God's existence from the works of nature, it was by no means certain that this God was benevolent or particularly good at world-making. For all we know, the past might have been full of his botched shots; God, like a bungling workman, might have created it by trial and error. We have no data for a cosmogony. Further, if the parts of an animal did not cohere, and were not 'curiously adjusted', then it would perish; such adjustments do not therefore prove that the animal has been designed, but are simply necessary conditions for its existence. Not only that, but nature is red in tooth and claw; perpetual war is kindled among all living creatures; necessity, hunger, and want stimulate the strong and courageous, while fear, anxiety, and terror agitate the weak and infirm. Curious artifices of nature embitter the life of every living being; and the whole world 'presents nothing but the idea of a blind nature, impregnated by a great vivifying principle, and pouring forth from her lap, without discernment or parental care, her maimed and abortive children'. There are here anticipations of *In Memoriam* and of *The Origin of Species*; but in fact Hume did not anticipate Darwin any more than did Empedocles, for Darwin – as T. H. Huxley pointed out – was essentially a teleological thinker, like most scientists. He made it respectable to think again about purposes and functions, admittedly in terms of a new conceptual system.

Before considering Paley, and the Bridgewater Treatises of the 1830s, we should return briefly to the earlier part of the eighteenth century. Hume was unsuccessful in persuading scientists to abandon Christianity; and the same seems to have been true of the Deists like Collins, Toland, and Tindal, who followed Locke in seeking a reasonable religion, and found French disciples. Though they may not always have been wholly orthodox, most men of science in eighteenth-century England were theists. Perhaps the power of Newton's example was responsible for this state of affairs; certainly Dr Johnson made much of this aspect of Newton. Butler's *Analogy of Religion* is in a sense a work of natural theology; but it is very famous, and as it does not set out to teach any of the sciences we shall not need to do more than mention it here. Perhaps one could describe it as a work on the philosophy of science, dealing as it does with the logic of analogy, and applying that to natural religion. The famous 'advertisement', remarking that many persons take it for granted that Christianity has been proved fictitious, must apply more to literary circles, and to wits of the coffee houses, than to scientists.

Among the various scientists who sought evidence for God's existence in the stars, one of the most interesting is Thomas Wright, an astronomer from Durham who is often credited with first correctly explaining the Milky Way. In his *Original Theory* there appears to be the suggestion we see the Milky Way as we do because we live in a disc-shaped galaxy. In the plane of the disc there are therefore many stars, which form a band of light; whereas at right angles to the plane there are far fewer. It is not clear whether this idea was really present in Wright's mind, for his *Original Theory* is written in a rhetorical style, and the main motive for writing it was theological rather than scientific. It has a place in the literature on plurality of worlds – Huygens' *Celestial Worlds* was one of the sources Wright used – but its main interest is the splendid plates, which make it, and his *Clavis Coelestis* (a descriptive work on the heavens) a collector's piece. Wright did not share the view of Huygens that there would be similar life on all planets. Rather he seems to have supposed that the universe was closed, with Heaven at the centre and Hell at the outside; and that after death one might hope to be reincarnated on

21 Stars and their planetary systems (Wright, T., *Original Theory*, 1750, pl 17)

a more central planet. The stars and planets were thus manifest mansions of eternal life; and the eye of God was put in many of the plates, at the centre. Wright was a better draughtsman than astronomer or theologian. He gave various lectures on astronomy, and in the beginning of the *Original Theory* he suggests, as no doubt many other popular lecturers did, that one's knowledge of God could not be said to be adequate unless one knew some of the facts of astronomy. He did himself have an observatory; but it seems unlikely that many of the subscribers to his handsome volumes would have been stimulated into making observations for themselves. A review of the *Original Theory* stimulated Immanuel Kant to produce his nebular hypothesis; but the nebular hypothesis of Laplace, and the explanation of the Milky Way by Sir William Herschel, both seem to have been independent of Wright's work, which was not generally read by scientists.

Paley's *Evidences* and *Natural Theology*, on the other hand, were read by scientists, partly because they became standard textbooks. Later editions – and editions went on appearing through most of the nineteenth century – even contained specimen examination questions appended by helpful editors. Darwin thought that the study of Paley was the most useful part of his formal courses at Cambridge. While he could not be described as an original scientist, Paley was an acute reasoner and he wrote well, if lengthily by modern standards. Because his books were so widely read by scientists and by laymen, they were among the most influential scientific books of the first half of the nineteenth century. Paley's example, that if on a walk one should come upon a watch one would not suppose that all the parts had been formed and come together by chance, and that the world exceeded a watch in the delicacy of adjustment of its parts, and must therefore have been designed, is justly famous, though it was far from new, and is only moderately convincing.

There are some remarks in the text at which one cannot but smile, such as: 'It is the most difficult thing that can be to get a wig made even; yet how seldom is the *face* awry', from the *Natural Theology*. Nevertheless, the reading of Paley would give anybody quite a good grasp of the biology of the day; and today we can read him to see how Darwin turned his arguments, even if we cannot take him altogether seriously in his own right. Most of the argument of the *Natural Theology* is derived from the biological sphere; and chapters such as those on 'prospective contrivances' (parts which are ready before the organism has need for them), on instincts, and on the relations of the various parts and the structure of complex organs like the eye, are very well done. The arguments for the benevolence of God are perhaps least convincing, being based upon the assertions that in most cases where contrivance is observed, it is beneficial, and that 'the Diety has superadded *pleasure* to animal sensations'. For Paley, nature was not primarily a battleground; insects, even greenfly, clearly enjoyed their life. Even more than eighteenth-century authors, natural theologians of the nineteenth century preferred to look on the bright side, though there were some sympathisers as well as enemies, who were ready to berate them for this easy optimism.

The next great surge of activity in the field of natural theology came with the publication of the Bridgewater Treatises, which again seem to have been fantastically popular, and from which a reasonable amount of science would have been absorbed by anybody who had managed to plough through the series. There were eight of them, and a ninth unauthorized one written by Charles Babbage. The eccentric eighth earl of Bridgewater, a prebendary of Durham who lived in Paris surrounded by dogs, left £8,000 to sponsor treatises on natural theology; the authors to be chosen by the President of the Royal Society, who was advised by the archbishop of Canterbury and the bishop of London. There was some adverse comment at the pious authors' getting £1,000 plus interest from the legacy on top of the profits from the sale of the books, all for arguing for the existence of God, when perhaps the case might have been made with more conviction by disinterested persons. But, nevertheless, competent authors were chosen and they did as good a job as could be expected.

The division into eight topics produced some ribald remarks at the time. Four authors – Thomas Chalmers, Peter Mark Roget, William Buckland, and William Kirby – required two substantial volumes to say all they felt to be necessary, so the set came to twelve volumes in all. Chalmers was a Scotsman, famous for his astronomical sermons; but his treatise was not specifically concerned with astronomy, which was alloted to William Whewell, but with 'the Adaptation of External Nature to the Moral and Intellectual Constitution of Man'. Whewell's book on astronomy and general physics was the only one specifically devoted to physical science; and one must admit that the arguments of Bentley and Derham were not greatly improved in Whewell's hands. In the stream of edifying reflections one misses the acuteness of his later books specifically on philosophy of science. Nevertheless, his was one of the best-selling of the Bridgewater Treatises, and seems to have been generally well received. It was criticized by Babbage, because of a passage in which Whewell implies that the more abstract branches of mathematical physics are less prone to lead to piety than are more experimental sciences. Babbage believed, as we shall see in a later chapter, that English science was weak because it was not sufficiently mathematical. To show that applied mathematics was not atheistic in tendency, he composed a fragmentary but very stimulating ninth Bridgewater Treatise, for which he got none of the legacy; it contains interesting remarks on Mossotti's theory of matter, and a discussion of induction in which the World is presented as a computer the programme of which we are attempting to guess. Certainly Babbage's is one of the few lively books on natural theology that appeared in the nineteenth century.

The main emphasis of the authorized Bridgewater Treatises was physiological. The earl of Bridgewater had been convinced, perhaps by watching the cats and dogs who dined at his table, that the human hand was a particularly good example of Design. Accordingly, one volume was devoted to *The Hand*; it was written by the great anatomist Sir Charles Bell, who is famous for his work on sensory and motor nerves. He had already written works on natural theology: *Essays on the Anatomy of Expression in Painting*, in which he attempted to show that various muscles had been designed for the expression of emotions and passions; and the brief *Animal Mechanics*. According to his brother, it was the former studies which led to his sorting out the nervous system of the human body. Darwin's *Expression of the Emotions* was written to show that muscles which now do no more than express emotions had other functions in our remote ancestors, and that Bell's arguments were therefore unsound. Bell's book on the hand is, as might have been expected, something of a curiosity; indeed all the physiological treatises seen from our position are odd because of what seems a wilful avoidance of any theory of evolution.

Roget, the polymath whose *Thesaurus* preserves his name, wrote the treatise on *Animal and Vegetable Physiology*; Kirby, a naturalist and entomologist, that on *The History, Habits and Instincts of Animals*; and Kidd, one of those responsible for getting science taught in nineteenth-century Oxford, that on *The Adaptation of External Nature to the Physical Condition of Man*. The geological treatise was the last to appear; geology was in an exciting state, Lyell's *Principles* having only recently been published. Buckland had,

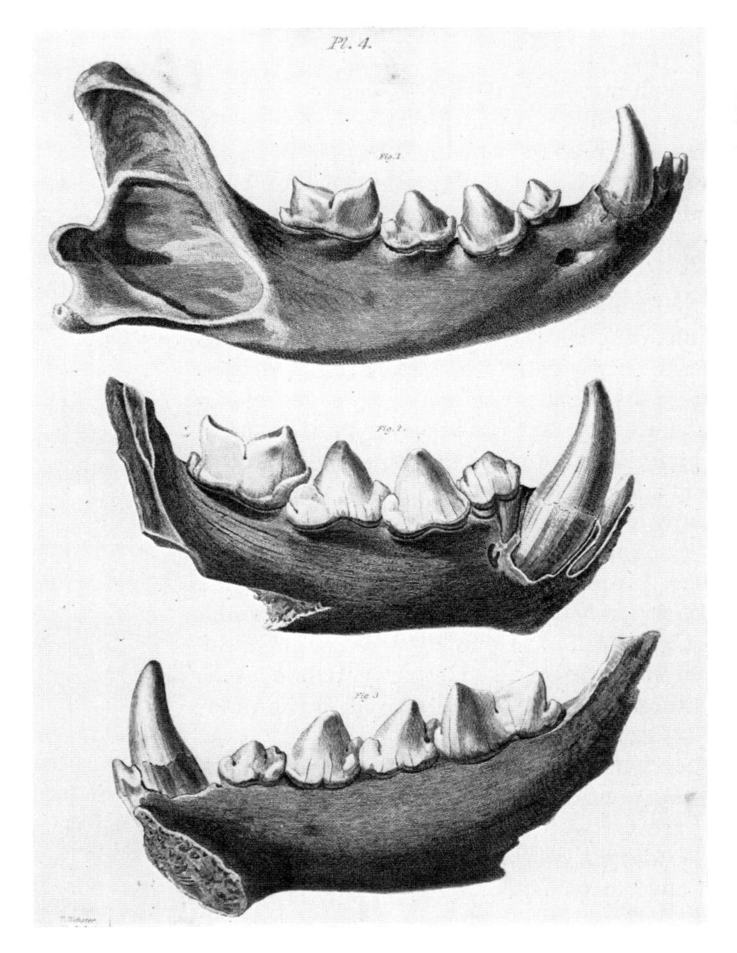

22 Jaws of ancient and modern hyaenas (Buckland, W., *Reliquae Diluvianae*, 1823, pl 4)

in his *Reliquae Diluvianae*, explained the bones found in the Kirkdale Cavern in Yorkshire as carried there by the Flood. It might have been expected that he would continue to propagate Mosaic geology, but in fact his treatise was a sound and up-to-date work, in which Buckland argued that the Bible was not there to teach us science. The eighth treatise was by William Prout, the author of Prout's Hypothesis – that hydrogen is the basic substance from which the world is built up – and the discoverer of hydrochloric acid in the stomach. He was given a strange medley of subjects, *Chemistry, Meteorology, and the Function of Digestion*, and it cannot be said that he had much success in the superhuman task of making a unity out of them. His approach in chemistry was vitalistic in that he asserted some immaterial agent or agents responsible for organizing organic substances into living beings. This seems rather similar to neo-Platonic and Helmontian speculation; but in reality Prout's treatise is fairly solid science.

Chemistry, which in the early decades of the nineteenth century was the most popular of the sciences, was relatively little represented and so were such active branches of physics as the study of light and electricity. In the eighteenth century, Bishop Watson's *Chemical Essays*, of which even Gibbon approved, fit into the general framework of natural theology. Various chemists in later years did write works on natural theology; the best-known are perhaps George Fownes' *Chemistry* of 1844 and George Wilson's *Religio Chemici* of 1862. Fownes held a Chair at University College, London. His book was written for the Actonian Prize, founded under the will of Mrs Acton who was moved by the example of the earl of Bridgewater. Fownes' book cannot be described as very inspiring; Wilson, who became the first Professor of Technology at Edinburgh, criticized its facile optimism. But he was an invalid. His book, written with Browne's *Religio Medici* in mind, is much more interesting: it consists of various essays, of which only one is concerned with chemistry and natural theology, the others including short biographies of Boyle, Wollaston, and Dalton, with discussion of their work. Other material is to be found in his essay 'Chemical Final Causes' in the *Edinburgh Essays*, 1857; and in his *Chemistry and the Electric Telegraph, together with the Chemistry of the Stars* in which he discusses the problem of why some elements should be so common, and others so rare; and why some should be so useful, and others apparently valueless.

The authors of the Bridgewater Treatises were, like Paley, concerned with evidences; and somehow eighteenth-century science was much more suitable to this kind of treatment than that of the nineteenth century, which developed in ways not obvious to most people in 1800. Among the more prescient was Humphry Davy, and his little book *Consolations in Travel* – the title must be modelled on that of Boethius – contains a very different view of the relations of science and religion. It was described by Cuvier as the work of a dying Plato. Davy was the friend of Coleridge, Wordsworth and Sir Walter Scott, and showed the distaste of Romantics for a faith based on 'evidences', and generally for that kind of religion in which reason plays an excessive rôle. Davy insisted that man must submit himself in faith to the inscrutable will of God. If a bird reasoned about the possibility of migration, it would starve in Europe; but relying on instinct, it gets safe to its destination. Davy did not deny that enormous intellectual satisfaction was to be derived from natural knowledge; but he asserted that God, and the phenomena of life, were incomprehensible. He believed that the scientist could only really make discoveries when he approached 'with reverence and awe the substantial majesty of nature'. And he preferred those sciences in rapid and revolutionary development – in his day, chemistry, electricity, and geology – to those – like astronomy – which appeared as sublime and completed edifices, to be passively admired and wondered at.

We shall look at books specifically devoted to geology or biology, but which helped to overturn the natural theology of the Bridgewater Treatises, in a later chapter. But we can conclude this one with a mention of the writings of Baden Powell, father of the defender of Mafeking, Savilian Professor of Geometry at Oxford, and a contributor to that explosive volume of theological papers, *Essays and Reviews*. There he welcomed *The Origin of Species*, published shortly before; and in his *Essays* of 1855 he had similarly welcomed pre-Darwinian evolutionary speculations. His reason for doing so was his fervent belief in the uniformity of nature: the phenomena of physics and chemistry were explained in terms of law, not of arbitrary interventions; and those of geology and biology must be similarly accounted for. Davy, like Wright, had suggested that upon the different heavenly bodies we might expect to be reincarnated, in some form or another, and that all were therefore rather different. Baden Powell's emphasis on uniformity led him to a view closer to that of Huygens. The tone of his writings is cool and Olympian; in the furore over *Essays and Reviews*, his essay was not one of those alleged to be heretical, though he went out of his way to undermine faith in miraculous occurrences. Another essay in the same volume, by C. W. Goodwin, disposed of the Mosaic Cosmogony; this was hardly a new enterprise, but it seemed very shocking at the time. And shortly afterwards the work of Bishop Colenso, proving that there were inaccuracies, to say the least, in quantitative passages in *Genesis* made it impossible for any but a fanatic to continue to believe in the verbal inerrancy of Scripture. The rather ridiculous fuss engendered by these arithmetical researches diverted attention, according to his recent biographer, from Colenso's more interesting moves in the direction of a kenotic Christology. It had become clear, therefore, by the 1860s that the tradition of natural theology which had extended from writers like Derham was barren; indeed that it had been rotten at the centre through the whole nineteenth century. No comparable edifice has been erected to replace it, but with a return of interest in phenomenology it might be possible to reconstruct something, linking man's search for natural knowledge to his deeper concerns.

3 THE RISE AND FALL OF NATURAL THEOLOGY

Babbage, C., *The Ninth Bridgewater Treatise . . .*, 1837.*

Baxter, R., *The Certainty of the World of Spirits*, 1691.

Bell, C., *Essays on the Anatomy of Expression in painting*, 1806.
Animal Mechanics, n.d. (1828?).
The Hand . . . as Evincing Design, 1833.*
Practical Essays, 2 vols., Edinburgh, 1841–2.

Bentley, R., *The folly and unreasonableness of atheism demonstrated*, 8 pts, 1693.

(Beverley, T.), *Reflections upon The Theory of the Earth . . .*, 1699.

Bridgewater Treatises; see Babbage, Bell, Buckland, Chalmers, Kidd, Kirby, Prout, Roget, and Whewell.

Buckland, W. E., *Geology and Mineralogy Considered with Reference to Natural Theology*, 2 vols., 1836.*

Burnet, T., *The Theory of the Earth*, 2 vols., 1684–90; the 4th ed., 1719, was called *The Sacred Theory. . . .*

Butler, J., *The Analogy of Religion to the Constitution and Course of Nature*, 1736.

* Bridgewater Treatises.

Chalmers, T., *On the Power, Wisdom, and Goodness of God . . . in the Adaptation of External Nature to . . . Man*, 2 vols., 1833.*
Church, H., *Miscellanea Philo-Theologica . . .*, 1637.
Cheyne, G., *Philosophical principles of natural religion . . .*, 1705.
Clarke, S., *A Demonstration of the Being and Attributes of God . . .*, 1705.
A Discourse concerning the unchangeable obligations of natural religion . . ., 1706.
A Collection of Papers, which passed between . . . Mr Leibnitz and Dr Clarke . . ., 1717.
Colenso, W., *St Paul's Epistle to the Romans . . .*, Cambridge, 1861.
The Pentateuch and Book of Joshua critically examined, 7 pts, 1862–79.
Collins, A., *A Discourse of Free Thinking*, 1713.
Croft, H., *Some animadversions upon . . . The Theory of the Earth*, 1685.
Cumberland, R., *A Treatise on the Laws of Nature . . .*, tr. J. Maxwell (1727).
Derham, W., *Physico-Theology . . .*, 1713.
Astro-Theology . . ., 1715.
Douglas, J., *A Dissertation on the Antiquity of the Earth*, 1785.
Fownes, G., *Chemistry as Exemplifying the Wisdom and Beneficence of God*, 1844.
Grew, N., *Cosmologia Sacra . . .*, 1701.
Howard, P., *The Scriptural History of the Earth and of Mankind . . .*, 1797.
(Hume, D.), *Philosophical Essays concerning Human Understanding*, 1748.
Dialogues concerning natural religion, 1779.
(Hutchinson, J.), *Glory or Gravity essential, and mechanical*, 1733.
Philosophical and Theological Works (ed. R. Spearman and J. Bate), 12 vols., 1748–9.
Keill, J., *An Examination of Dr Burnet's Theory of the Earth . . .*, Oxford, 1698; 2nd ed., 1734, has annexed to it a translation of Maupertuis' *Dissertation on the different figures of the celestial bodies.*
Kidd, J., *On the Adaptation of External Nature to the Physical Constitution of Man*, 1833.*
Kirby, W., *On the . . . History, Habits, and Instincts of Animals*, 1835.*
Locke, J., *The Reasonableness of Christianity*, 1695.
Malebranche, N., *Treatise concerning the search after truth*, tr. T. Taylor, 2 vols., 1694.
Mather, C., *The Christian Philosopher . . .*, 1721.
More, G., *A Demonstration of God in his Workes . . .*, 1597.
Paley, W., *A View of the Evidences of Christianity . . .*, 2 vols., 2nd ed., 1794; 1st ed. not seen.
Natural Theology . . ., 1802.
P(atrick), S., *A Brief Account of the new Sect of Latitude-men Together with some reflections upon the New Philosophy*, Cambridge, 1662.
Prout, W., *Chemistry, Meteorology, and the Function of Digestion*, 1834.*
Ray, J., *The Wisdom of God manifested in the Works of the Creation*, 1691.
Miscellaneous Discourses concerning the Dissolution and Changes of the World, 1692.
Roget, P. M., *On Animal and Vegetable Physiology Considered with Reference to Natural Theology*, 2 vols., 1834.*
Steno. N., *The Prodromus to a dissertation concerning solids Naturally Contained within Solids*, tr. H.O., 1671.
Temple, F., *et al.*, *Essays and Reviews*, 1860.
Tindal, M., *Christianity as old as the Creation . . .*, 1730.
Toland, J., *Christianity not Mysterious . . .*, 1696.
Warren, E., *Geologia: or, a Discourse concerning the Earth before the Deluge . . .*, 1690.
Whewell, W., *Astronomy and General Physics Considered with Reference to Natural Theology*, 1833.*
Whiston, W., *A New Theory of the Earth . . .*, 1696.
Astronomical Principles of Religion . . ., 1717.
Whitehurst, J., *An Inquiry into the Original State and Formation of the Earth*, 1778.

* Bridgewater Treatises.

Wilson, G., *Religio Chemici*, 1862.
(Wollaston, W.), *The Religion of Nature delineated*, 1722; another ed., 1724; under Wollaston's name, 1725.
Woodward, J., *An Essay towards a natural history of the earth . . .*, 1695.
Wright, T., *Clavis Coelestis . . .*, 1742.
An Original Theory or new hypothesis of the universe . . ., 1750.
Second Thoughts, ed. M. A. Hoskin, 1968.

Recent Publications

Barbour, I. G., *Issues in Science and Religion*, 1966.
Bentley, R., *A Bibliography*, by A. T. Bartholomew, Cambridge, 1908.
Chadwick, O., *The Victorian Church*, 1966–70.
Cockshutt, A. O. J. (ed.), *Religious Controversies of the Nineteenth Century: Selected Documents*, 1966.
Cragg, G. R. (ed.), *The Cambridge Platonists*, New York, 1968.
Dillenberger, J., *Protestant Thought and Natural Science*, 1961.
Hinchliff, P., *John William Colenso*, 1964.
McAdoo, H. R., *The Spirit of Anglicanism: A Survey of Anglican Theological Method in the Seventeenth Century*, 1965.
Newsome, D., *Godliness and Good Learning*, 1961.
The Parting of Friends, 1966.
Reardon, B. M. G., *Religious Thought in the Nineteenth Century . . .*, Cambridge, 1966.

Additional Reading

Brock, W. H., *From Protyl to Proton*, Bristol, 1985.
Cosslett, T. (ed.), *Science and Religion in the Nineteenth Century*, Cambridge, 1984.
Crowe, M.J., *The Extraterritorial Life Debate*, Cambridge, 1986.
Cannon, S. F., *Science in Culture*, New York, 1978.
Davies, P., *Other Worlds*, 1980.
Dawkins, R., *The Blind Watchmaker*, 1985.
Dick, S. J., *Plurality of Worlds*, Cambridge, 1982.
Jay, E., *Faith and Doubt in Victorian Britain*, 1986.
Knight, D. M., *The Age of Science*, Oxford, 1986.
Kohn, A., *False Prophets*, Oxford, 1986.
Oppenheim, J., *The Other World*, Cambridge, 1985.
Polkinghorne, J. C., *One World*, 1986.
Redoni, P., *Galileo Heretic*, tr R. Rosenthal, 1988.
Russell, C. A., *Science and Social Change 1700–1900*, 1983.
Cross Currents, 1985.
Thomas, K., *Religion and the Decline of Magic*, 1971.
Man and the Natural World, 1983.

4 *The Reception of Newtonian Physics*

Newton's stature was so great that eighteenth-century scientists had perforce to live in his shadow; and in a sense all can be described as Newtonians. Optical and chemical discussions in the early years of the nineteenth century were still dominated by the remarks, or *obiter dicta*, of the illustrious Sir Isaac. By the time of Newton's death, his physics was generally accepted in Britain, and in Holland; but it did not prevail in France until Voltaire and Maupertuis succeeded in destroying Cartesian physics. Newtonians fell into various groups. First came those who sought simply to popularize the Newtonian system; to ensure that everybody knew how the celestial motions were the result of a single law, and that white light was composed of rays differently refrangible. Well-written examples of this *genre*, especially if illustrated, are still agreeable and interesting to read; for they were not dry textbooks – such things hardly existed – but were intended for an educated general readership. Some of these, such as Colin Maclaurin's *Account of Sir Isaac Newton's Philosophical Discoveries*, went beyond exposition of Newton and contained some original matter in the way of developments of his physics. Other groups of Newtonians sought to extend the boundaries of science by applying what they conceived to be the methods of the master; quoting liberally from his writings, they ventured into regions which he had not explored.

Newton's optical researches appeared first in the *Philosophical Transactions*. Not until 1704, after Hooke's death, was the *Opticks* published. The experiments described in the book are masterly; particularly those with the prism, establishing the composite nature of white light. The treatment reveals Newton as a great experimentalist; and it reveals how much progress could be made in understanding the phenomena of light and colours without explicit commitment to one hypothesis as to the nature of light. But in fact the attention of the reader, both then and now, is perhaps most engaged by the queries which are appended to the book. The first edition contained sixteen of these, all relating to light and colours. All of them were put in the form of a question, but were naturally taken to be expressions of Newton's own hypothetical views. In later editions the number and scope of the queries was widened; and one of the tasks which the Newtonians set themselves was to try to answer them. In the Latin translation of the *Opticks* of 1706 there appeared queries relating to the ether; presenting a theory not unlike that of Newton's famous letter to Boyle of 28 February 1678/9, published in Birch's edition of Boyle's *Works*, and of the second optical paper which may be found in Birch's *History of the Royal Society*, vol III. By the fourth edition of the *Opticks*, which appeared in 1730, not long after Newton's death, the number of queries had risen to thirty-one. They contain Newton's views on the constitution of matter, discussions on the cause of gravity, and remarks on the nature of light; and indicate that he was not averse to hypotheses provided they were kept distinct from the facts.

The *Opticks*, as a superb example of the union of theory and experiment, of mathematical presentation and hypothetical queries, was very widely read in the eighteenth century. It provided a useful paradigm for workers in other fields; appropriately so in the case of electricity, but less usefully perhaps in psychology. The popular appeal of the *Principia* was considerably less. Anybody who cared to make the effort could read the *Opticks*, but the *Principia* demanded a course of mathematics first. Moreover, it was in Latin; and three Latin editions had appeared, with such important additions as Roger Cotes' preface directed against Leibniz and the Cartesians, and Newton's General Scholium concerned *inter alia* with the rôle of God in the Universe, before an English translation by Andrew Motte was published in 1729, after Newton's death. The important correspondence between Newton and Cotes concerned with the preparation of the second edition of the *Principia* was published by Edelston in 1850; Cotes' *Hydrostatical and Pneumatical Lectures* was published in 1738, long after his premature death. The *Principia* was a daunting work to anybody; its conclusions needed to be popularized before they could be generally accepted. Bentley's Boyle lectures represent one of the earliest attempts at popularization, though Bentley, like John Locke, could not follow all the mathematics himself. In his review of the *Principia* in the *Philosophical Transactions*, Edmond Halley tried to present its conclusions in non-mathematical terms; and in the eighteenth century there appeared a series of works by authors of various nationalities, attempting to spread knowledge of the Newtonian natural philosophy to those who lacked the mathematics necessary for reading Newton himself.

The first biography of Newton was Fontenelle's obituary, which appeared in English in 1728. This is much more interesting than one expects an *éloge* to be – though the Académie des Sciences had high standards in this respect – because while appreciating Newton's stature as a mathematician and an experimentalist, Fontenelle remained to the end a Cartesian. To him, the 'action at a distance' which seemed a postulate of Newtonian physics represented a return to the occult qualities of the Aristotelians. He demanded of explanations in physics that they be clear and mechanical, as can be seen in his own *Plurality of Worlds*. To such folk it appeared that Newton, in producing the equations which simply and correctly predicted the positions of the heavenly bodies, had not explained anything. Only when the mathematics was linked to a mechanical account of gravitation could Newtonian physics be said to be complete. It is amusing that this attitude is in fact Aristotelian. It was generally taken for granted before the Scientific Revolution that the mathematician might make any assumptions he chose; all that mattered was that he should produce the right figures in the end. The physicist, on the other hand had to try to tell a true story, and his postulates must be examined with due care. Our cautious talk about models and their applicability, instead of truth or falsehood, indicates a further change in outlook, and the story of the reception of the Newtonian-Galilean physics is therefore to some extent a history of the replacement of one view of physics and mathematics by another.

In France, Cartesian physics, with its clear mechanisms but poor fit with the facts, survived well into the eighteenth century, until Maupertius and Voltaire succeeded in replacing it by the Newtonian system. In England Newton was by the end of his life generally recognized as the greatest of scientists. He received that accolade reserved for mathematicians in the form of anecdotes, recorded by Joseph Spence, about his inability to add and subtract. Voltaire's remark, in his *Letters on England*, that in Paris the world was full (a plenum) and in London empty (largely void) is familiar. But it would be a mistake to suppose that the Cartesian system had been completely displaced in this country. The syllabus at Cambridge remained Cartesian although Newton was the university's most famous son.

The textbook used there was the *Physics* of Jacques Rohault, which had been translated into Latin by John Clarke. It had first appeared in 1671. There was little in it that was new, except a defective theory of capillary attraction, but as a textbook it was an enormous success. It went through numerous editions, and when an English translation was prepared, including Samuel Clarke's notes, it saw three editions. The notes were drawn from Newton's philosophy; at first they appeared at the back of the book, but by the third edition of 1737 they had been moved to the bottom of the page to which they refer. They had also been expanded, to such an extent that on some pages there is more footnote than text. Frequently the note flatly contradicts the passage to which it refers. Thus readers of what appeared to be a Cartesian textbook would have found themselves exposed willy-nilly to the physics of Newton.

Clarke argued for atomism; for matter being essentially massy, against the Cartesian view that it was simply extension. And, in the same vein as Cotes in his Preface to the *Principia*, he showed the futility of assuming vortices and subtle matter. In particular, the motion of comets was according to Descartes random, but Newtonians had shown that they moved regularly. Their very eccentric orbits, inclined to the ecliptic, would not fit into the scheme of vortices, and if the depths of space were not void, or nearly so the comets would not return with such predictability. The vortices were therefore incompatible with the cometary motions, with Kepler's laws, and with the

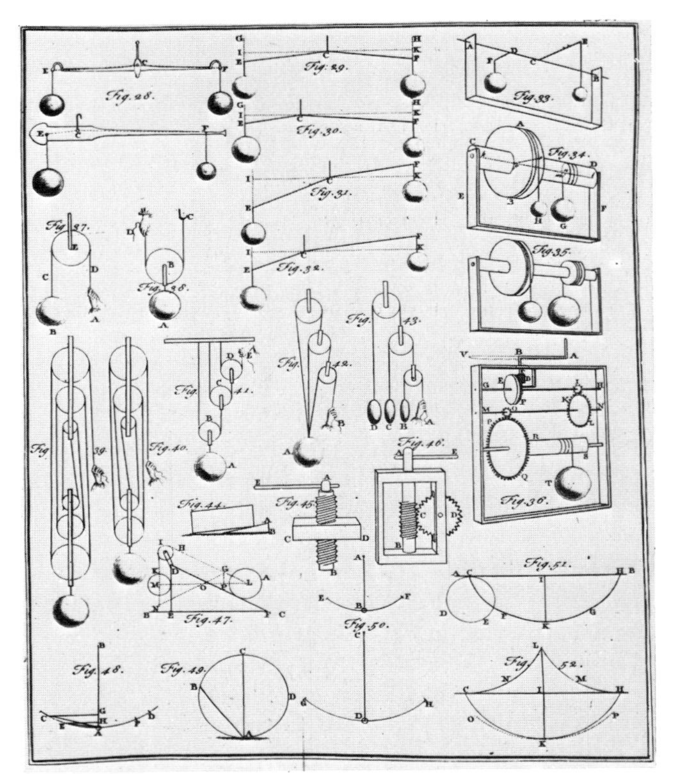

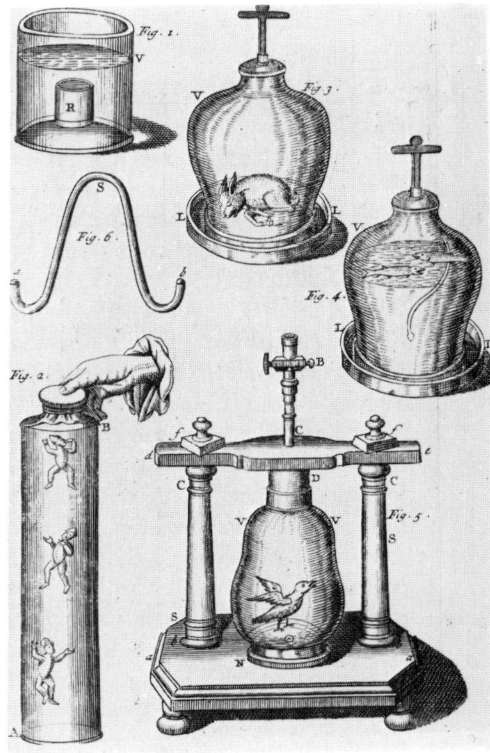

23 (*left*) Illustrations of the general principles of mechanics (Pemberton, H., *A View of Sir Isaac Newton's Philosophy*, 1728, pl facing p 94)

24 (*right*) Experiments on air-pressure ('s Gravesande, W. J., *Mathematical Elements of Natural Philosophy*, 1754, 2nd ed, vol I, pl 31)

rates of rotation of the planets; and moreover if there were vortices they would soon come to rest. We must accept instead, Clarke declared, that the planets 'are so placed in the most free and open spaces, as to revolve about certain centres by a force compounded of gravity and a projectile motion in straight lines'. It is worthy of remark that the biological parts of Rohault's treatise are much less annotated than the physical sections. The Cartesian programme of searching for mechanical explanations in biology was evidently more acceptable than the astronomical mechanisms.

By 1737 it was no longer necessary, unless demanded by some syllabus, to learn Newtonian physics from the footnotes of Rohault; for if the *Principia* itself, in English or Latin, was too demanding, there were good popularizations already available. Probably the most famous is that by Henry Pemberton, who had helped Newton to prepare the third Latin edition of the *Principia*. His *View of Sir Isaac Newton's Philosophy* appeared in 1728; the preface contains some anecdotes of Newton, including the story that he hit upon the idea that the Moon was subject to the Earth's gravitation while sitting in the garden. Voltaire's *Letters on England* added the information about the falling apple.

Pemberton's book is an excellent summary. It contains Newton's Rules of Reasoning in Philosophy, and in the Introduction a discussion of what is involved in the mathematical method of pursuing science. Book I deals, like the first book of the *Principia*, with the motions of bodies in general; Book II with the planetary motions; and Book III with light and colours. So the reader of Pemberton would have emerged with a good idea of the whole Newtonian achievement, in the realms of physics. The presentation is relatively non-mathematical, and no doubt many readers skipped the mechanics of Book I. Because Pemberton had been associated with Newton, his remarks on what Newton believed are of great interest and the anecdotes are as well founded as can be expected from an old man recollecting the events of half a century before. Pemberton tells us that Newton admired Huygens' presentation, although he did not accept his theory of light, and that he distrusted the algebraic methods of Descartes, preferring the rigorous demonstrations of the ancient geometers. Newton's own *Treatise of the System of the World* appeared in English in 1728, with a second edition in 1740; Cajori believed that the translator was Andrew Motte, and he appended the work to his revision of Motte's translation of the *Principia*. It presents the conclusions of Book III of the *Principia* in a rather easier form.

After Pemberton's probably the best-known popularizations of Newtonian science are those by Voltaire and by Count Algarotti, both of which were translated into English. Neither is on the same level as Pemberton's. Voltaire had previously, in his *Letters on England*, described some general features of the Newtonian system; his *Elements of Sir Isaac Newton's Philosophy* is clear and amusing, and well calculated to wean away from Cartesian views those who were interested in science but had no intention of becoming scientists. Voltaire's Newton is a figure of the Age of Reason, rather than the full Newton wrestling with metaphysical problems, absorbed in alchemical and chronological speculation, and leaving loose ends in the queries for his successors to pick up. Voltaire tells us how planets having satellites can be 'weighed'; that is, how their mass relative to that of the Earth, can be determined. The relative distances of planets from the Sun could also be easily computed, from Kepler's Third Law; to convert these relative figures into pounds and miles was one of the major tasks which confronted astronomers in the eighteenth century. The Earth was weighed by Henry Cavendish, in a famous experiment described in the *Philosophical Transactions* for 1798, in which he employed a torsion balance, and by Nevil Maskelyne's experiments to see how much the mountain of Schiehallion in Perthshire pulled a plumb-line out of true. The distance between the Earth and the Sun was determined, with considerably less accuracy, from observations on the transits of Venus across the Sun in 1761 and 1769; this will be discussed in more detail later.

Algarotti's popularization of Newton was written for ladies, after the example of Fontenelle's *Plurality of Worlds*; children could learn Newtonian physics from the works of Wells, Tom Telescope, or Benjamin Martin. Serious scientists learned their physics from authors such as W. J. 's Gravesande,

P. van Musschenbroek and J. T. Desaguliers. 's Gravesande's lectures on the Newtonian system, delivered in the university of Leyden, were translated into English; and so was his *Mathematical Elements of Natural Philosophy* the 'authorized' translation of which appeared in two volumes in 1720–1, and went through six editions. It contains a number of handsome plates illustrating scientific apparatus which was then less austere in appearance than it is today. Musschenbroek's handsome textbook, which appeared in 1744, is also pleasantly illustrated. Desaguliers, who translated 's Gravesande, himself wrote *A Course of Experimental Philosophy*, which is an excellent source of information on the state of the physical sciences in the first half of the eighteenth century. Particularly interesting are his investigations of the weight a man can carry if properly loaded. Desaguliers was well known for his public lectures with demonstration experiments, a field in which scientists from this country were very successful throughout the eighteenth and most of the nineteenth century, by which time they had fallen behind their Continental colleagues in providing the proper complement of laboratory training.

Galileo had used the telescope essentially to make qualitative astronomical observations; but by the mid-seventeenth century it was evident that accurate telescopic observations of the heavens would be required. When Greenwich observatory was set up after the Restoration one of its major tasks was the making of such observations; but Flamsteed, the first Astronomer Royal, did not get on well with Newton and Halley. The first series of observations were extracted from Flamsteed and published by Newton and Halley, who thought that Flamsteed had been unfairly keeping his data to himself. But he argued that the observations were not yet sufficiently complete and accurate to be published, and got the edition suppressed so that copies are now very rare. One of the reasons for founding Greenwich observatory was to solve the greatest problem facing navigators – the determination of longitude.

To determine latitude is a relatively simple matter; one observes the altitude of the Sun at noon, or of a fixed star. Determination of longitude depends most simply upon comparing local time with Greenwich Mean Time; every hour's difference corresponds to 15° of longitude. At the start of the eighteenth century there were no clocks which would keep sufficiently accurate time. This problem was eventually solved by John Harrison, with his marine chronometer. Other attempts at finding longitude had been made by astronomers; observation of an eclipse, with the local time noted, provides the data for elucidating longitude; but eclipses happen infrequently. In the eighteenth century, astronomers tried to work out methods of deriving longitude from lunar observations, with some success. But the methods were not easy, and could not compete in practicability with the perfected chronometer; but Henry Salt, in his embassy to Abyssinia in 1810, determined longitude there by 'lunars', and navigators checked their chronometers by such observations. In 1713 Whiston had, with Humphrey Ditton, published a proposal that lightships be stationed in mid-ocean, and should fire rockets at midnight so that passing ships could correct their watches. In the following year they published their ideas in book form. This impractical proposal led to an Act of Parliament of 1714 promising a reward of up to £20,000 for a satisfactory method of determining longitude. Harrison's chronometers are described in his publications; Nevil Maskelyne, who became Astronomer Royal in 1765, supported the method of lunar observations. In 1763 he produced, under authority from the Board of Longitude, *The British Mariner's Guide*; out of which developed *The Nautical Almanac*, which began to appear in 1767.

Before the end of the eighteenth century, therefore, navigators could determine their position on the Earth's surface with considerable accuracy, as the voyages of Captain Cook proved. Of less practical importance, perhaps, was the establishing of the form of the Earth. That the Earth was a sphere had been known since antiquity, and various estimates of its diameter had been made, of which Columbus adhered to the smallest. Pemberton, explaining why Newton did not publish the theory of universal gravitation until twenty years after he had hit upon it, declares that because his value for an arc of a degree was inaccurate, the figures did not quite tally as they should have done. The 'arc of a degree' is that distance on the Earth's surface corresponding to one degree of longitude; Newton's value in 1666 was apparently sixty miles, whereas it should have been about 69·5. The accurate measurements were made by Norwood in England – though his determination did not become generally known – and by Picard in France.

During the debate between Newtonians and Cartesians, it became very important to measure the exact shape of the Earth. According to Newtonian theory one would expect the Earth to be an oblate spheroid, that is a sphere flattened at the Poles; whereas the Cartesian doctrine would predict a prolate spheroid, that is, one compressed around the Equator. Expeditions were sent therefore by the French Academy of Sciences, one to the Equatorial, and the other to the Polar, regions; to measure the arc of a degree in both places. That near the Pole proved to be longer, and hence the Newtonian prediction was fulfilled. Maupertuis wrote an account of the expedition to Lapland, of which he was the leader. His book, translated into English in 1738, was one of the sources for Coleridge's 'Ancient Mariner'. These expeditions were the forerunners of others, determining various physical constants, in succeeding centuries.

When the Earth's shape and size were known, the next step was to determine the distance between the Earth and the Sun. Kepler's laws enabled the relative distances of planets from the Sun to be calculated from the length of time they take to complete their orbits. If the actual distance of any one planet from the Sun were known, then the distances of all the rest could readily be found. It was pointed out in the *Philosophical Transactions* of 1716, by Halley, that this distance could be determined from observations of the transit of Venus across the disc of the Sun. Because of the inclinations of the orbits of the Earth and Venus, these transits can only be observed about twice in a century, when they occur with an eight-year interval; in the eighteenth century these were to be transits in 1761 and 1769. If the transit is observed from two widely separated places on the Earth, and the path of Venus seen from both places plotted, then, because the relative distances from the Earth to Venus and from Venus to the Sun are known, and the actual distance between the observation points known, the diameter of the Sun can be calculated. It is then a simple matter of trigonometry to calculate how far away the Sun is, since its angular diameter can easily be measured.

There were numerous expeditions to observe the transits; the crucial observations required were the exact times at which Venus entered the Sun's disc, and left it. Unfortunately it became clear that these moments cannot be determined unequivocally; and that the exactitude of the method was therefore limited. Enough observations were made for quite a good value to be calculated from them when scientists had made more progress. The expeditions ran into difficulties because Britain and France were at war, and sea captains did not always take much notice of the safe-conducts given to the scientists. Most of the accounts of the expeditions appeared in scientific journals; for the transit of 1769 the greatest number of observations were recorded by English-speaking astronomers. There was considerable activity in the American colonies; and John Winthrop and Benjamin West published accounts of observations in Boston and in Providence. Captain Cook observed the transit of 1769 in Tahiti, during his voyage on the *Endeavour*; indeed, this was the major object of the voyage, although its most important results turned out to be in the realms of geography and natural history. J. Chappe d'Auteroche was sent to observe the two transits in Siberia and in California; his journals were translated into English in 1774 and 1778. His results were among the best. In California the expedition was struck by a disease to which all but one eventually succumbed, Chappe himself being the last to die. In Russia he had been shocked by the backwardness and barbarity of the country, and said so in his journal, in which he included passages on the progress of arts and sciences in that country. A refutation of these passages was published anonymously, but apparently by Catherine II, and appeared in English in 1772 under the title *The Antidote*.

Authors describing the theory and practice of transit observation were numerous. William Whiston published a broadside in 1723 plotting predicted paths of Venus and Mercury across the solar disc; he recommended observing transits of Mercury, which happen much more often, but it was found that this planet moved too rapidly for accurate observations to be made. James Ferguson, a Scottish autodidact, described in the fifth edition of his *Astronomy Explained* how the distance of the planets had been compared from transit observations. E. Stone published in 1768 *The Whole Doctrine of Parallaxes*; and M. Stewart, in 1763, *The Distance of the Sun from the Earth Determined*; both these set out the general principles involved. Nevil Maskelyne, who had gone to St Helena in 1761 but had been unable to see the transit because of cloudy skies, published in 1768 *Instructions* for those who were intending to observe the transit of 1769.

In 1761 Benjamin Martin, an instrument-maker, had published *Venus in*

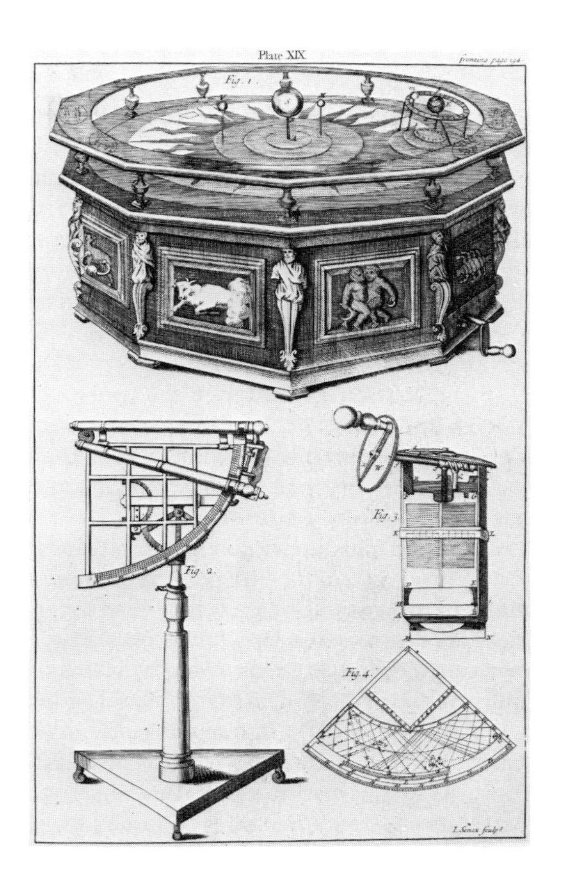

25 An orrery and astronomical instruments (Bion, N., *Mathematical Instruments*, 1723, pl 19)

the Sun; another instrument-maker, James Short, wrote a number of papers on the transits for the *Philosophical Transactions*, and also supplied a great number of the telescopes used by observers of many nationalities. Early eighteenth-century instruments are shown in the handsomely illustrated translation of Nicholas Bion's *Mathematical Instruments*, the second edition of which contains more plates in an appendix. John Dollond, who first produced achromatic lenses, also supplied instruments to observers. The output of telescopes and other instruments from the workshops of London makers at this period is surprisingly large; and correspondingly the incomes of the best-known instrument-makers were high, by comparison, for example, with academic salaries. Newton had suggested that because of chromatic aberration – coloured fringes around the image – reflecting telescopes would always be superior to refractors. The perfection of the achromatic lens, which in its simplest form consists of two lenses composed of different kinds of glass, changed this. The best-equipped transit expeditions, particularly in 1769, used refracting telescopes.

Reflectors came back into use in the latter years of the eighteenth century with William Herschel, whose papers appeared in the *Philosophical Transactions*, but were collected into a handsome set of two volumes in 1912. He built giant telescopes with which to study the nature and distribution of the fixed stars and nebulae, whereas most eighteenth-century astronomers had been mainly concerned with the solar system. To him we owe the discovery of Uranus, and of the proper motion of the Sun relative to the fixed stars, and the elucidation of the shape of our galaxy. Herschel was the first to observe double stars; he also suggested, following Whiston, that the Sun might be inhabited: that beneath fiery clouds there might be a temperate globe, the sunspots being glimpses through to the cool interior. Spectroscopic observations in the second half of the nineteenth century disposed of this engaging hypothesis. Herschel's work also initiated the controversy as to whether the nebulae were 'island universes', other galaxies an enormous distance away; or stars and planets in process of formation, and therefore immersed in nebulous matter; or collections of solid objects which appeared indefinite only because the telescope had not yet resolved them.

These were problems to engage astronomers of the nineteenth century. As far as the eighteenth century is concerned, it seems to be unquestionable that mathematicians in Britain fell behind those on the Continent, especially those in France. Only Maclaurin can be said to have shone in the field of mathematical astronomy, whereas on the Continent one finds such men as Clairaut, Euler, D'Alembert, Lagrange, and Laplace, who, applying the methods of analysis, were able to demonstrate the great power of the theory of universal gravitation in accounting for the phenomenon of the solar system without recourse to Divine interventions. An English translation of Euler's popular *Letters to a German Princess* appeared in 1795, and L'Hôpital's standard works had appeared much earlier. We shall meet Laplace later in discussing the 'decline of science'. In Britain Newton's *Opticks* rather than his *Principia* was taken as the model; and experimental scientists, particularly in the fields of electricity and chemistry, tried to achieve in those regions what Newton had done in his optical researches.

We shall examine their publications in the next chapter; there are other aspects of the Newtonian tradition worth glancing at in this one. The austere mathematical methods of the *Principia* found their culmination in France; close experimental reasoning in the chemists; and speculations about the rôle of the ether in one of the most influential books of the eighteenth century, David Hartley's *Observations on Man*. This appeared in two volumes: the first is an account of the hypothesis, and the second is devoted to showing that its theological consequences are not as alarming as might at first sight have appeared. Though the first volume is set out in the geometrical manner of Propositions, Corollaries, and Scholia, there is no mathematics in the book, except for some attempt to calculate the probability of past events, and of testimony generally; a favourite eighteenth-century exercise, taken up by Hume in his discussion of miracles. The works of Newton on which Hartley mainly drew were the *Opticks*, the letter to Robert Boyle concerning the ether, the *Chronology*, and the *Commentary on Daniel*. These latter works form the subject of a book by F. E. Manuel, and there is no need to say much about them here except that they show that Newton's interests were not confined to physics for its own sake. In the *Chronology* he tried to fix the dates of events recorded in ancient Greek histories and in the Old Testament; his work, published posthumously, gave rise to furious controversy. The rise of the science of archaeology in the nineteenth century may be said to have superseded this kind of erudition.

Hartley is famous for his doctrine of association of ideas, which he made the basis of his system. He accepted the corpuscular philosophy of Boyle and Newton; and the empiricist view that at birth the mind was a *tabula rasa*, containing no innate ideas. He believed that external objects caused vibrations in the ether and in the nerves, and ultimately in the medullary substance of the brain. The magnitude of each sensation was to be estimated from these hypothetical vibrations in the brain; the importance of vibrations in the biological sphere being demonstrated by the important rôle that they played in explanations of heat, light, and electricity in Newtonian physics. Hartley was able to account for pleasure and pain, learning and memory, on the doctrine of vibrations; he supposed certain vibrations to become associated together so that one started the other automatically. It was this *necessity* in the system which made it alarming; though in the eighteenth century there seems to have been greater readiness to accept determinism than at other periods. In the second volume Hartley argued that religion does not presuppose free-will in the sense that one could do different things in the same circumstances. For Hartley motives were causes: 'each Action results from the previous circumstances of Body and Mind, in the same manner, and with the same Certainty, as other Effects do from their Mechanical Causes.' His was an orderly world, from which caprice was excluded.

We no longer find impressive the doctrine of vibrations, and Hartley's psychological observations seem shallow when compared with, say, *The Anatomy of Melancholy* or numerous later works. We can read the book for its historical importance – especially for its influence on Wordsworth and Coleridge – and as an illustration of the attitude towards science of the mid-eighteenth century. The second volume, for example, is full of reasonable religion; Hartley was tolerant, he was impressed by natural religion and by 'evidences', he urged that general happiness was in fact found to prevail, and he exhorted his readers to eschew licentiousness and pursue benevolence. Like Mr Moffat in *Humphry Clinker* he did not believe that sinners would be punished for ever, but merely for a prolonged period; ultimately all mankind would be made happy. Whereas we tend to suppose that the pursuit of pure research is self-evidently good, and the foundation of academic freedom, Hartley warned his readers that in this region, as in that of the pleasures of sense, 'our Appetites must not be made the Measure of our Indulgences.' The pursuit of truth should be guided by temperance; and entered upon with a view to the glory of God and the good of mankind.

In the first volume we find a discussion of philosophy of science. Davy, in the early nineteenth century, referred to Bacon, Locke, and Hartley as the representatives of the empiricist tradition; and it would seem that Hartley's views on method are again representative of his period. We notice first that natural philosophy and natural history are merely branches of 'science', which was a word then used to embrace any organized body of knowledge. Hartley includes among sciences philology, mathematics, logic, history, and religion. The section on Propositions and the Nature of Assent contains a discussion of statistical evidence, mostly drawn from A. de Moivre's book, *The Doctrine of Chances*, a pioneering mathematical work chiefly devoted to calculating the odds in games, and premiums for life insurances. Hartley applied these ideas, as de Moivre had suggested might be done, to the inducing of laws of nature from a collection of facts. His discussions of simplicity and analogy, and of the importance of independent claims of evidence for laws, still make interesting reading. Hartley was by no means against hypotheses, and argued that the hypothetico-deductive method of proceeding was the right one. The problem was to generate hypotheses which could be tested, and abandon those which failed the test; those who tried to avoid all hypothesis in fact succeeded, he believed, only in holding confused ideas in place of clear and testable ones.

Hartley's remarks on method were taken note of, but it was his doctrine of necessity which aroused most interest. His book appeared, without change, in a number of editions; and a shorter account of it, with the doctrine of vibrations omitted, was published by Joseph Priestley in 1775. This gave rise to a considerable amount of controversy, mostly of a theological kind. Coleridge began as a disciple of Hartley, and named his eldest son after him; but in reaction against what he later believed the superficiality of the Hartleyan system he named his next child Berkeley. The most important follower of Hartley in the field of systematic associationist psychology and empiricist philosophy was James Mill, whose *Analysis of the Phenomena of the Human Mind* appeared in 1829. He hoped to make the human mind as plain as the road from Charing Cross to St Paul's. But the work displays the worst characteristics of Hartley, and one turns in relief from the aridities of James Mill to the wider sympathies, and less tidy mind, of his son, John Stuart Mill. James Mill's book does not seem to have been of great importance to scientists, who appear to have read chiefly the younger Mill, John Herschel, and William Whewell, in working out their philosophy of science.

Of other philosophers writing in English, the rôle of Locke in making generally acceptable the way of thinking of the corpuscularian scientists is too well known to need any comment. Berkeley's criticisms of Newton, and of the atomistic world-view in general, are also far from unknown. His *New Theory of Vision* appeared in 1709 and is one of the most important critiques of the mathematical treatment of light and colours which had culminated in Newton. Berkeley argued that we do not estimate distances by measuring angles but by using our judgement. This emphasis on what we perceive rather than on the models used by the applied mathematicians reminds us of Goethe's *Theory of Colours* of a century later. Berkeley also criticized the Newtonian doctrine of absolute space and time; in his Latin treatise *de Motu* of 1721, and in English in his *Analyst* of 1734, addressed to an infidel mathematician, and *Siris*, concerned with the virtues of tar-water, of 1744. This book begins with case studies, but includes inquiries as to whether the ancients really knew as much physics as we do – speculations which Newton also indulged in – and remarks on absolute space, and on the doctrine of the Trinity. The *Further Thoughts on Tar Water* of 1752 contain only further clinical trials of the efficiency of this nasty nostrum. Professor Popper has pointed out similarities between Berkeley and Ernest Mach as philosophers of science; but unlike Mach, Berkeley does not seem to have made any great impact on contemporary scientists. The same would seem to be true of Hume, whose *Treatise*, raising sceptical doubts over the justification of induction and the nature of causality, appeared in 1739. Eighteenth-century scientists were working under Newton's shadow, and had a sufficient number of problems to tackle with his methods; they did not need to take much notice of philosophers.

At the end of the eighteenth century, the Scottish 'common sense' school of philosophy, particularly Thomas Reid and Dugald Stewart, seem to have exercised some influence on scientists. In particular, Thomas Young and John Dalton were responsive to this school; and in his early days Davy also read them, when he was still apprenticed to an apothecary-surgeon in Penzance. Davy also read at that period W. Enfield's *History of Philosophy*, which appeared in two volumes in 1791. Enfield had given courses of physics lectures at the Dissenting Academy at Warrington, which were published as *Institutes of Natural Philosophy*, dedicated to Priestley, in 1785. His *History of Philosophy* is a spacious survey, based on Brucker's *Historia Critica Philosophiae*, which gives an account of all the various philosophical schools. Although Enfield found Plotinus romantic, mystical, obscure, and fanatical, and wrote that 'this philosopher made it the main scope and end of his life to dazzle his own mind, and the minds of others, with the meteors of enthusiasm, rather than to illuminate them with the clear and steady rays of truth', nevertheless he did devote a considerable amount of room to an exposition of the systems of the neo-Platonists, of the Cambridge Platonists, and of modern 'eclectics' including Bruno, Cardan, and Campanella as well as Bacon, Hobbes, Descartes, and Leibniz. Enfield's book is essentially a translation of a compilation; neither in physics nor in philosophy was he original, and his books are valuable as a guide to the general beliefs and sympathies of the period. Taken with the translations of Plotinus, Porphyry, and Proclus by the autodidact Thomas Taylor, they show that it was not difficult in the late eighteenth century to find non-empiricist philosophy. And in 1798 there was published the first account in English of the Kantian philosophy – *Elements of the Critical Philosophy*, edited and translated by A. F. M. Willich, a physician who had attended Kant's lectures in 1778–81 and again in 1792. The book is too close a translation from the German to be easy to read, but the glossary of Kantian terminology is of some interest, and the criticism of the 'common sense' solution to Hume's problems is quite clear. There will be

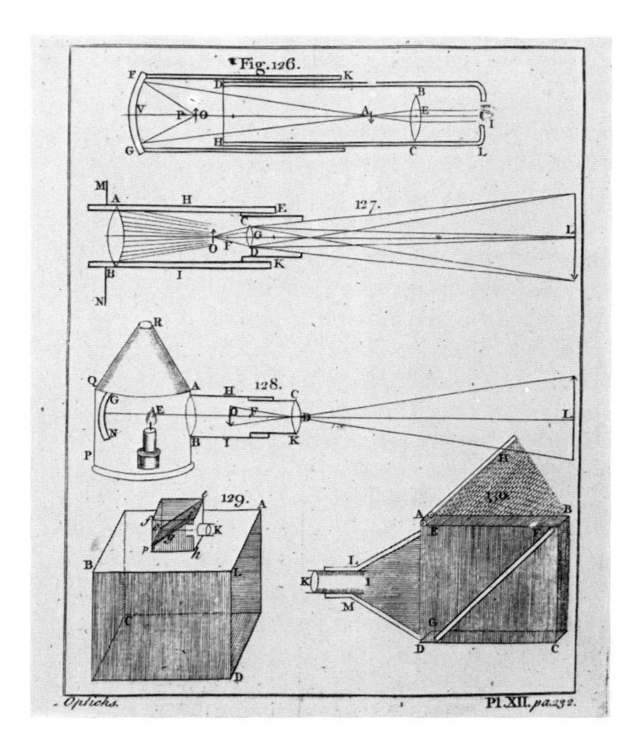

26 Telescope, magic lanthorn and camera obscura (Emerson, W., *The Elements of Optics*, 1768, pl 12)

more to say about the philosophy of Kant when we come to discuss early nineteenth-century chemistry.

Another lecturer in physics in the late eighteenth century was George Atwood, whose *Description of Experiments* to illustrate his course came out in 1776. A synopsis of his lectures at Cambridge on natural philosophy was published in 1784. These publications are of some theoretical interest because the two-fluid theory of electricity which they contain was taken up by Faraday while he was still a bookbinder's apprentice, in preference to the then orthodox one-fluid theory. Otherwise the books illustrate the very competent level of lecture-demonstrations at this time. Other standard eighteenth-century textbooks were William Emerson's *Principles of Mechanics*, with many plates; and Robert Smith's *Opticks*, which includes instructions for grinding and polishing lenses as well as sections on the properties of light, and on the eye.

At a more popular level comes 'Sir' John Hill's *Urania*, published in 1754, which sought to convey in dictionary form both descriptive and theoretical astronomy, and gave particular emphasis to heavenly bodies mentioned in Scripture and in the literature of antiquity. A most useful work for biographical information and for showing the state of science about 1800 is Charles Hutton's *Mathematical and Philosophical Dictionary*, the first edition of which appeared in 1796. The second edition, in three volumes instead of two and containing more material, came out in 1815. Hutton was one of those responsible for the *Abridgement* of the *Philosophical Transactions*, from the beginning of the series up to 1800; this is extremely valuable for those who have not easy access to the original volumes, and has a good index. The *Biographia Britannia* is also invaluable for the lives of scientists; and so is Benjamin Martin's *Biographia Philosophia* of 1764, which is chronologically arranged, and contains articles on foreign scientists as well as British ones. In addition to this and his publication on the transit of Venus, Martin published *The Philosophical Grammar*, a successful popularization of physics; and edited *The General Magazine of Arts and Sciences*, which appeared from 1755 to 1763, and formed an encyclopedia of science illustrated with plates. Martin was one of the most dedicated and competent popularizers of the Newtonian physics.

Anti-Newtonians writing in English were rare, but besides Berkeley we may note the theologian John Hutchinson, whose *Glory or Gravity essential and mechanical* appeared in 1733; and William Jones of Nayland, who published an *Essay on the First Principles of Natural Philosophy* in 1762, in which he supported Leibniz against Clarke, and argued that the Newtonians raised obscurities in seeking to talk of laws rather than mechanical causes. But these authors seem to have had little influence; as earlier had the pious Robert Green of Clare College, Cambridge, who argued for a plenum and a dynamical physics.

4 THE RECEPTION OF NEWTONIAN PHYSICS

Algarotti, F., *Sir Isaac Newton's Philosophy explain'd for the use of ladies* (tr. E. Carter), 2 vols., 1739.

(Berkeley, G.), *An Essay towards a new theory of vision*, Dublin, 1709.

The Analyst; . . . addressed to an infidel mathematician . . ., 1734.

Siris: a chain of . . . inquiries concerning the virtues of tar water, Dublin, 1744.

Bion, N., *The Construction and Principal uses of Mathematical Instruments*, tr. E. Stone, 1723; 2nd ed. (reprint + supplement), 1758.

[Catherine II, of Russia], *The Antidote*, 1770.

Chappe d'Auteroche, J., *A Journey into Siberia*, 1774.

A Voyage to California, 1778.

Clarke, J., *A Demonstration . . . of Sir Isaac Newton's Principles of Natural Philosophy*, 1730.

Cotes, R., *Hydrostatical and Pneumatical Lectures*, ed. R. Smith, 1738.

de Moivre, A., *The Doctrine of Chances . . .*, 1718.

Desaguliers, J. T., *A System of Experimental Philosophy . . .*, 1719; a pirated edition.

A Course of Experimental Philosophy, 2 vols., 1734–44.

Emerson, W., *Principles of Mechanics*, 2nd ed., 1758; 1st ed. not seen.

The Elements of Optics, 1768.

Enfield, W., *Institutes of Natural Philosophy*, 1785.

The History of Philosophy . . ., 2 vols., 1791.

Euler, L., *Letters to a German Princess . . .*, tr. H. Hunter, 2 vols., 1795.

Elements of Algebra . . ., 2 vols., 1797.

Ferguson, J., *Astronomy Explained upon Mr Isaac Newton's principles . . .*, 1756; 5th ed., 1772, discusses Transits of Venus.

Life of J. F., ed. E. Henderson, Edinburgh, 1867.

Flamsteed, J., *Historia Coelestis . . .*, ed. E. Halley, 1712.
Historia Coelestis Britannica . . ., 3 vols., 1725.
Atlas Coelestis . . ., 1729.
Fontenelle, B. le B. de, *The Elogium of Sir Isaac Newton*, 1728.
Green, R., *A Demonstration of the truth and divinity of the Christian Religion . . .*, Cambridge, 1711.
The Principles of natural philosophy . . ., Cambridge, 1712.
The Principles of the philosophy of the expansive and constructive forces . . ., Cambridge, 1727.
Gordon, G., *A Compleat Discovery of . . . the Longitude at Sea*, 1724.
(Harrison, J.), *An Account of the Proceedings, in order to the Discovery of the Longitude at Sea . . .*, 1763.
Hartley, D., *Observations on Man . . .*, 2 vols., 1749.
Helsham, R., *A Course of Lectures in Natural Philosophy*, 1739.
Herschel, W., *The Scientific Papers*, ed. J. L. E. Dreyer, 2 vols., 1912.
Hill, J., *Urania*, 1754.
L'Hôpital, G. F. A. de, *An Analytick Treatise of Conic Sections . . .*, tr. E. Stone, 1723.
The Method of Fluxions, tr. E. Stone, 1730.
(Hume, D.), *A Treatise of Human Nature . . .*, 3 vols., 1739–40.
Hutton, C., *A Mathematical and Philosophical Dictionary*, 2 vols., 1795–6 (vol. II appeared first).
Jones, W., *An Essay on the first principles of Natural Philosophy . . .*, Oxford, 1762.
Keill, J., *An Introduction to Natural Philosophy . . .*, 1720.
An Introduction to the true Astronomy . . ., 1721.
Maclaurin, C., *An Account of Sir Isaac Newton's Philosophical Discoveries . . .*, ed. P. Murdoch, 1748.
Martin, B., *The Philosophical Grammar*, 1735.
Philosophia Britannica : . . . the Newtonian Philosophy, 2 vols., Reading, 1747.
Venus in the Sun . . ., 1761.
Biographia Philosophica . . ., 1764.
Maskelyne, N., *The British Mariner's Guide containing . . . Instructions for the Discovery of Longitude . . .*, 1763.
An Account of the Going of Mr John Harrison's Watch . . ., 1767.
Maupertuis, P. L. M. de, *The Figure of the Earth . . .*, 1738.
An Essay towards a history of the principal comets . . ., Glasgow, 1770; see also Keill, J., in bibliography of ch. 3.
Mill, J., *Analysis of the Phenomena of the Human Mind*, 2 vols., 1829.
Musschenbroek, P. van, *The Elements of Natural Philosophy . . .*, tr. J. Colson, 2 vols., 1744.
Newton, I., *Opticks*, 2 pts, 1704; 4th ed., 1730, has 31 Queries.
Universal Arithmetick, tr. J. Raphson, 1720.
The Chronology of Ancient Kingdoms amended, 1728.
A Treatise of the System of the World, 1728.
The Mathematical Principles of Natural Philosophy, tr. A. Motte, 2 vols., 1729.
Observations upon the Prophecies of Daniel and the Apocalypse of St John . . ., 1733.
Four Letters from Sir Isaac Newton to Dr Bentley . . ., 1756.
Pemberton, H., *A View of Sir Isaac Newton's Philosophy*, 1728.
Rohault, J., *System of Natural Philosophy*, tr. J. Clarke, notes by S. Clarke, 2 vols., 1723; later eds., 1728–9; 1735.
A Treatise of Mechanics, tr. T. Watts, intr. W. Whiston, 1716.
Salt, H., *A Voyage to Abyssinia . . .*, 1814.
's Gravesande, W. J., *An Explanation of the Newtonian Philosophy . . .*, 1735.
Mathematical Elements of Natural Philosophy, tr. J. T. Desaguliers, 2 vols., 2nd ed., 1726; 1st ed. not seen.

Smith, R., *A Compleat System of Opticks*, Cambridge, 1738.
Harmonics, or the philosophy of musical sounds, Cambridge, 1749.
Spence, J., *Anecdotes, Observations and Characters of Books and Men*, ed. S. W. Singer, 1820.
Stewart, M., *Tracts Physical and Mathematical*, 2 pts, Edinburgh, 1761–3; pt 2 is concerned with the Solar Parallax.
Stone, E., *The Method of Fluxions* . . . , 1720; 1st pt tr. from L'Hôpital.
A New Mathematical Dictionary . . . , 1726.
Stone, Edw., *The Whole Doctrine of Parallaxes* . . . , Oxford, 1768.
The Whole Doctrine of Parallaxes . . . , Oxford, 1763.
Telescope, Tom (Newbery, J.), *The Newtonian System of Philosophy, Adapted to the Capacities of Young Gentlemen and Ladies*, 1761.
Vince, S., *A Complete System of Astronomy*, 3 vols., Cambridge, 1797–1808.
Voltaire, F. M. A. de, *Letters concerning the English Nation* . . . (tr. J. Lockman), 1733.
The Elements of Sir Isaac Newton's Philosophy . . . , tr. J. Hanna, 1738.
Wells, E., *The Young Gentleman's Course of Mathematicks* . . . , 3 vols., 1712–14.
West, B., *An Account of an Observation* . . . , Providence, R. I., 1769.
Whiston, W., *The Longitude and Latitude found by the Inclinatory or Dipping Needle*, 1721; appended to it, R. Norman, *Newe Attractive*, with title page dated 1720, and Ditton, H., *A New Method for discovering the longitude*, 1714.
Winthrop, J., *A Relation of a Voyage* . . . , Boston, 1761.
Two Lectures on the Parallax . . . , Boston, 1769.

Recent Publications

Bush, D., *Science and English Poetry*, New York, 1950.
Cohen, I. B., *Franklin and Newton* . . . , Philadelphia, 1956.
Dugas, R., *Mechanics in the seventeenth century*, tr. J. R. Maddox, Neuchâtel, 1958.
Halley, E., *Correspondence and Papers*, ed. E. F. MacPike, 1932.
Horrox, J., *The Transit of Venus across the Sun*, tr. A. B. Whatton, 1859.
Hume, D.: T. E. Jessop, *A Bibliography of David Hume and of Scottish Philosophy from Francis Hutchinson to Lord Balfour*, 1938.
Jones, W. P., *The Rhetoric of Science* . . . , 1966.
Koyré, A., *Newtonian Studies*, 1965.
Manuel, F. E., *Isaac Newton, Historian*, Cambridge, 1963.
Newton, I., *Papers and Letters on Natural Philosophy* . . . , ed. I. B. Cohen, Cambridge, 1958.
Correspondence, ed. H. W. Turnbull, Cambridge, 1959–.
Unpublished Scientific Papers, ed. A. R. and M. B. Hall, Cambridge, 1962.
Mathematical Works, ed. D. T. Whiteside, New York, 1964–.
Mathematical Papers, ed. D. T. Whiteside, Cambridge, 1967–.
A Bibliography, by G. J. Gray, 1907.
R. B. Webber and H. P. Macomber, *A Descriptive Catalogue of the Grace K. Babson Collection of the Works of Sir Isaac Newton* . . . , 2 vols., New York, 1950–5.
Nicolson, M., *Newton demands the Muse* . . . , Princeton, 1946.
Quill, H., *John Harrison*, 1966.
Royal Society, *Newton Tercentenary Celebrations*, Cambridge, 1947.
Wheatland, D. P., *The Apparatus of Science at Harvard, 1765–1800*, Cambridge, Mass., 1968.
Woolf, H., *The Transits of Venus*, Princeton, 1959.

Additional Reading

Cantor, G., *Optics After Newton*, Manchester, 1983.
& Hodge, M. J. S., *Conceptions of Ether*, Cambridge, 1981.
Wallis, R. V. & P. J., *Bibliography of British Mathematics*, Newcastle, 1986.
Westfall, R. S., *Never at Rest: a Biography of Isaac Newton*, Cambridge, 1980.

5 Science in the Eighteenth Century

It seemed that in the *Principia* Newton had announced the great law governing the motions of the planets, so that all that remained for his successors was to determine such constants as the Solar Parallax and the density of the Earth. As Maclaurin put it, Newton had 'begun and carried this work so far, that he left to posterity little more to do, but to observe the heavens, and compute after his models'. The work of William Herschel indicated that the inverse square law of gravity applied even to double stars far beyond the limits of our system. In the second half of the eighteenth century French mathematicians were engaged in completing the edifice of Newtonian physics; but their English-speaking contemporaries were for the most part content to present the Newtonian system as an object for contemplation rather than as something requiring further labours. Newton's *Opticks*, on the other hand, with its experimental approach and its wide-ranging Queries, could not be so regarded; and most English-speaking chemists, electricians, and even biologists can be viewed as experimental Newtonians.

Newton had, in the tradition of Boyle, attempted to give chemical explanations in terms of corpuscles and of the forces between them. In the orthodox corpuscularian view, all the different kinds of substances were composed of particles simply of matter, in different arrangements; the particles might also differ in shape. There was thus a hierarchy; the ultimate corpuscles formed the simplest bodies, and these in their turn were the building blocks of which complex substances were composed. Transmutation was therefore not impossible; but it would be a process at a deeper level than ordinary chemical change. In the Preface of John Harris' *Lexicon Technicum* of 1704–10 appeared a paper of Newton's, *de Natura Acidorum*, with an English translation, in which Newton briefly expounded this doctrine. He also suggested, here and in the Queries to the *Opticks*, that the particles of bodies were held together not by hooks but by 'attraction' that is, by a force akin to gravitation. Chemists of the eighteenth and early nineteenth centuries tried to achieve a theoretical chemistry based upon this idea. Newton suggested that the actual solid matter in the universe might be a very small proportion of the total; that the spaces between particles might be very large indeed in proportion to their size. His disciples took this up with enthusiasm, and Priestley suggested that all the solid matter in the solar system might only fill a nutshell. It was not matter but forces which filled space; and chemistry would therefore become a quantitative science, as astronomy had, when enough data had been assembled for the laws of these forces to be calculated by some Newton of chemistry. In the event, as we know, chemistry did not develop along these lines, although the theory of phlogiston can perhaps be seen as a precursor of that of chemical energy; Lavoisier and Dalton made chemistry a science of weights, and the study of chemical energy was not put on a firm quantitative basis until the middle of the nineteenth century.

Newton's chemical writings were brief, and suggestive rather than systematic; and it was left to such disciples as John Keill, James Keill, John Freind, and Francis Hauksbee to develop them. James Keill's *Animal Secretion* of 1708 is an attempt to apply these ideas in biochemistry; he even attributed to Hippocrates the doctrine that the small particles of matter attract one another. His brother's *Introductio . . .* a Latin treatise of 1702, conveyed similar doctrines; and so did Harris' introduction to his *Lexicon.* The doctrine that matter was chiefly empty space ran counter to the Cartesian and Leibnizian doctrine, which ruled out atoms and the void; its supporters were therefore chiefly to be found, in the early years of the century, in Britain and Holland. Friend's *Praelectiones Chymicae* ... appeared in 1709, and an English translation in 1712; it represents a sustained and almost comic attempt to apply the doctrine of corpuscles and the attraction between them to explaining the facts of chemistry. It aroused the ire of the Leibnizian *Acta Eruditorum*, but was favourably reviewed in the *Philosophical Transactions.* Freind took 'attraction' between particles as an established fact, and explained distillation, sublimation, and dissolution in acids in terms of it. In the eighteenth and nineteenth centuries English chemists in the Newtonian tradition tended to be more interested in explaining chemical change in principle than in thinking in specifically chemical terms, or performing arduous series of quantitative chemical experiments. Thus Priestley's discoveries were systematized by Lavoisier; Davy's insights by Berzelius; and Faraday's electrochemical work by Hittorf, Kohlrausch, and Arrhenius.

One of the works in which Newtonian natural philosophy and chemistry was popularized was Benjamin Martin's *Philosophical Grammar* of 1738. Its form of questions and answers remained popular for such books well into the nineteenth century, Mrs Marcet's *Conversations on Chemistry* being perhaps the best-known because it first attracted Faraday to science. In Martin's book the Newtonian corpuscularian doctrines are clearly expressed, and their implications made clear; but it makes no claim to being a work of originality. Statements of a dynamical corpuscular view of chemistry appeared in textbooks well into the nineteenth century, when this view was already being superseded by Daltonian atomism; the problem was that the corpuscular doctrines made very little difference to the chemistry which filled the body of the book. The ultimate particles were very small, inaccessible to observation, impenetrable, hard, and massy; and chemists were really only concerned with the higher levels in the hierarchy of arrangements. Peter van Musschenbroek in his *Elements of Natural Philosophy* of 1744 discussed in some detail the way in which the ultimate particles formed 'particles of the first order', which might differ in size, shape, and weight; and these particles in their turn composed others of the second order, having still more widely different properties.

Musschenbroek was one of the Dutch Newtonians, of whom the most important from the chemical point of view was Hermann Boerhaave. His *Method of Studying Physick* appeared in English in 1719; his *New Method of Chemistry*, based upon students' notes of his lectures translated by Shaw and Chambers, appeared in 1727: and an authorized edition under the title of *Elements of Chemistry* was translated from the Latin by Dallowe in 1735. Chemistry was in the eighteenth century generally taught as part of a medical course; but under Boerhaave the science began to acquire greater autonomy, and his is one of the greatest of chemical textbooks. The first volume dealt with theoretical chemistry, giving it a Newtonian emphasis; and the second with practical chemistry. The book had a great influence on the teaching of chemistry in the Scottish universities in the eighteenth century, and Boerhaave was one of the two authors recommended to his students by Joseph Black in 1767–8; the other was Macquer, whom we shall meet later.

The most original English Newtonian chemist of this period was Stephen Hales, perpetual curate of Teddington, friend of Alexander Pope and of the young Gilbert White. His *Vegetable Staticks* of 1727 contains numerous measurements of the sap pressure in plants, and represents an important contribution to plant physiology. But it was his researches into the air 'inspired by Vegetables' which were to prove so fruitful in chemistry. William Wotton had written after Boyle's experiments 'that there is scarce any one Body, whose Theory is now so near being compleated, as is that of the Air'. Hales' work showed how far from true this was; that the chemical nature and rôle of air was very little understood. He inaugurated the pneumatic chemistry which led ultimately to Lavoisier's chemical revolution. Hales invented the 'pneumatic trough' for collecting gases over water; to Priestley we owe the refinement of using a mercury trough for water-soluble gases. Hales burnt candles and placed small animals in vessels full of ordinary air, and in various 'airs' obtained by heating animal, vegetable, or mineral substances in a retort; and found that in time the air was made bad thereby. He attributed this to a loss of elasticity in the air. When this elasticity was lost, and the particles became 'fixt' by the strong attraction of acid or sulphurous particles, the air became vitiated, and unfit to support combustion or respiration. Air, this 'now fixt, now volatile Proteus', should be restored to its rightful place among the chemical elements or principles; and if instead of trying to make gold, chemists were to investigate air their researches would be 'rewarded with very considerable and useful discoveries'.

Hales' book is full of tributes to Newton, and is a splendid example of research in the Newtonian manner. But it became clear to at least some chemists that the particles of the corpuscular philosophy had no real importance in detailed chemical explanations, and that chemical affinity differed from gravitational attraction just as much as it resembled it. Stahl, who made the phlogiston theory of combustion the central tenet of chemistry, declared that the corpuscularians could only give a surface account of the phenomena. His *Philosophical Principles of Universal Chemistry* appeared in English in 1730; he distinguished the 'physical principles' of the corpuscularians from the 'chemical principles' which formed the limits of analysis. After this, chemists and physicists took little account of one another's theories of matter until about the 1870s, and even beyond that prominent chemists have urged the autonomy of their science. Stahl's book oddly does not contain his theory of phlogiston, that combustible substances contain this substance, which they lose on burning. With this theory, Stahl and his successors were able to impose some order on the facts of chemistry; and it was not until the last two decades of the century that Lavoisier and his associates exposed its incoherence in the face of new discoveries.

Peter Shaw, the translator of Boerhaave and Stahl, published his own *Chemical Lectures* in 1734. He distinguished between the atoms of the philosophers, which were 'metaphysical speculations', and the 'grosser

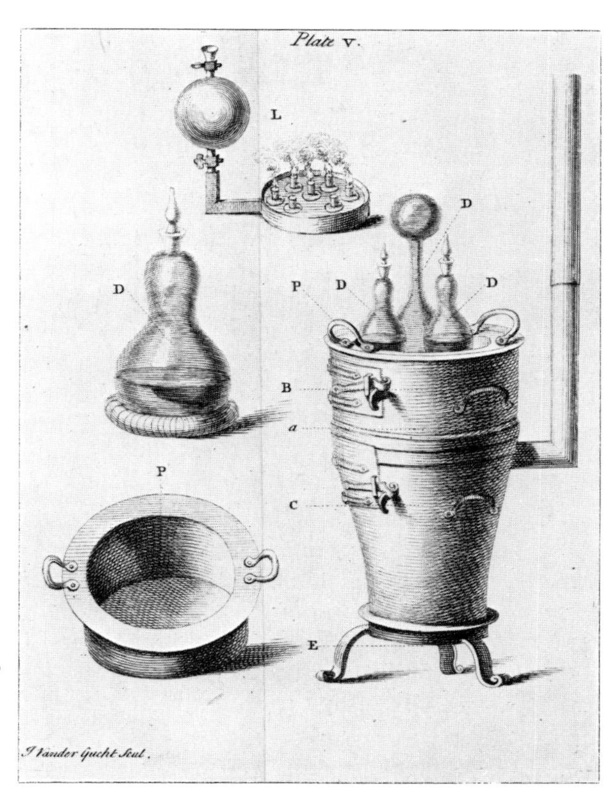

27 A portable laboratory in use (Shaw, P., and Hauksbee, F., *Portable Laboratory*, 1731, pl 5)

principles' of the chemists, which were evident to the senses. William Cullen and Joseph Black in Glasgow and Edinburgh were probably the most important teachers of chemistry in Britain in the mid-eighteenth century; but Cullen's course was never published, and Black's only appeared in print in 1803, after his death. But in 1756 in *Essays and Observations, Physical and Literary, read before a Society in Edinburgh* had appeared his account of his quantitative researches on magnesia alba, our magnesium carbonate. He found that on heating, this substance lost weight, and that the product did not effervesce with weak acid. The loss in weight he attributed to loss of air; and he found that if magnesia were heated and then dissolved in acid, it required about as much acid as if it had been dissolved without previous heating. From the solution, magnesia alba could again be precipitated on addition of a mild alkali. He interpreted these reactions as proving that the difference between magnesia alba and its ignition product, and between mild and caustic alkalies, was that the former contained 'fixed air' and the latter did not. The paper was one of the most important contributions to chemistry made in the eighteenth century, and it was reprinted in later editions of the *Essays*, and also by itself.

One of the problems of applying the Newtonian paradigm to chemistry is that while gravitational attraction is universal – all matter attracts all other matter according to the inverse square law – chemical affinity is elective. Some substances hardly react at all; and more reactive metals will displace less-reactive ones from their salts. In Black's paper and in his lectures we find references to the doctrine of elective attractions. Tables of elective attractions were introduced into chemistry by Geoffroy in 1718: at the top of vertical columns he put the name of an acid or alkali, and below it the names of the substances which reacted with it, the most reactive being at the top. Any substance named would displace any other below it in the list. The name 'double elective attractions' was given in the eighteenth century to what we call double decompositions; reactions between two salts in which the metallic components are exchanged. The word 'elective' has an anthropomorphic flavour, and indeed Goethe wrote a novel with the title *Elective Affinities*; but this is not apparent in the writings of chemists.

The most important work on elective attractions was Tobern Bergman's *Dissertation* which appeared in English translation in 1785; the translator was Thomas Beddoes, who also edited an English translation of the *Chemical Essays* of Bergman's protégé Charles William Scheele. Bergman's book contains diagrams with symbols and brackets to illustrate chemical reactions; Black had used a notation with symbols in circles; and it is surprising that chemical equations went out of favour in the early nineteenth century, to reappear in modern form in about the 1830s. Bergman's *Dissertation* with its view of the invariability of the elective attractions, represents an important step towards the doctrines of definite and multiple proportions, and hence towards Daltonian atomic theory. One of its more interesting successors was William Higgins' *Comparative View* of 1789, in which he drew diagrams attempting to quantify affinities; this book has been claimed – notably by

Higgins – as a precursor of Dalton's *New System* because of its advocacy of an atomic theory; but the atomic theory is that of the corpuscularians, not that of nineteenth-century chemistry.

Scheele's *Essays* are chiefly devoted to experiments on a number of bodies, including the dye Prussian Blue: and researches on manganese (our manganese dioxide) which led to the discovery of 'dephlogisticated muriatic acid', which we call chlorine. The experiments are very clearly reported, and give us a good idea of the procedures of one of the greatest discoverers of new substances in the history of chemistry. Other essays in the book include the recipe for 'Scheele's Green', a pigment; estimates of the quantity of 'pure or fire air' (oxygen, of which Scheele was a discoverer) in the atmosphere; and studies on ether, on lactic acid, and on vinegar. Scheele believed erroneously that sulphuric acid was a constituent part of the ether evolved from a mixture of it and alcohol; the theory of this reaction was first clearly worked out by A. W. Williamson in 1850. Scheele's actual account of the discovery of oxygen was translated in 1780.

The other independent discoverer of oxygen was Joseph Priestley, whose *Experiments and Observations on different Kinds of Air* appeared between 1774 and 1786, in six volumes; and in an 'abridged and authorized' form in three volumes in 1790. Priestley believed in the rapid publication of results, and thought it misleading to produce polished and highly organized papers. His reports are therefore more like a laboratory notebook; he recorded his hypotheses and his fruitless experiments as well as his successes. This makes his work more exciting to read than that of more cautious authors; but perhaps it has meant that Priestley has been taken less seriously than he deserves. The discovery of oxygen comes in the second volume, of 1775; Priestley interpreted his discovery in terms of the prevailing phlogiston theory, to which he adhered until his death in 1804, although his experiments provided some of the most important evidence used by Lavoisier in overthrowing phlogiston. Priestley also worked on ammonia, on the oxides of nitrogen, and on sulphur dioxide; and made pioneering studies on photosynthesis. Priestley's quantitative volumetric observations command respect, but he was at his best in qualitative researches.

In Priestley's *Disquisitions relating to Matter and Spirit* of 1777 there appeared an account of Roger Boscovich's theory of the atom, in which the hard massy particles of orthodox atomism were transformed into mere points – 'mathematical points' for Boscovich, and 'physical points' for Priestley – surrounded by forces which were at different distances attractive and repulsive. All the properties of the atom depended upon the forces or powers, and not upon the nucleus. This model exerted a fascination upon natural philosophers in England for more than a century. Priestley employed it to destroy the opposition between matter and spirit. Whereas Cudworth, Newton, and Maclaurin, for example, believed matter to be essentially inert, on the hypothesis of Boscovich, as interpreted by Priestley, matter was characterized by force and power rather than by inertia, and spirit was therefore redundant.

The Unitarian Priestley believed that the acceptance of materialism would be of assistance to religion, for the pagan hypothesis of the Immortality of the Soul could be thereby exploded in favour of the truly Christian doctrine of the Resurrection of the Body. Chemists of the next generation resisted this attempt to use their science to forward materialism, and chemistry and theology did not finally separate until well into the nineteenth century.

Perhaps the best standard textbook of the later eighteenth century was Richard Watson's *Chemical Essays*, a work which is quoted by Gibbon. Watson was elected Professor of Chemistry at Cambridge, and set about learning the subject; and these volumes were the result. He was later appointed to the see of Llandaff, and is frequently quoted as Bishop Watson. This amateurism continued in England; but in France the new professional spirit began to be visible, and with it came a new chemistry. Macquer had been the most important French chemist of the generation before Lavoisier's; his *Dictionary of Chemistry* and *Elements of Chemistry* are particularly useful for showing what was known and believed just before the chemical revolution. Of particular interest is his discussion of chemical elements, which he defined as the limits of analysis, and then added that they were fire, air, earth, and water. Lavoisier did not first use the modern definition of a chemical element, but he did make it generally accepted. Lavoisier's *Essays Physical and Chemical*, translated by Thomas Henry, appeared in 1776; and his *Essays on Atmospheric Air* in 1783. The latter contains his doctrine of combustion which he opposed to the phlogiston theory; but it is in his *Elements of Chemistry*, translated hurriedly and not very elegantly by Robert Kerr in 1790, that we find his view of chemistry systematically presented. The work of the French nomenclators, Guyton de Morveau, Lavoisier, and others, who introduced new names to replace those which referred to phlogiston, was translated in 1788; it was attacked by S. Dickson in his *Essay* of 1796. Richard Kirwan, of Dublin, wrote an *Essay on Phlogiston* supporting the old view. This was thought to be sufficiently important to be translated into French by Madame Lavoisier with notes by the French chemists; and in the second English edition these notes were translated by William Nicholson, so that one finds both sides of the question expressed. Higgins' *Comparative View* was also intended as a detailed refutation of Kirwan.

We shall return to the history of chemistry after the triumph of the anti-phlogistic system in a subsequent chapter. We should now turn to another science which involved action at a distance; the study of electricity. Volta's discovery that when two different metals are dipped in water and connected, a current flows, was published in 1800, and electricity in the eighteenth century meant electrostatics. 'An electric battery', similarly, meant a series of Leyden-jar condensers, and not a voltaic, or galvanic cell. The most important author in English on the subject of electricity was Benjamin Franklin, whose *Experiments and Observations on Electricity* went through numerous editions. The book is very readable, being experimental rather than mathematical in approach; the famous experiment with the kite in the

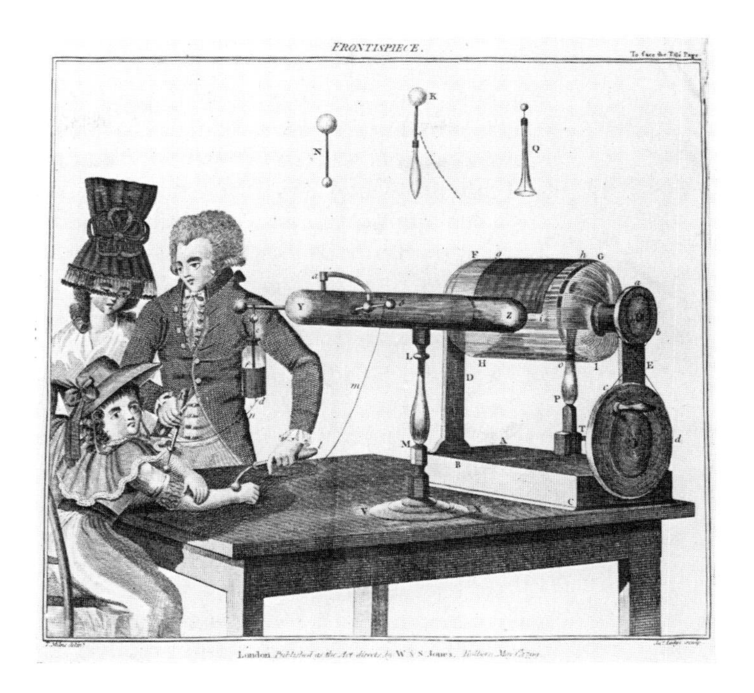

28 A girl being electrified (Adams, G., *Essay on Electricity*, 1799, 5th ed, frontispiece)

thunderstorm is described. We may notice that Franklin's advocacy of lightning conductors was not very effective in England, and W. Snow Harris as late as 1843 had to urge their adoption, listing damage caused to church steeples in the previous years. Franklin favoured the view that electricity was a weightless fluid; the mathematics of this were worked out by Aepinus, and by Henry Cavendish, whose electrical papers were published by James Clerk Maxwell in the nineteenth century. Cavendish's treatment is noteworthy for his separation of fact and theory; he sets up the 'model' of an electric fluid, deduces consequences from it, and then sees whether these conclusions hold in the real world. The chief objection to the one-fluid theory of electricity was that when a discharge is passed through a piece of paper, a burr is observed on both sides; and the view that electricity was the effect of two fluids, one positive and the other negative, had gained ground by 1800.

It was in electricity that Priestley first made his reputation, with the writing of his *History and Present State of Electricity*, published in 1767. Before he had finished writing the history – which is a most useful work – he had himself begun to make experiments, which he duly recorded. The success of the book was such that Priestley hoped to produce similar works in other fields of science, and wrote his *History of Light and Colours*; but this was generally felt to be less successful, perhaps because Priestley lacked the mathematics necessary for treating parts, at least, of optics, and for making discoveries in that science. Also on the subject of electricity, we should notice the *Complete Treatise* of Tiberius Cavallo, with many details of experiments, and handsome plates; and a popular *Essay* by G. Adams, an instrument-maker, with an alarming frontispiece of a girl about to be given a shock from an electrical machine. Cavallo also wrote a *Treatise on Magnetism*; in general, magnetism was a less active science than electricity in the eighteenth century; a lot of Cavallo's book is taken up with advising sailors how to make observations.

Franklin's chief adversary was Jean Nollet, whose theory of electricity involved the postulation of a subtle matter which penetrated the pores of bodies; he has been described as a Cartesian rather than a Newtonian. His *Lectures in Experimental Philosophy* appeared in English in 1752, and make interesting reading to those who do not care only to peruse the production of the winning side. Other electrical writings which deserve mention are those of Benjamin Hoadly and Benjamin Wilson, whose *Observations* appeared in 1756; ten years earlier, Wilson had tried to explain the phenomena of electricity in terms of Newton's ether. In the same year, Benjamin Martin, the indefatigable popularizer, had also published an *Essay* in which he tried to account for electricity 'on the principles of Sir Isaac Newton's theory of vibrating motion, light and fire'. These authors were drawing on the same kind of Newtonian model as David Hartley; eighteenth-century electricity illustrates the way in which a simple and powerful theory, that of Franklin, gave coherence to a collection of otherwise mystifying experimental results, and led to new discoveries and practical applications.

It was from Italy that the next great advance – that of Galvani and Volta – was to come, and the works of Italian electricians appeared in English in the eighteenth century. J. B. Beccaria's studies of atmospheric electricity appeared in 1776; in the previous year his memoirs on phosphorescent matters had been translated. And in 1793 Eusebius Valli's *Experiments on Animal Electricity* were published; Galvani's own book did not find an English translator until the twentieth century. Electricity was already being invoked to explain the mechanism of the nervous system; and great interest was aroused by the various fishes, such as the electric eel and the torpedo, which appeared to stun their enemies or their prey by an electric shock. John Walsh tried to explain the effects, in a paper in the *Philosophical Transactions*, and Cavendish made model torpedoes out of wood and leather, incorporating a number of Leyden-jars; and found that these contraptions did give shocks rather as the fish does. But it was not until well into the nineteenth century that complete proof was available of the identity of 'animal' and 'common' electricity; and electricity did not play any other than a speculative rôle in physiology until later still.

Chemistry and medicine had had a long and intimate association: but the first major application of chemistry in the biological sciences was the explanation of animal heat, independently and in somewhat different terms, by Lavoisier and Adair Crawford. Previous authors had written of the vital flame, but respiration had been generally believed to cool the heart; now it was seen as supplying oxygen for processes in the body akin to slow combustion. Crawford's explanation was in terms of the theory of phlogiston, whereas Lavoisier's involved the modern view of the rôle of oxygen. Their publications marked the culmination of a series of attempts to explain the phenomenon, usually in mechanical terms; thus Archibald Pitcairne supposed animal heat to arise from attrition of the parts of the blood caused by its rapid motion through the arteries. He rejected the chemical view that the heat arose through fermentation. Robert Douglas also favoured friction as the cause of animal heat; but his essay is set out in Euclidean form, though hardly with Euclidean self-evidence. Leslie in 1778 took up the theory advanced by Andrew Duncan in lectures at Edinburgh, that animal heat was generated by the evolution of phlogiston from the blood, and published it in his *Philosophical Inquiry*.

Crawford was able to improve upon this theory using the researches of Black on latent and specific heats, and his book had a much firmer experimental basis than those of his predecessors. Thus he endeavoured to show experimentally that 'dephlogisticated air' (oxygen) contained more heat than the 'fixed air' (carbon dioxide) breathed out; and that arterial blood had a greater capacity for heat than venous blood, and could therefore absorb heat in the lungs. With the triumph of Lavoisier's chemistry, this explanation needed rephrasing; and the chemical explanation was open to attacks like that of Hugh Moises, who remarked in his *Treatise on the Blood* that we had no authority to suppose that combustion could take place at body temperature. Similar criticisms were made in the nineteenth century but, in general, the explanation was accepted, and led to new questions for physiologists to answer.

Hales' researches on sap pressure, recorded in *Vegetable Staticks*, were intended to test the view that the sap in vegetables circulated like the blood in animals, and the methods he used he had already employed to measure the blood-pressure in animals. But these experiments were not published until 1733, in his *Haemastaticks*, which is a disagreeable volume full of painful experiments on animals, into whose veins tubes were pushed. Hales believed that the experiments would be valuable to humanity, and presumably shared the Cartesian view that animals were soulless mechanisms; we would agree with his friend Pope, who asked: 'How do we know, that we have a right to kill creatures that we are so little above as dogs, for our curiosity, or even for some use to us?' Nevertheless, the book does contain a thorough study of the phenomena of the circulation, in a quantitative form. The *Haemastaticks* formed the second volume of *Statical Essays*, of which *Vegetable Staticks* was the first; and the index to both volumes is to be found at the back of *Haemastaticks*.

Much more agreeable are works of natural history in the tradition of Ray and Willughby. It is difficult, and probably mistaken, to try to separate serious biological works from handsome picture books at this period, because biology and geology only really became fields for professionals in the nineteenth century. To take an interest in animals, plants, and minerals, and perhaps to fill a cabinet with a collection of handsome specimens, was expected of a person of culture; and contributions to natural history were made by those whom we would describe as amateurs. Probably the most famous work on natural history by an Englishman of this period was Gilbert White's *Selborne*: and the most handsome, George Stubbs' *Anatomy of the Horse*, which

29 (*left*) Horse (Anon., *The English Farrier*, 1649, frontispiece)
30 (*right*) Horses (Snape, A., *Anatomy of an Horse*, 1683, frontispiece)

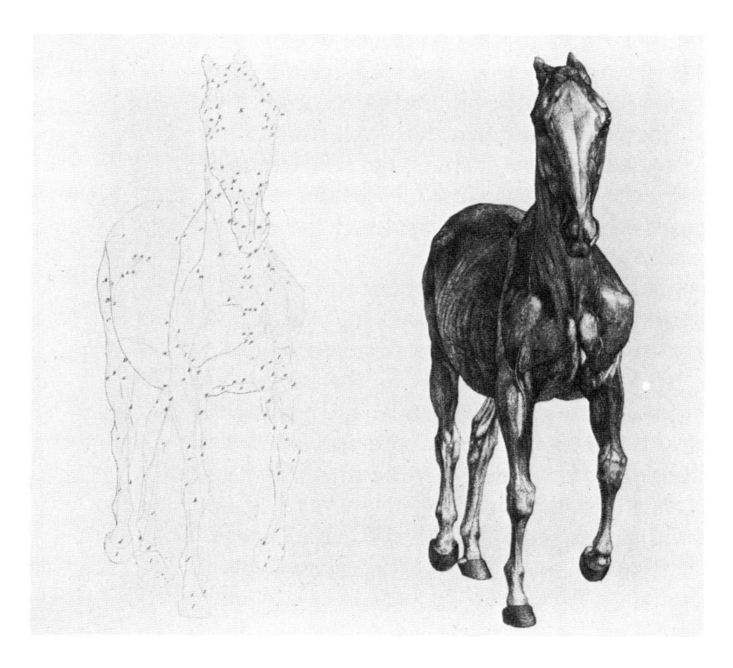

31 Anatomical study of a horse (Stubbs, G., *Anatomy of the Horse*, 1766, 6th anatomical table)

appeared in 1766. The eighteen splendid plates show the horse in various stages of dissection, seen from the side, back, and front; he could find nobody to engrave the drawings, and had to do them himself. Not only does the work reveal the knowledge behind Stubbs' famous paintings, but it also stands in its own right as a work both of science and of art. He dissected 'a great number of horses' and injected their blood vessels with wax; he worked on each horse for a number of weeks, untroubled by the smell, in an isolated farmhouse in Lincolnshire over a period of eighteen months. Some of his working drawings have been recently discovered and published with a facsimile of the book. It is interesting to compare Stubbs' book with the crude early seventeenth-century works: and with the handsome volume by Snape.

White's *Selborne* needs no recommendation, and its text has inspired numerous illustrators. But the most important naturalists of the century were not British. Tournefort's *Complete Herbal* appeared in English in 1719–30, with over 250 plates; Tournefort was one of the great botanists of the pre-Linnean period, who travelled in the Levant collecting species with the great botanical artist Aubriet. G. L. le Clerc, comte de Buffon, translated Hales' *Vegetable Staticks* into French; selections from his own enormous *opus* appeared in English in various versions in the last quarter of the eighteenth century. Historians have argued over whether Buffon really believed in some kind of evolution, or in a theory of degeneration from a golden age. To his contemporaries the system behind his work was not easy to see; a reviewer in the *Philosophical Transactions* remarked that such an enterprise must collapse under its own weight. Buffon's definition of a species, that members of the same species mate together and produce fertile offspring, was useful until relations between extinct species began to demand elucidation in the nineteenth century. But apart from considerations of theory, Buffon's *Natural*

32 Crocodiles (Buffon, *Oviparous Quadrupeds*, 1802, vol 1, pl 12)

History was a splendid and well-illustrated account, particularly of the animal kingdom, and was very widely used and copied from; it illustrated animals from remote parts of the world which were by the eighteenth century beginning to be explored by men with an interest in natural history.

In vertebrate zoology the number of species to be classified is relatively manageable, but among invertebrates and in the realm of botany the situation is much less clear. Neither the Aristotelian idea, recommended by Ray, of placing living things in natural groups on a basis of multiple criteria, nor the Great Chain of Being or Ladder of Nature so popular in the eighteenth century, proved very valuable to workers in these fields. The artificial system of Linnaeus, in which species of plants were distinguished according to the characteristics of their sexual parts, enabled much more rapid progress to be made in the ordering and naming of the new species which came flooding in as botanists explored the Tropics, and also looked more carefully at the productions of their own countries. By the end of the century, botany had reached a stage at which it was possible to try to achieve a natural classification; and Antoine Laurent de Jussieu, a member of a famous family of botanists, and A. P. de Candolle brought the science back into the tradition of Aristotle and Ray.

The earliest work of Linnaeus to be translated was an essay on travelling in one's own country, in a collection of tracts the translator of which wanted to show 'how far all mankind is concerned in the study of natural history.' The book has the typographical peculiarity of using a lower-case 'i' for capital 'I'. The Linnean system was expounded in *An Introduction to Botany* by J. Lee, a nurseryman, who had refrained from actually translating any of Linnaeus' writings because they were 'interspersed with philosophical and critical Remarks that are of less general Use'. His book was therefore a summary; and it was followed by a flood of works expounding the Linnean system. One of the pleasantest was by Jean-Jacques Rousseau, translated by T. Martyn; it took the form of letters addressed to a lady. He believed that cultivated plants were 'monsters'; while domesticated trees produced fruits with more pulp, wild ones were larger and more vigorous.

William Withering, who is chiefly remembered for the introduction of digitalis for the cure of dropsy, and who had already translated Bergman's *Outlines of Mineralogy*, published in 1776 his *Botanical Arrangement of Vegetables*, which is essentially a translation from Linnaeus of the descriptions of plants indigenous to Britain. Linnaeus' *System of Vegetables* was translated by a society in Lichfield in 1783; this appears to have been edited by Erasmus Darwin, as was the translation of Linnaeus' *Families of Plants*. This prepared him for the poetical effusions which earned him such a reputation in his own day, and which now give him a claim to be counted a precursor of his grandson in propounding a theory of evolution. *The Botanic Garden*, popularizing the Linnean System, appeared in two parts, 'The Economy of Vegetation' and 'The Loves of the Plants'. It was followed by *Zoonomia*, of which the first volume contains an account of Darwin's natural philosophy, including the idea that all warm-blooded animals are derived from one living filament, and had diverged in character to meet different circumstances. *Phytologia*, concerned with agriculture and gardening, appeared next; Professor Schofield remarks that it contains 'much shrewd observation, experiment, and common sense overlaid by a fantastic theory which did much to conceal the value of the practical material'. In this case the theory was that a plant was a low form of animal; an analogy which had been abandoned by most biologists early in the century. And finally Darwin produced *The Temple of Nature*, which also stresses the progressive character of nature, and has notes dealing with electricity, chemistry, and spontaneous generation. These poems were widely read and influenced, among others, Shelley; but it is hard to take them really seriously. Darwin was a man with an immense range of interests, but hardly a great original scientist.

Sir James Smith, who became President of the Linnean Society, the first society in England devoted to a particular branch of science, was himself a botanist, responsible for the text of James Sowerby's *English Botany*, which appeared in thirty-seven volumes between 1790 and 1814, containing over 2,500 drawings of plants. Sowerby also produced *English Fungi* and *British Mineralogy*, which are illustrated with splendid plates. The *English Botany* became a standard work as well as an object of beauty, and James Sowerby was the first of a family several of whom became distinguished illustrators of natural history in the nineteenth century. Smith translated a number of works of Linnaeus, including his *Correspondence* and his *Reflections on the Study of Nature*; and with Sowerby produced the splendid volumes of *Exotic Botany*, and also a *Botany of New Holland*. In 1821 he published his *Grammar of Botany*, which contains an exposition of the natural system of de Jussieu. Sinclair's *Hortus Gramineus Woburnensis* contains the results of experiments carried out at Woburn, with advice from scientists including Davy, on growing different grasses in different situations: it contains plates drawn on stone by Parez. Strutt's *Sylva Britannica* describes splendid trees of different species, with engravings; but is not really a work of botanical science.

Among other translators and popularizers of Linnaeus were Sibly, whose major interest appears to have been astrology, and Donovan, who produced a superbly illustrated and famous series of works dealing with British insects, birds, fishes, shells, and quadrupeds. By the end of the century, botany had become thoroughly respectable and indeed fashionable; but the study of insects or spiders still appeared repulsive, if one may judge from prefaces to books on the subject. Réaumur's *Natural History of Bees* appeared in English in 1744, and another account of his work in 1800. Swammerdam's *Book of Nature*, with a life of the author by Boerhaave, was published in 1758; and T. Martyn, the Professor of Botany at Cambridge, produced *The English Entomologist* in 1792. This contains a number of plates of beetles, some very splendid; they were produced by Martyn's academy for drawing and printing natural history, in Great Marlborough Street, London. Martyn had previously produced *The Universal Conchologist*, with handsome plates of shells,

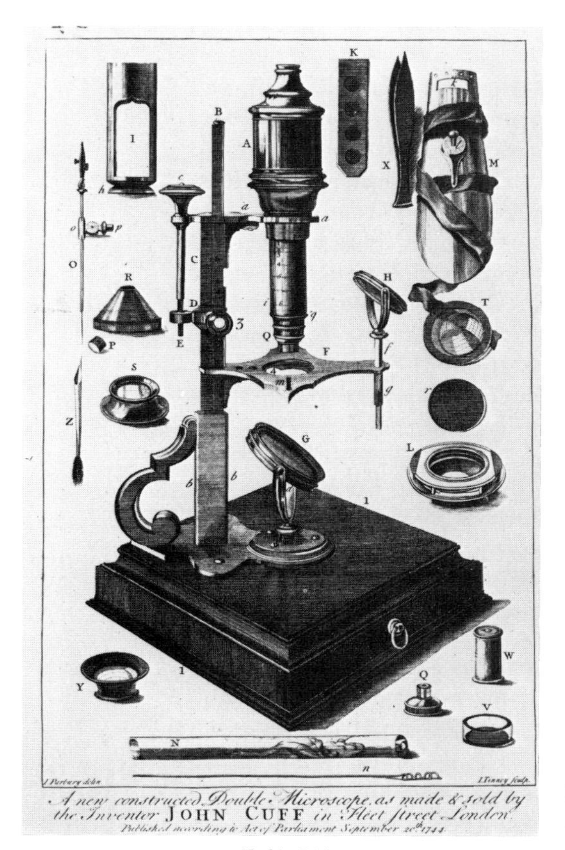

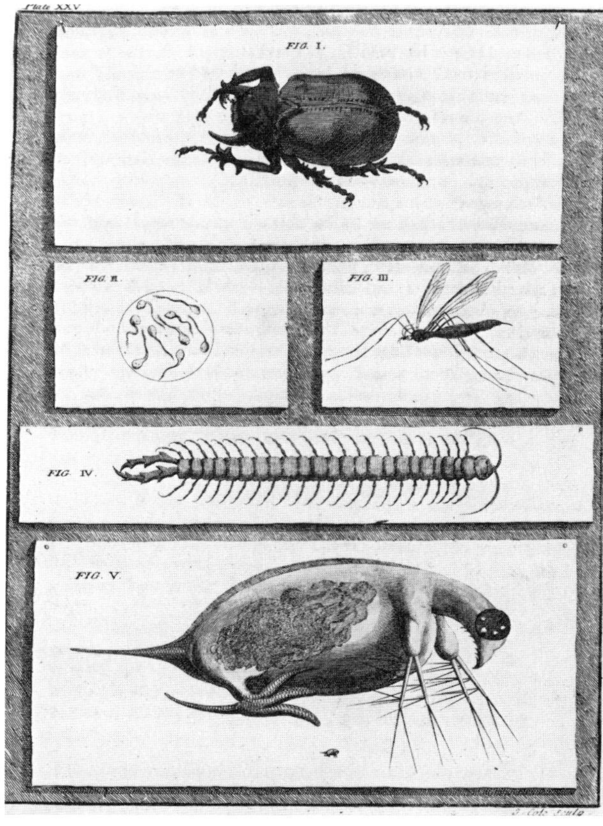

33 (*left*) Microscope (Baker, H., *Employment for the Microscope*, 1764, 2nd ed, pl facing p 422)

34 (*right*) Insects and animalculae from Thames water seen through the microscope (Bradley, R., *Philosophical Account of the Works of Nature*, 1721, pl 25)

and a text in French and English, which was not uncommon in works of natural history of this period. Martyn also edited, in 1793, Eleazar Albin's *Natural History of Spiders*, the first edition of which had appeared in 1736. Wildman's *Treatise on the Management of Bees* was an essentially practical treatise; a more systematic work on insects was Kirby and Spence's *Introduction to Entomology* which contains a number of coloured plates but is a work of science rather than of fine art. By the early nineteenth century, this transition from natural history towards the sterner science of biology was under way, though splendidly illustrated volumes continued to be, and still are being, produced. While the beginning of the nineteenth century can be seen as the beginning of a new era of uniformitarian geology, and thus, ultimately, of evolutionary biology, there was no discontinuity in natural history. Some eighteenth-century works will be mentioned at the beginning of the ninth chapter; and some from the early nineteenth century here.

Studies of invertebrates include Albin on insects as well as on spiders; and

35 Life-cycle of the ephemera (Swammerdam, J., *The Book of Nature*, 1758, pl 13)

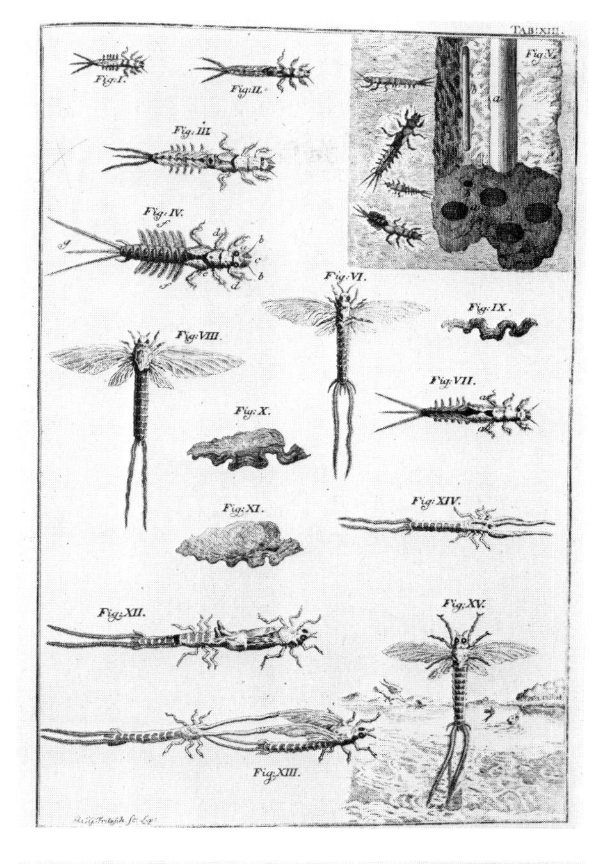

36 (*below left*) Stag beetle (Martyn, T., *The English Entomologist*, 1792, pl 5)

37 (*below right*) Spider (Albin, E., *Natural History of Spiders*, 1736, pl 34)

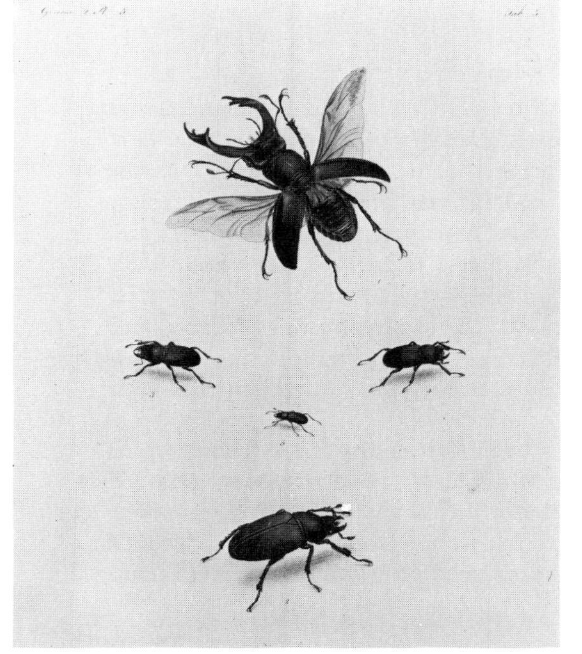

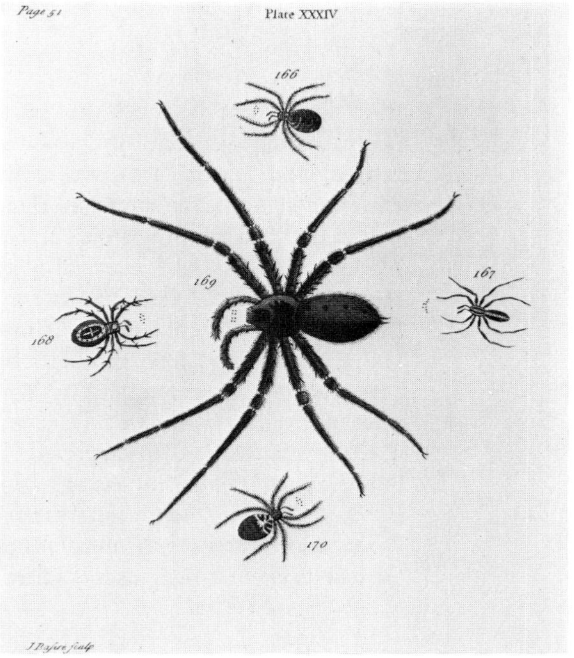

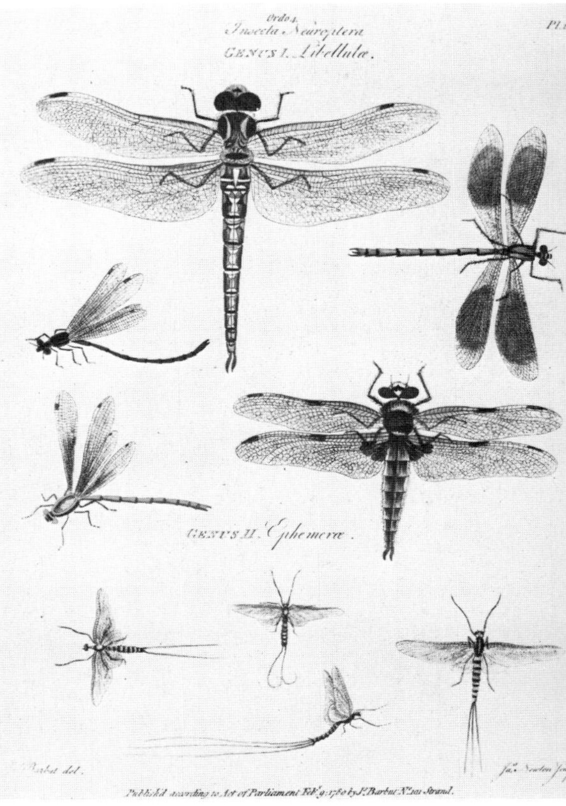

38 (*left*) Corals at low tide (Ellis, J., *Natural History of Corallines*, 1755, frontispiece)
39 (*right*) Dragonflies (Barbut, J., *Genera Insectorum of Linnaeus*, 1780, pl 11)

Ellis' handsome volumes on corals and on zoophytes. This latter appeared posthumously, edited by Daniel Solander. Drury's plates of insects deserve notice; and Barbut's volumes illustrating the 'worms' and insects of the Linnean system, with text in English and French, are very splendid. Later works include Montagu on shells, of 1803, with pretty little drawings; and Miller's *Crinoidea* of 1821 with splendid plates. The *Rare and Remarkable Animals of Scotland* described by J. G. Dalyell in 1847–8 are all zoophytes, illustrated in colour; and the contemporary *British Mollusca* of Forbes and Hanley had nice plates of shells by G. B. Sowerby. The Huber's studies of ants and of bees were standard works in their day, and can still be read with pleasure.

Of works devoted to larger creatures, Albin's was one of the first fine bird books, with beautiful though rather stiff plates. Its successors included Brown's *Zoology*, again with somewhat stiff birds; and Edwards' *Uncommon Birds* and *Natural History*, which include some lively parrots. Latham's *General Synopsis of Birds* has some delightful plates too; though none of these

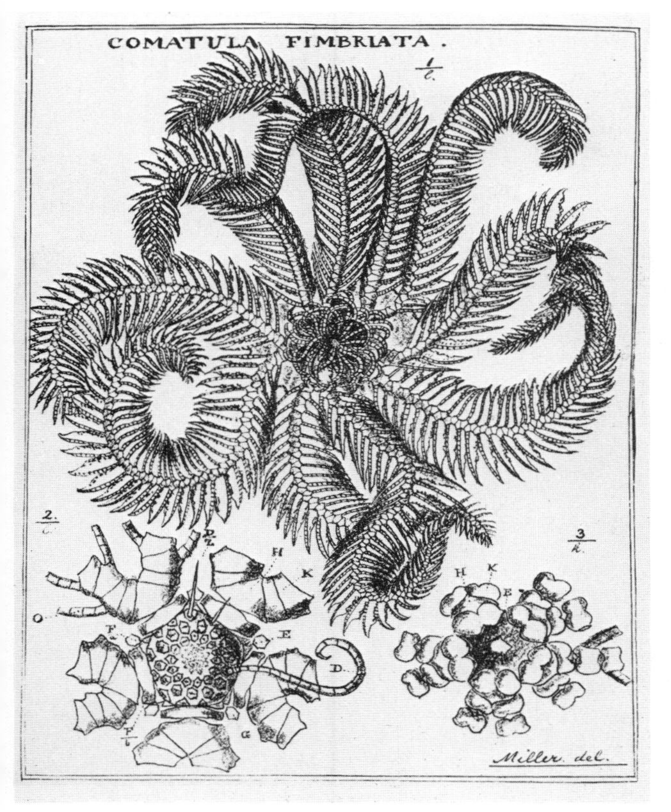

40 (*left*) Starfish (Miller, J. S., *Crinoida*, 1821, frontispiece)
41 (*right*) Seashells (Forbes, E., & Hanley, S., *British Mollusca*, 1848–53, vol 1, pl 23)

have the vivacity which Audubon was to bring into this field. The dodo was illustrated in a treatise of 1848 by Strickland and Melville. Exotic quadrupeds were agreeably depicted in the volumes of Thomas Pennant, and Horsfield's *Java* of 1824.

In the realm of botany, the egregious 'Sir' John Hill produced a handsome *Exotic Botany* and a herbal, of which versions went on appearing well into the nineteenth century. He urged that since the end of the sciences was utility, Culpepper was to be preferred before Linnaeus. Fossil flora in coal measures are beautifully depicted in Mammatt's *Geological Facts*; and delicate, exquisitely coloured, drawings of fungi were published by Bolton in 1788. Barton published a *Flora of North America* in 1821–3; and the *Ferns* of J. E. Sowerby and of Moore enable us to compare the approximately contemporary plates of a very competent artist with 'nature-prints' taken directly from actual specimens. Nature-printing was also used for the beautiful *Sea-weeds* of Johnstone and Croall; they and Moore give copious descriptions of varieties as well as species.

42 Brazilian bittern (Brown, P., *New Illustrations of Zoology*, 1776, pl 34)

43 (*below left*) Parrot (Edwards, G., *Uncommon Birds*, 1747–51, vol IV, pl facing p 175)

44 (*below right*) Penguin (Latham, J., *Synopsis of Birds*, 1781–90, vol III, pt 2, pl 103)

45 (*above*) Moose (Pennant, T., *Arctic Zoology*, 1784–7, vol I, pt 2, pl 7)

46 (*above right*) Malay tapir (Horsfield, T., *Zoological Researches in Java*, 1824)

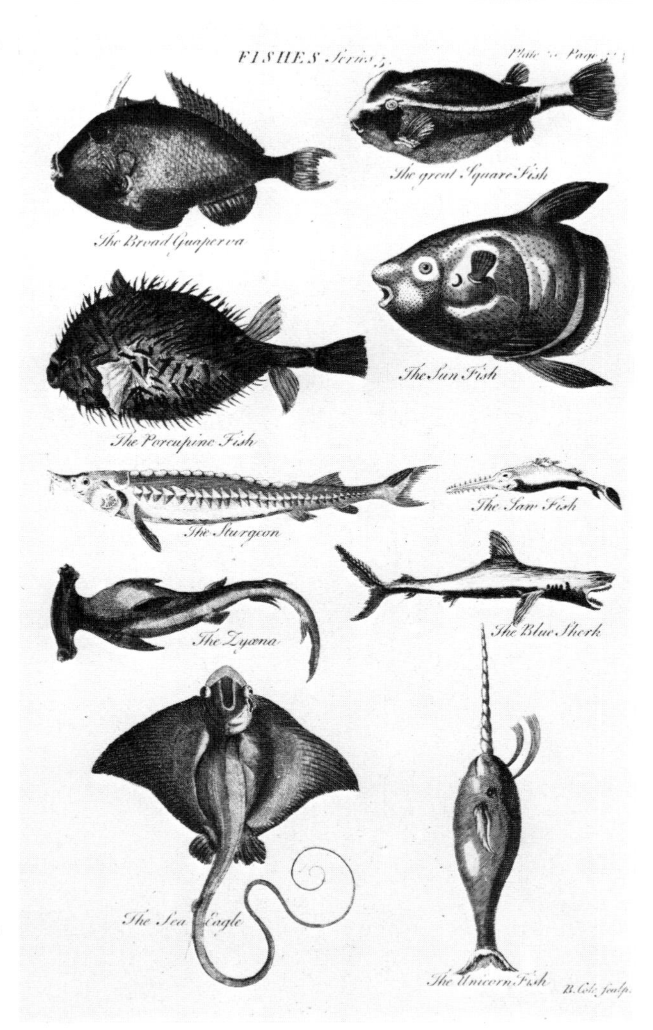

47 Fishes, from 'Sir' John Hill's compilation, *General Natural History*, 1748–52, vol III, pl 16, facing p 314

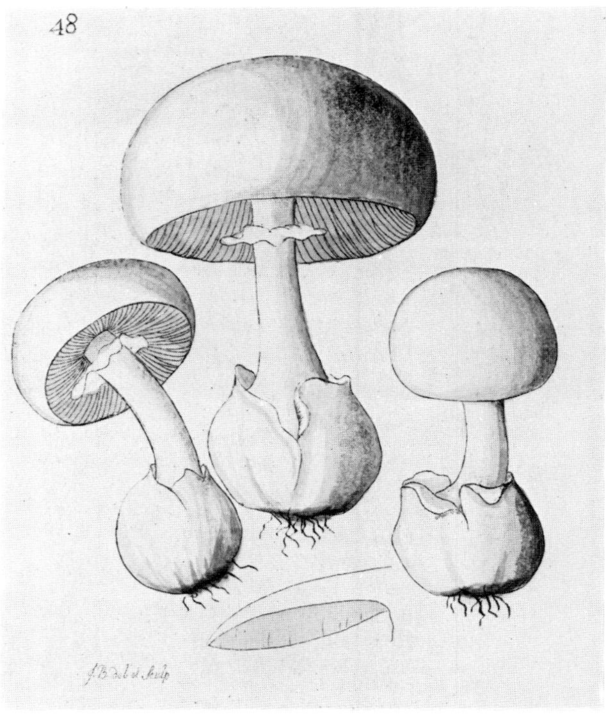

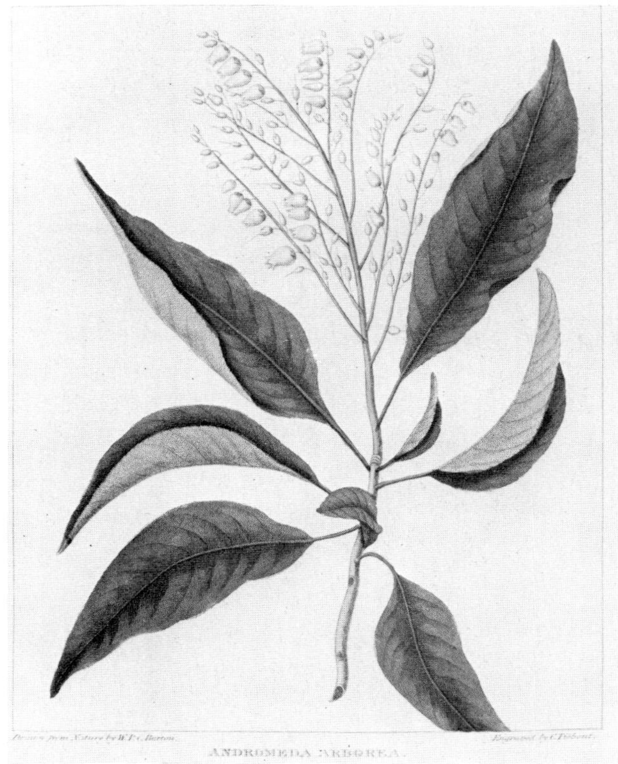

48 (*above left*) Fossil ferns (Mammatt, E., *Geological Facts*, 1836, pl 29)

49 (*above right*) Poisonous toadstool (Bolton, J., *History of Fungusses*, 1788, vol II, pl 48)

50 (*left*) Sorrel tree (Barton, W. P. C., *Flora of North America*, 1821–3, vol I, pl 30)

51 (*opposite*) Black maidenhair spleenwort (Moore, T., *Ferns of Great Britain and Ireland*, 1855, pl 36)

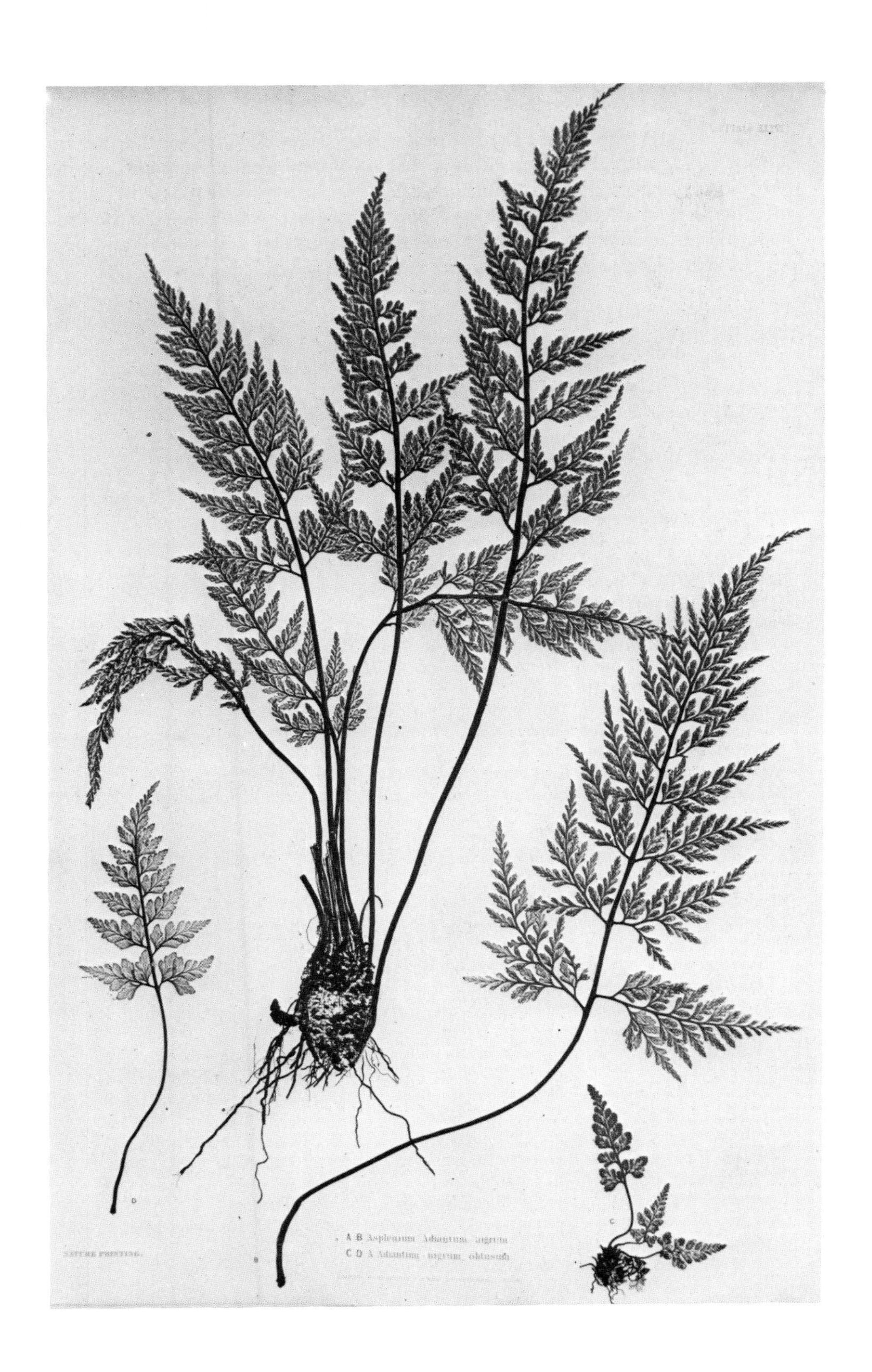

A B Asplenium Adiantum nigrum
C D A Adiantum nigrum obtusum

Both natural history and natural philosophy were readably popularized by Oliver Goldsmith in the 1770s; his *Animated Nature* went through numerous editions. More original contributions to natural history were made by Lazaro Spallanzani, whose works were translated in the last two decades of the century. His experiments refuting spontaneous generation are famous; and lead us towards the researches of Pasteur in the nineteenth century.

5 SCIENCE IN THE EIGHTEENTH CENTURY

Adams, G., *An Essay on Electricity . . .*, 1784; 5th ed., ed. W. Jones, 1799.

Albin, E., *A Natural History of English Insects . . .*, 1720.
A Natural History of Spiders . . ., 1736.
A Natural History of English Song-birds . . ., 1737.

Anon., *The English Farrier . . .*, 1649.

Baker, H., *Employment for the Microscope*, 1753, 2nd ed., 1764.

Barbut, J., *The Genera Insectorum of Linnaeus . . .*, 1780; *Genera Vermium*, 2 vols., 1783–8, English/French.

Beccaria, J. B., *A treatise upon artificial electricity . . .*, 1776.

Bergman, T., *Outline of Mineralogy . . .*, tr. W. Withering, Birmingham, 1783.
An Essay on the Usefulness of Chemistry (tr. J. Bentham and F. X. Schweidaur), 1784.
Physical and Chemical Essays . . ., tr. E. Cullen, 2 vols., 1784.
A dissertation on elective attractions . . ., tr. T. Beddoes, 1785.

Black, J., in: *Essays and Observations, Physical and Literary, read before a Society in Edinburgh*, 1756.
Lectures on the Elements of Chemistry . . ., ed. J. Robison, Edinburgh, 1803.

Boerhaave, H., *A New Method of Chemistry*, tr. P. Shaw and E. Chambers, 1727; 2nd ed., 2 vols., 1741; a pirated edition.
Elements of Chemistry, tr. T. Dallowe, 2 vols., 1735.

Bolton, J., *A History of Fungusses growing about Halifax . . .*, 3 vols., 1788; supplement, 1791.

Bradley, R., *A Philosophical Account of the Works of Nature . . .*, 1721.

Brown, P., *New Illustrations of Zoology . . .*, 1776; English/French.

Buffon, G. L. le Clerc, comte de, *Natural History general and particular . . .*, tr. W. Smellie, 2nd ed., 9 vols., 1785; 1st ed. not seen.
The Natural History of Birds . . ., 9 vols., 1792.
Barr's Buffon: Buffon's Natural History, tr. W. Barr, 10 vols., 1807–12; there seem to be earlier editions.
Natural History of Birds, Fish, Insects and Reptiles, 6 vols., 1808–16.
after Buffon, *A History of the Earth and Animated Nature . . .*, Alnwick, 1810.
The Natural History of Oviparous Quadrupeds . . ., tr. R. Kerr, 4 vols., Edinburgh, 1802.

Candolle, A. P. de, *Vegetable Organography . . .*, tr. B. Kingdon, 2 vols., 1839

Cavallo, T., *A Complete Treatise of Electricity . . .*, 1777; 4th ed., 3 vols., 1795.
A Treatise on Magnetism . . ., 1787.

Chenevix, R., *Remarks upon Chemical Nomenclature . . .*, 1802.

Cowper, W., *Myotomia Reformata: . . . to which is prefix'd an introduction concerning muscular motion*, 1724.

Crawford, A., *Experiments and Observations on Animal Heat . . .*, 1779.

Darwin, E., *The botanic garden . . .*, 2 pts., 1789–91.
Zoonomia . . ., 2 vols., 1794.
Phytologia . . ., 1800.
The Temple of Nature . . ., 1803.
Dickson, S., *An Essay on Chemical Nomenclature . . .*, 1796.
Donovan, E., *The Natural History of British Insects*, 16 vols., 1793–1813.
Instructions for collecting and preserving various subjects of natural history, 1794.
The Natural History of British Birds, 4 vols., 1794–7.
The Natural History of British Shells, 5 vols., 1799–1803.
The Natural History of British Fishes, 5 vols., 1804–8.
Descriptive Excursions through South Wales and Monmouthshire . . ., 1805.
The Natural History of British Quadrupeds, 3 vols., 1820.
Douglas, R., *An Essay concerning the generation of heat in animals*, 1747.
Drury, D., *Illustrations of Natural History . . .*, 3 vols., 1770–82.
Illustrations of Exotic Entomology . . ., ed. J. O. Westwood, 3 vols., 1837; a new edition of the preceding.
Dryander, J., *Catalogus Bibliothecae historico-naturalis Josephi Banks*, 5 vols., 1796–1800.
Edwards, G., *A Natural History of Uncommon Birds . . .*, 4 vols., 1747–51; English/French.
Gleanings of Natural History . . ., 3 vols., 1758–64; English/French.
Ellis, J., *An Essay towards a Natural History of the Corallines . . .*, 1755.
The Natural History of . . . Zoophytes, ed. D. Solander, 1786.
Fordyce, G., *Elements of Agriculture*, Edinburgh, 1765.
Franklin, B., *Experiments and observations on electricity . . .*, 2 pts, 1751–3; 5th ed., 1774.
Freind, J., *Chymical Lectures . . .*, tr. J. M., 1712.
Hales, S., *Vegetable Staticks*, 1727; later eds. formed vol. I of *Statical Essays*; of which vol. II was *Haemastaticks*, 1733.
Harris, J., *Lexicon Technicum . . .*, 2 vols., 1704–10.
Harris, W. S., *On the Nature of Thunderstorms . . .*, 1843.
Hauksbee, F., *Physico-mechanical experiments on various subjects . . .*, 1709.
Hauksbee, F., the younger, *A Course of mechanical, magnetical . . . experiments*, *c.* 1730; a syllabus, illustrated, of lectures by J. J. Whiteside, Whiston, and Hauksbee.
Higgins, B., *Minutes of the Society for Philosophical Experiments and Conversations*, 1795.
Experiments and Observations relating to Acetous Acid . . ., 1796.
Higgins, W., *A Comparative View of the Phlogistic and Antiphlogistic Theories*, 1789.
Experiments and Observations on the Atomic Theory . . ., 1814.
Hill, J., *A General Natural History*, 3 vols., 1748–52.
Essays in Natural History and Philosophy . . ., 1752.
The British Herbal, 1756.
Exotic Botany illustrated . . ., 1759.
Hoadly, B., and Wilson, B., *Observations on a series of electrical experiments*, 1756.
Huber, F., *New observations on the natural history of bees*, 1806.
Huber, P., *The natural history of ants*, tr. J. R. Johnson, 1820.
Johnstone, W. A. and Croall, A., *The Nature-printed British Sea-weeds*, 1859–60.
Jussieu, A. de, *The Elements of Botany*, tr. J. H. Wilson, 1849.
Kirby, W., and Spence, W., *An Introduction to Entomology . . .*, 4 vols., 1815–26.
Kirwan, R., *An Essay on Phlogiston . . .*, 1787; 2nd ed., 1789.
Kunkel, J., Stahl, G. E., and Fritschius, J. C., *Pyrotechnical Discourses . . .*, 1705.
Latham, J., *A General Synopsis of Birds*, 3 vols. in 10, 1781–90.
Lavoisier, A. L., *Essays, Physical and Chemical*, tr. T. Henry, 1776.
Essays, on the Effects produced by various processes on Atmospheric Air, tr. T. Henry, Warrington, 1783.
Elements of Chemistry, tr. R. Kerr, Edinburgh, 1790.

et al., Method of Chymical Nomenclature . . ., tr. J. St. John, 1788.
Lee, J., *An Introduction to Botany*, 1760.
Lemery, N., *A Course of Chymistry*, tr. W. Harris, 1677; another ed., tr. J. Keill, 1698.
Leslie, P. D., *A Philosophical Inquiry into the Cause of Animal Heat* . . ., 1778.
Linnaeus, C., in: *Miscellaneous tracts relating to natural history* . . ., tr. B. Stillingfleet, 1762.
Institutes of Botany . . ., tr. C. Milne, 1771; never completed.
A generic and specific description of British Plants . . ., tr. and ed. J. Jenkinson, Kendal, 1775.
A System of Vegetables . . ., 2 vols., Lichfield, 1783.
Reflections on the Study of Nature . . . (tr. J. E. Smith), 1785.
A Dissertation on the Sexes of Plants, tr. J. E. Smith, 1786.
The Families of Plants . . . (tr. E. Darwin), 2 vols., Lichfield, 1787.
The Animal Kingdom . . ., vol. I, pts 1 and 2, tr. R. Kerr, 1792.
Elements of Natural History; being an introduction to the Systema Naturae of Linnaeus, 2 vols., 1801–2.
A General System of Nature, tr. W. Turton, 7 vols., 1802–6.
Lachesis Laponica, or a tour in Lapland, tr. J. E. Smith, 2 vols., 1811.
A Selection from the Correspondence . . ., ed. J. E. Smith, 1821.
Macquer, P. J., *Elements of the Theory and Practice of Chemistry* (tr. A. Reid), 2 vols., 1758.
A Dictionary of Chemistry . . ., 2 vols., 1771.
(Mandeville, B. de), *The Fable of the Bees* . . ., 2 pts, 1714–29; the 2nd pt has more relation to science; ed. F. B. Kaye, 2 vols., Oxford, 1924.
Markham, G., *Markham's Maister-peece, or, what doth a Horse-man lacke* . . ., 1610.
The English Husbandman . . ., 3 pts, 1613–15.
Martin, B., *An Essay on Electricity* . . ., Bath, 1746.
Martyn, T., *The Universal Conchologist* . . ., 4 vols., 1784; English/French.
The English Entomologist . . ., 1792.
Miller, P., *The Gardener's and Florist's Dictionary*, 2 vols., 1724.
Mills, J., *A Treatise on Cattle*, 1776.
Moises, H., *A Treatise on the Blood* . . ., n.d. (1794).
Nollet, J. A., *Lectures in Experimental Philosophy*, tr. J. Colson, 1752.
Pemberton, H., *A Course of Chemistry* . . ., ed. J. Wilson, 1771.
Pennant, T., *British Zoology*, 4 vols., 1768–70; earlier version, by T. P., 2 vols., 1766.
Synopsis of Quadrupeds, 1771.
History of Quadrupeds, 2 vols., 1781.
Introduction to Arctic Zoology, 2 vols., 1784–7.
Pitcairne, A., *The Whole Works*, 1715.
The philosophical and mathematical elements of physick, tr. J. Quincy, 2nd ed., 1745; 1st ed. not seen. A satire on Pitcairne is *Apollo Mathematicus: or the art of curing diseases by the mathematics*, 2 pts, 1695.
Price, R., *A Free Discussion of the doctrines of materialism and philosophical necessity*, 1778.
Priestley, J., *The History and Present State of Electricity*, 1767.
The History and Present State of Discoveries relating to Vision, Light, and Colours, 1772.
Hartley's Theory of the Human Mind, 1775.
Experiments and Observations on different Kinds of Air . . ., 6 vols., 1774–86; three-volumed, Birmingham, 1790.
Disquisitions relating to Matter and Spirit, 1777.
Pulteney, R., *Historical and Biographical Sketches of the Progress of Botany in England*, 2 vols., 1790.
Réaumur, R. A. Ferchault de, *The Natural History of Bees* . . ., 1744.
The Art of Hatching and bringing up domestic fowls by means of Artificial Heat, 1750.

after Réaumur: *A Short History of Bees . . .*, London, 1800.
Rigby, E., *Chemical Observations on Sugar*, 1788.
Rousseau, J.-J., *Letters on the Elements of Botany . . .*, tr. T. Martyn, 1785.
Martyn, T., *Thirty-eight plates . . . adapted to the Letters on the Elements of Botany*, 1788.
Scheele, C. W., *Chemical observations and experiments on air and fire . . .*, tr. J. R. Forster, 1780.
Chemical Essays . . . (tr. F. X. Schwediauer and T. Beddoes), 1786.
Seward, A., *Memoirs of the Life of Dr Darwin*, 1804.
Shaw, P., *Three Essays in Artificial Philosophy . . .*, 1731.
Chemical Lectures . . ., 1734.
and Hauksbee, F., *An Essay for introducing a Portable Laboratory*, 1731.
Sibly, E., *An Universal System of Natural History*, 14 vols., 1794–1807; vols. 8–14 tr. from Linnaeus.
Sinclair, G., *Hortus Gramineus Woburnensis*, 1816.
Smith, J. E., *A Sketch of a Tour on the Continent*, 3 vols., 1793.
A Specimen of the Botany of New Holland, vol. I, 1793.
The Natural History of the Rarer Lepidopterous Insects of Georgia, 2 vols., 1797; English/French.
Tracts relating to Natural History, 1798.
Exotic Botany . . ., 2 vols., 1804–5.
An Introduction to Physiological and Systematical Botany, 1807.
A Grammar of Botany . . ., 1821.
Snape, A., *The Anatomy of an Horse . . .*, 1683.
Sowerby, J., *English Botany*, text by J. E. Smith, 37 vols., 1790–1814.
Coloured Figures of English Fungi or Mushrooms, 3 vols., 1797–1809.
British Mineralogy . . ., 5 vols., 1804–17.
A New Elucidation of Colours . . ., 1809.
Sowerby illustrated J. E. Smith's *Exotic Botany*, and *Botany of New Holland*.
Spallanzani, L., *Dissertations relative to the natural history of animals and vegetables . . .* (tr. T. Beddoes), 2 vols., 1784.
Tracts on the nature of animals and vegetables (tr. J. G. Dalyell), Edinburgh, 1799.
Stahl, G. E., *Philosophical Principles of Universal Chemistry*, tr. P. Shaw, 1730.
Strutt, J. G., *Sylva Britannica, or Portraits of Forest Trees*, 1822.
Deliciae Sylvarum, or grand and romantic forest scenery . . ., 1828.
Stubbs, G., *The Anatomy of the Horse*, 1766.
Swammerdam, J., *The Book of Nature . . .*, tr. T. Flloyd, rev. J. Hill, 1758.
Taylor, T. (tr.), *The Philosophical and Mathematical Commentaries of Proclus . . .*, 2 vols., 1788–9.
The Works of Plato, 5 vols., 1804.
The Works of Aristotle, 10 vols., 1806–12.
Tournefort, J. P. de, *The Complete Herbal*, tr. T. Martyn, 2 vols., 1719–30.
Trembley, A., in: Adams, G., *Micrographia Illustrata*, 1746.
Wakefield, P., *An Introduction to Botany*, 1796.
Excursions in North America, 1806.
An Introduction to Entomology . . ., 1815.
Watson, R., *Chemical Essays*, 5 vols., 1781–7.
White, G., *The Natural History and Antiquities of Selborne . . .*, 1789.
Wildman, T., *A Treatise on the Management of Bees . . .*, 1768.
Wilson, B., *An essay towards the explication of the phænomena of electricity, deduced from the æther of Sir Isaac Newton*, 1746.
A Treatise on Electricity, 1750.
A Series of Experiments relating to Phosphori . . . with two memoirs . . . by J. B. Beccari, 1775.

An Account of experiments made at the Pantheon on the nature and use of conductors . . ., 1778.
A Short View of Electricity, 1780.
Wilson, G., *A Compleat Course of Chymistry*, 1699.
Withering, W., *A Botanical Arrangement of all the vegetables naturally growing in Great Britain*, 2 vols., Birmingham, 1776.
A Botanical Arrangement of British Plants, 3 vols., Birmingham, 1787–92.
A Systematic Arrangement of British Plants, 1801.
Miscellaneous Tracts . . ., 2 vols., 1822.

Recent Publications

Banks, J., *The* Endeavour *Journal, 1768–71*, ed. J. C. Beaglehole, 2 vols., Sydney, 1962; this edition supersedes that of J. D. Hooker, 1896.
Bergman, T., *Foreign Correspondence*, ed. G. Carlid and J. Nordström, Stockholm, 1965–.
Black, J., *Notes from Dr Black's Lectures on Chemistry*, by T. Cochrane, ed. D. McKie, Wilmslow, Cheshire, 1966.
Clark-Kennedy, A. E., *Stephen Hales, D.D., F.R.S.*, Cambridge, 1929.
Crosland, M. P., *Historical Studies in the Language of Chemistry*, 1962.
Darwin, E., *The Essential Writings*, ed. D. King-Hele, 1968.
Kent, A., *An Eighteenth Century Lectureship in Chemistry*, Glasgow, 1950.
Lavoisier, A. L., *A Bibliography*, by D. I. Duveen and H. S. Klickstein, 1954; *Supplement*, 1965.
Lindeboom, G. A., *Herman Boerhaave*, 1968.
Lovejoy, A. O., *The Great Chain of Being*, Cambridge Mass., 1936.
Mendelsohn, E., *Heat and Life*, Cambridge, Mass., 1965.
Meyer, G., *The Scientific Lady in England, 1650–1760*, Los Angeles, 1965.
Priestley, J., *A Bibliography*, by R. E. Cook, 1966.
A Scientific Autobiography, ed. R. Schofield, Cambridge, Mass., 1966.
Ritterbush, P. C., *Overtures to Biology; the Speculations of Eighteenth Century Naturalists*, 1964.
Schierbeek, A., *Jan Swammerdam*, Amsterdam, 1967.
Schofield, R., *The Lunar Society of Birmingham*, Oxford, 1963.
Thackray, A., *Atoms and Powers*, Cambridge, Mass., 1970.
Wheeler, T. S., and Partington, J. R., *The Life and Work of William Higgins, Chemist*, Oxford, 1960.
White, G., *A Bibliography*, by E. A. Martin, 1897; rev. ed., 1934.
Woodward, B. B., Wilson, W. R., and Soulsby, B. H., *A Catalogue of the Works of Linnaeus*, 2nd ed., 1933.

Additional Reading

Hufbauer, K., *The Formation of the German Chemical Community 1720–95*, Berkeley, 1982.
Joppien, R., & Smith, B., *The Art of Captain Cook's Voyages*, 2 vols., New Haven, 1985.
Jordanova, L., & Porter, R. (eds.), *Images of the Earth*, 1977.
King-Hele, D., *The Letters of Erasmus Darwin*, Cambridge, 1981.
Knight, D. M., *Zoological Illustration*, 1977.
The Transcendental Part of Chemistry, 1977.
Porter, R., *The Making of Geology*, Cambridge, 1977.
Sepper, D. L., *Goethe Contra Newton*, Cambridge, 1988.
Smith, B., & Wheeler, A., *The Art of the First Fleet*, New Haven, 1988.
Stansfield, D. A., *Thomas Beddoes 1760–1808*, Dordrecht, 1984.
Underwood, E. A., *Boerhaave's Men*, Edinburgh, 1977.

6 *The Industrial Revolution*

Although chemistry, electricity, and natural history all made these great strides in the eighteenth century, as far as England is concerned the period is perhaps chiefly exciting on account of the beginning of the transformation of the economy in the Industrial Revolution. The nature of the relationship between science and technology remains unresolved; it seems that whereas few scientists, until the rise of nuclear physics, were not happy to see their work applied in technology, this was in most cases not their chief motive. That scientific discoveries would be beneficial in a material sense remained in most cases a pious hope from Bacon's time until the nineteenth century was well advanced. Benjamin Franklin's retort, 'What is the use of a baby?' was typical; and so was the comparison between a new discovery and a recently explored territory which had not yet been exploited. Some discoveries were rapidly applied; the pressure cooker came within a generation of the discovery that the atmosphere exerted a pressure, and Newcomen's atmospheric steam engine within another generation. The use of coal-gas for lighting followed hard upon the isolation of distinct gases. On the other hand, the practical use of electric power came long after the discoveries of Oersted and Faraday which made it possible. The reasons would seem to be partly technical and partly economic; and technologists usually seem to have applied the science of a previous generation in their efforts to do something better or more cheaply than their predecessors. The study of the Industrial Revolution is more the province of the economic historian than of the historian of science.

On the other hand, science and engineering did not become professionalized in England until well into the nineteenth century, and the line between scientists and technologists is hard to draw. To suggest that all scientists were mandarins pursuing harmony and intellectual beauty alone is as misleading as to suppose them all utilitarians. Recent studies of the Lunar Society of Birmingham has shown how close were relations between some scientists and some industrialists, and how the talents of the discoverer and of the entrepreneur were fruitfully combined. Even those who sought in the sciences light on revealed religion, or data for a cosmology, could not escape from technology, for they needed to use apparatus of one kind or another. Thus in the biological field, discoveries in embryology and bacteriology became possible only as microscopes were improved: this has been described as the existence of a technical frontier. In pure technology, Watt's improvements to the steam engine required the much lower tolerances which only Boulton could provide. In chemistry, W. H. Wollaston's discovery of a process for getting platinum into a metallic (instead of grey powdery) form, made available to chemists crucibles in which reactions, hitherto very difficult or impossible to follow, could be readily made to occur. The use of this metal in jewels came much later, fortunately for nineteenth-century chemists. Similarly, towards the end of the century, the study of the phenomena of discharge tubes, which led to an understanding of spectra and of cathode rays, was made possible by the development of pumps with which very high vacua could be obtained.

An important early work on mechanics was John Wilkins' *Mathematical Magick; or the Wonders that may be performed by Mechanical Geometry*. The title is deceptive because the book contains nothing of magic, as a Renaissance mathematical book might well have done, but is concerned with the science of mechanics. Mechanical devices such as screws, pulleys, levers, and gear-wheels are explained, and Wilkins makes it clear that in mechanics we cannot get something for nothing; that by arrangements of devices we can move a greater weight, but it will be moved more slowly. He is somewhat equivocal on the question of perpetual-motion machines, which he is unwilling to exclude as a possibility; but speculations on this topic persisted among men of science well into the eighteenth century. Wilkins showed that one could contrive circular motions as fast as the *primum mobile* had been supposed to go; or, correspondingly, immeasurably slowly, which seemed to him more amazing. He described the siege-engines of antiquity, with diagrams, and concluded that those designed by Archimedes exceeded in power the largest cannon in use (by the Turks) in the seventeenth century.

Since the days of Hero of Alexandria, mechanical devices had given delight essentially as toys; and Wilkins devotes a section to automata. One of these is the smoke-jack, a useful device which consisted of a propeller placed in the chimney which, being turned by the smoke, caused a spit to rotate: it could also, he adds, chime little bells or rock a cradle. Another device is a sailing chariot; and there follow descriptions of clocks and of dolls in animal or human form which execute various intricate motions. The most dramatic of these were the toy birds that flew; Archytas of Tarentum had made one in antiquity, and Regiomontanus had made a fly and an eagle at Nuremberg – which occasioned Sir Thomas Browne's remark that the fly was to be preferred. Wilkins suggested that a flying machine could be constructed which would hold man; and that ultimately they might even fly to the moon in such a contraption. Wilkins' book is not an important original work, but it did make available, in English, knowledge of mechanics, and it proved popular.

52 Boring elms for water-mains (Caus, I. de, *Water Works*, 1659, pl 12)

Thus it is referred to in the earliest work by an English writer devoted to the raising of water: D'acres' *Art of Water Drawing* which appeared in 1659. His machine consisted of a vessel immersed in cold water, connected to a furnace by a tube with a stop-cock. When the cock is opened the vessel fills with hot gases; then the cock is closed, the gases cool and contract, and the atmospheric pressure forces the water up a pipe from the place it is desired to drain into the vessel. Then the cock from the furnace is again opened, this water and the cool gases are driven out and the cycle begins again. D'acres urged that among the advantages of his machine were that it took up little room, had few moving parts, and would have a high value as scrap at the end of its life. He also discussed perpetual motion, and the various forms of pumps that were available; but his book contained no diagrams, and appears not to have made any impact in its own day. Better known is the handsome illustrated volume on water-works by Caus; and the work of Edward Somerset, marquis of Worcester, whose *Century of . . . Inventions* of 1663 contains, among numerous other matters, a description of a kind of pump depending upon the condensation of steam. Worcester secured a patent for his 'water commanding engine': but the serious prehistory of the steam engine is usually held to begin with Thomas Savery. His *Miner's Friend* was published in 1702, containing pictures of his machine, which he had patented in 1698 and began to manufacture himself in London. Steam was admitted into a vessel which was then chilled with cold water. As the steam condensed, the pressure of the atmosphere pushed water up from the mine or well into the vessel, as in D'acres' device, and the water was driven out or further upwards by a fresh ingress of steam. The stop-cocks were turned by hand, and the device was neither efficient nor rapid in operation; but it proved the possibility of harnessing steam, or rather heat, to make it do mechanical work in place of the animals, rivers, and winds which had been the only prime movers hitherto available.

Wilkins' book was concerned with theoretical mechanics, and the others with water-pumping: they do not tell us in detail how the mechanic of the

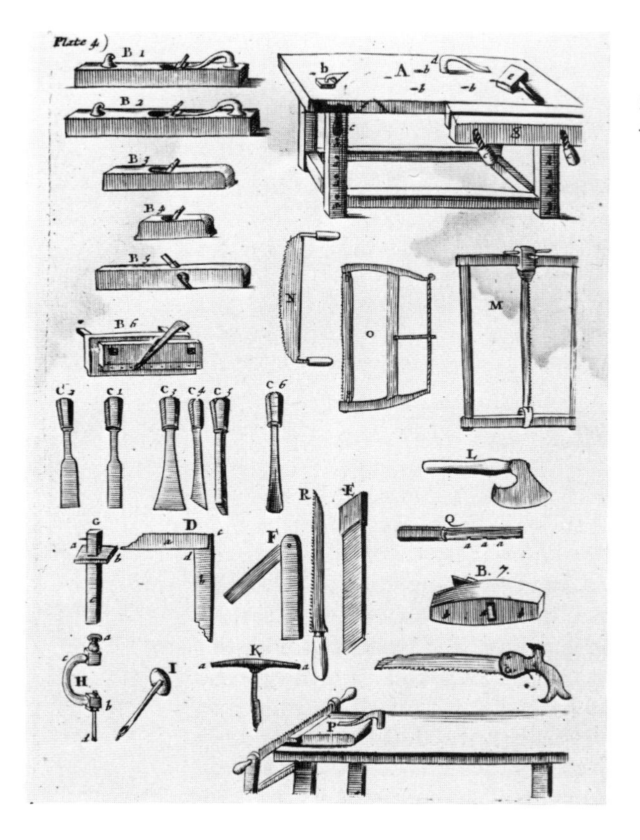

53 Workbench and tools (Moxon, J., *Mechanick Exercises*, 1678, pl to pt 4)

seventeenth century made, for example, his gear-wheels. We can find these technical details in the *Mechanick Exercises* of Joseph Moxon, hydrographer to the king, which came out in monthly parts beginning in 1677, and eventually formed two volumes. The second is entirely devoted to printing, and the first to teaching the basic operations of wood and metal work. In his preface he declared that there was nothing sordid about manual labour; and that that of the smith formed the best introduction to the rest of the skilled trades. Each part was illustrated with a copperplate engraving showing the various tools and operations described. The apprentice engineer would find most of these familiar today. Such processes as riveting and forging are described; small screws could be made using a plate through which the screw-pin was turned to acquire its thread. But large screws were cut individually, using a cold chisel. We find how to case-harden iron, and how to temper steel; how to make a level plain of iron, how to use the various tools of the joiner and the smith, and how to make glue. As a description of how the seventeenth-century craftsman set about his labours, and of what means were at his disposal, Moxon's account is invaluable.

Although the engineer of today would recognize these tools and techniques, his subject was completely transformed during the eighteenth century by the development of the steam engine. It is to Thomas Newcomen, about 1712, that we owe the true steam engine, of which such machines as Savery's were but precursors; though it should be noted that it was not until the closing years of the eighteenth century that steam engines were applied to drive machinery directly, and not simply to pump water. They were thus crucial for the extractive industries rather than for manufacturing; the factory system was well under way before steam engines were applied to driving the machinery. Newcomen's engines were of very low efficiency and it was, again, not until about 1800 that the power available from steam engines began to exceed greatly that obtained from windmills or water-wheels. But a device had been put in the hands of engineers which, given improvements in theory and in technical practice, was capable of immense development. The question of whether the steam engine gave more to science than vice versa has often been raised; but it would certainly be a mistake to suppose that such men as Newcomen, Smeaton, and Watt were ignorant craftsmen not in close touch with leading men of science. Newcomen's engine represents a synthesis of different devices coming ultimately from both Europe and Asia; it was self-acting in that the valves were opened and closed automatically. Newcomen's first engine was built at Dudley Castle, Staffordshire, in 1712: by 1715 several were at work in the coalfields of Northumberland and Durham; and in 1725 the York Buildings Company in London installed one to pump water, in place of Savery's pump which had not been satisfactory, but it proved too expensive to run. The cost of advanced technology was as hard to forecast in

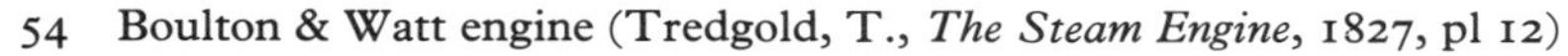
54 Boulton & Watt engine (Tredgold, T., *The Steam Engine*, 1827, pl 12)

the eighteenth century as it is today. The heavy cost in fuel of these engines was a reason for using them chiefly to pump out coal-mines. Steam engines were assembled *in situ*, and the brick and wood parts were locally obtained. The beam arrangement worked very well for pumping, and meant that it could be erected in sites that were not accurately levelled. One of the differences between the histories of science and technology is that in the latter to find descriptions and pictures of machines one usually has to go to books not written by the inventor himself. Thus one of the best-known plates of a Newcomen engine is to be found in Desagulier's *Experimental Philosophy*; and books of natural philosophy and political economy between them provide numerous illustrations of machinery. But there are standard early histories of steam engines – by Thomas Tredgold, an engineer; by John Farey; and by Rigg – which are illustrated and valuable.

There were projects to illustrate trades in the eighteenth century, and the plates of Diderot's *Encyclopédie* are justly famous as showing a great range of crafts as practised on the eve of the Industrial Revolution. Rather surprisingly, a similar compilation, *T'ien-Kung K'ai-Wu* had appeared in China in the seventeenth century, giving us the opportunity to compare crafts in two very different regions before those of Europe were transformed by the advent of steam. Encyclopedias published in Britain followed the French in giving a considerable amount of space to technical matters, and the plates of the *Britannica* in its various editions, of the *Metropolitana*, and of Rees' *Cyclopedia*, are excellent as illustrating machinery from the end of the eighteenth century. There had been technical dictionaries from a good deal earlier: Harris' *Lexicon Technicum* has been mentioned in connection with Newton's views on chemistry, but it was chiefly concerned with techniques and is a valuable source of information. Much chemistry and physics, technical as well as theoretical, could be learned from the dictionaries of such men as Charles Hutton, William Nicholson, and Thomas Brande, but we shall examine them more fully in discussing the popularization of science in the nineteenth century.

We must be careful not to let the advent of the steam engine blind us to other achievements of the eighteenth century. Agriculture passed through a revolution almost as dramatic as that in manufacturing, though it was not until about 1800 that scientists began to have much to do with it, and not until the mid-nineteenth century, with the work of Justus von Liebig, that the chemist had anything very useful to contribute to the farmer. New forms of plough and harrow were invented in the eighteenth century; and in 1701 Jethro Tull invented the first practicable seed-drill, which he described in his *Horse-houghing Husbandry* of 1731. An advantage of thus sowing seeds in neat rows instead of broadcasting them was that they could be easily weeded; and Tull's horse-hoe was designed to dig up the weeds between the rows. It was apparently particularly successful with turnips, and helped to encourage their adoption into crop rotations. Tull seems to have believed that the mere breaking up of the soil made it easier for the roots of plants to absorb particles of nourishment, and he was sceptical of the use of manure. Duhamel de Monceau was an advocate in France of Tull's methods; his books were translated into English and admired by Arthur Young. In 1785 Robert Ransome invented the self-sharpening ploughshare, in which one side was made harder than the other by more rapid cooling; and in 1808 he designed an iron frame for ploughs from which broken parts could easily be removed and replaced. Agricultural machinery available by the mid-nineteenth century is described and illustrated in J. A. Ransome's *Implements of Agriculture* of 1843, in Wilson's *Farmer's Dictionary* of 1850–52, and in the encyclopedias of agriculture of J. C. Loudon and of J. C. Morton, which describe general farming practice as well, in considerable detail. With new forms of plough, with seed-drills, reapers, and threshing machines, the up-to-date farm of the nineteenth century was very different from that of two or three generations earlier. By 1850 steam engines, stationary or on wheels, were beginning to supply power on farms. A good idea of rustic life and trades in this period can be got from the delightful illustrations of William Pyne's *Microcosm* of 1845, though no doubt the reality was less Arcadian; and, a little later, urban life and industry are depicted with greater grimness in Gustave Doré's famous drawings for his *London: A Pilgrimage* of 1872.

While improvements of machinery were one aspect of the Agricultural Revolution, the changes in crop rotation and the selective breeding of farm animals, both associated with enclosure of the common fields, were really more important. The works of Arthur Young, the tireless advocate of improved agriculture, need no recommendation, for they are enjoyable to read as well as being standard sources of information on social history and on agricultural practice. The *General Views* or reports on different counties, some by Young, which were published by the Board of Agriculture – of which he was Secretary – about 1800, give an exceedingly valuable picture of the state of farming of the period; a time when such events as the sheep-shearings at Woburn and Holkham were attended by a distinguished company of landowners and men of science. A five-volume *Review and Abstract* of the

55 Woodstock waggon (Young, A., *General View of the Agriculture of Oxfordshire*, 1809, pl 23)

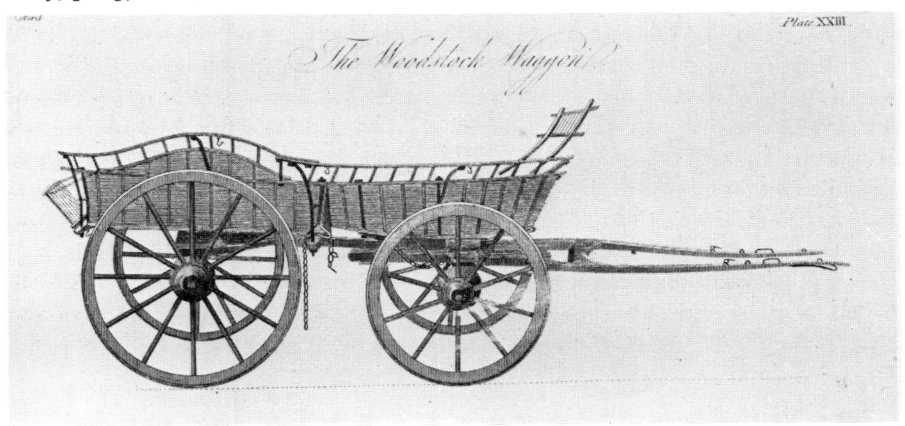

Reports was made by William Marshall and is a most useful survey of current practices. The improved strains of farm animals developed in the late eighteenth century by such breeders as Robert Bakewell are depicted in Thomas Bewick's *History of Quadrupeds*, where they appear with amusing but less accurate depictions of such creatures as the giraffe and the rhinoceros.

Boyle had hoped that chemistry might be able to contribute to agriculture, and some of the ideas of the French potter Bernard Palissy on the marling of soil were advocated in seventeenth-century England, notably in Plat's *Jewell House of Art and Nature*, and in Blith's *English Improver* of 1649, which describes early 'field trials'. Marl is a soil consisting of calcium carbonate and clay, which is a valuable fertilizer. The diarist, and fellow of the Royal Society, John Evelyn wrote, as well as his famous *Sylva* and *Pomona* dealing with forest trees and fruit trees, *A Philosophical Discourse of Earth*, published in 1676. This remained the standard work on soil science for more than a century, and was very widely read. Evelyn adhered to the view, supported by experiments made by Nicholas of Cusa, van Helmont, and Boyle, that plants grew by transmuting water, which was all that they apparently took in. This hypothesis was only overturned when in the late eighteenth century Priestley, Lavoisier, and Ingenhousz isolated oxygen and elucidated the process of photosynthesis, by which plants in sunshine derive their carbon from the carbon dioxide of the air, and release the oxygen. Evelyn gave recommendations for digging over the soil, because exposure to the air increased its fertility, and for preparing compost, using various kinds of dung. Suitable arrangements for collecting and conserving dung are illustrated, *inter alia*, in Hale's *Compleat Body of Husbandry* of 1756. Evelyn had previously, in 1664, published the first gardening calendar, *Kalendrium Hortense*. In 1757 there appeared Francis Home's *Principles of Agriculture and Vegetation*, which was an attempt to marry chemistry and agriculture, and discussed the nature of soils and composts, and their effects on plants.

In the last quarter of the eighteenth century the Bath and West and the Highland and Agricultural Societies were founded, and began to promote an interest in agricultural science, as did the Board of Agriculture set up in 1793: and the Royal Society of Arts took a great interest in agricultural innovations. The Royal Society of Agriculture received its charter in 1810. The Board of Agriculture arranged for Humphry Davy, who had recently been appointed, in his early twenties, Professor of Chemistry at the Royal Institution, to give courses of lectures annually, beginning in 1802. These lectures were published, as *Elements of Agricultural Chemistry*, in quarto format in 1813. There was a second, octavo, edition in the following year, followed by others in England and America, and the work was translated into German and French; so that although the book received somewhat tepid reviews it was very influential. Davy was an unusually talented lecturer, and the style of the book is easy, although the treatment does not rise to the level of his writings on electrochemistry. Davy stressed the importance of the absorption by the roots of plants of fluid matter from the soil, believing that solid substances must be liquefied by fermentation before they could nourish the plant. He advocated soil analysis, and described and illustrated apparatus for doing such analyses in which gases were collected, in an old-fashioned manner, in bladders.

In 1819 there was published William Grisenthwaite's *New Theory of Agriculture*, which attracted little attention, being written by an obscure apothecary from Wells; but which anticipated some of Liebig's conclusions. For Grisenthwaite, agriculture was a means of transforming manure into vegetable matter; soil analysis was therefore of relatively little importance, compared to attention to fertilizers, the ideal composition of which could be determined by analysing plants to see what they were composed of. In 1838 there appeared another book on soil, John Morton's *Nature and Property of Soils*, which introduced geological considerations as well as the emphasis on textures of soils which had hitherto preoccupied investigators; although Davy, who claimed that he had given the first public geology lectures in London, devoted half a chapter of his book to elementary geology. Morton was not a chemist, but a pioneer of soil physics; his advice was directed to improving the texture of the soil, by the addition of various substances to make it lighter if it were too thick, and heavier if it were too loose.

Neither of these books seems to have made any contemporary impact, and the next scientist whose work modified agricultural science was Charles Daubeny, Professor of Chemistry and of Botany (simultaneously) in the University of Oxford, and one of those chiefly responsible for the introduction of honours degrees in the sciences at Oxford. Daubeny's experiments of growing plants in rotations on some plots and consecutively on others were published in the *Philosophical Transactions*; but some essays on agriculture appeared in his entertaining *Miscellanies* of 1867, along with others on the difficulties of convincing other academics of the importance of the sciences. From the point of view of agriculture, Daubeny's most important pupil was John Bennet Lawes, the founder in 1843 of the Rothamsted Experimental Station. Lawes was the inventor of superphosphate fertilizer, the secret of which was to get the phosphate into a soluble form in which it could be used by plants. This fertilizer proved particularly valuable for turnips. Lawes' experiments were conducted along the lines of Daubeny's, but on a much more extensive scale; they were published by A. D. Hall in 1905, after appearing in journals.

Davy's book was still selling in 1839, ten years after his death, when his *Collected Works* in nine volumes began to come out; and in order that they should not compete, the *Agricultural Chemistry* was split so that it fills two half-volumes of the *Works*. But in 1840 there appeared Justus von Liebig's bold essay, *Organic Chemistry in its application to Agriculture and Physiology*, which revolutionized the subject; after this the importance of chemistry to farming could no longer reasonably be doubted. Liebig's agricultural ideas can also be found in his more popular *Familiar Letters on Chemistry*, the first edition of which was a slim volume, which grew and was modified rapidly through successive editions, remaining always interesting, controversial, concerned with general problems, and readable. Liebig believed that organic matter was unnecessary for the growth of plants. The ashes of a given normal crop should be analysed; the soil should then also be analysed and anything lacking which the crop would require should be made up by adding fertilizer. Although in the first edition of the *Agricultural Chemistry* Liebig suggested that nitrogenous fertilizer should be added to the soil, in later editions he came down in favour of purely mineral fertilizers. His book gave rise to controversy on this question, and also over whether fertilizers were food or stimulants for plants. Lawes and Gilbert, at Rothamsted, published experiments in the *Journal of the Royal Agricultural Society*, a very important vehicle, which showed that plants do benefit from nitrogen; but the existence of crops like clover which 'fix' atmospheric nitrogen, and thereby enrich the soil, complicated research on this topic. Of less originality than Liebig's writings but of some importance are the books of J. F. W. Johnston, a pupil of Berzelius who was one of the founders of the British Association for the Advancement of Science, and became Reader in Chemistry at the new University of Durham in 1832. His *magnum opus* was *Elements of Agricultural Chemistry and Geology*; his shorter *Lectures on Agricultural Chemistry*, and *Contributions to Scientific Agriculture*, are noteworthy; but his most successful book was his *Catechism of Agricultural Chemistry and Geology* which passed through an enormous number of editions. His last book, *The Chemistry of Common Life*, is deservedly a classic. He visted America, and made important contributions to the beginning of experimental farming there.

We have in our discussion of agriculture got far beyond the point at which we left engineering. Newcomen's steam engine held out the promise of supplies of power on a scale not previously dreamed of; but perhaps more important for ordinary life was the revolution in communications which began with the building of canals and great bridges in the eighteenth century, and culminated in the building of the main railway network in the 1830s and 1840s. The men responsible include Brindley, Rennie, Smeaton, Telford, Metcalf, McAdam, the Brunels, and the Stephensons. Still one of the best sources of information is Samuel Smiles, in his *Lives of the Engineers*, which are amply illustrated. Smiles is notorious for his philosophy of self-help, but this seems to have influenced him in his choice of engineers – and in *Industrial Biography*, of toolmakers – and not led to serious distortions in his narrative. Thus Isambard Kingdom Brunel must be one of the most original and versatile engineers ever, but neither he nor his father, who built the Thames Tunnel, was written up by Smiles; we should be careful not to accept his list of heroes as complete. He takes his engineers and their problems seriously and there is no mere antiquarianism in Smiles; he succeeds in demonstrating the importance and the profitability of opening up inland communications as the first step in industrialization. Regions which had been brutal, backward and isolated had been transformed into flourishing communities buzzing with industry. By comparison with the engineers, the toolmakers' story is less dramatic, and the biographies are all more fragmentary; so that while *Industrial Biography* is a useful volume, it was never as popular as his other writings.

John Smeaton's experiments which led to improved designs of water-wheels and windmills were published in the *Philosophical Transactions*. It is a commonplace in the history of technology that the invention of a new device – in this case the steam engine – may lead to rapid improvements in its seemingly obsolescent competitors; but although the first textile mills were driven by water-wheels, the steam engine soon triumphed. Smeaton improved the Newcomen engine; but he believed that James Watt's great advance, the use of a separate condenser in place of a spray of cold water in the cylinder, was technically impossible. It would indeed have been so, had it not been for Matthew Boulton. There is a two-volume joint life of Boulton and Watt by Samuel Smiles, which formed the fourth and fifth volumes of the later editions of the *Lives of the Engineers*. Desaguliers and Smeaton had both been interested in comparing the power available from engines and horses; and the modern standard figure for horse-power was defined by Boulton and Watt in 1783. For Telford, we have an autobiography, principally concerned with his works and accompanied by a handsome folio volume illustrating his achievements, of which the best-known is probably the suspension bridge across the Menai Straits. Further illustrations of Telford's bridges, and of Rennie's – which included London, Southwark, and Waterloo bridges – may be found in Cresy's *Encyclopedia of Civil Engineering*, which appeared in 1847. Of the engineers selected by Smiles, only Rennie had a university education; and it has been remarked that at this period Britain produced great engineers but no great theologians or classical scholars despite the bias towards these studies in the universities.

As well as Smiles' account, we have two earlier biographies of Watt, described by H. W. Dickinson as 'standard works for all time'; that by George Williamson, which appeared in 1856 and contains much material on Watt's early life, and that by J. P. Muirhead, which came out in three volumes in 1854, and a shorter one-volume version in 1858. Whereas as the *Compleat Collier* by J. C. shows, Newcomen's engine had come at a most opportune time for the coal-mines, as working penetrated into strata so wet that existing methods could not dry them, Watt's innovations gave new life to the tin-mining industry in Cornwall, where economy of fuel was obviously more important than in a coalfield. But it was in the coalfields that the next major innovation in the use of steam took place, namely its application to haulage on railways. Wooden railways, and trucks with flanged wheels to run upon them, had been used from the seventeenth century to carry coals from the mines to the Tyne or Wear to be loaded on to boats; and at the end of the eighteenth century cast iron began to replace wood as the material for the rails. As an alternative, wrought-iron strips were sometimes laid along the top of the wooden rails. When there was no higher ground between the pithead and the river, the full trucks would run down under gravity, slowed by a brake, and the

56 Entrance to Liverpool railway-station (Booth, H., *Liverpool and Manchester Railway*, 1830, frontispiece)

empties could be hauled back by a horse. Less fortunately situated pits, and those further from rivers – and these began to be opened up as engineers began to penetrate the magnesian limestone of County Durham – cried out for some powerful system of haulage. Steam locomotives soon became possible with the development of high-pressure steam engines, which dispensed with any condenser. These were developed particularly in Cornwall in the first years of the nineteenth century, and applied by Trevithick in 1804. With the discovery that a rack and pinion was not necessary, since the friction of the wheels on the track was sufficient – a rack arrangement had been used by Blenkinsop in Leeds in 1811 – the steam locomotive was launched. Various railways used them in the second and third decades of the nineteenth century; but their triumph was not assured until Stephenson's Rocket demonstrated its potentialities on the Liverpool and Manchester Railway, which opened in 1830. The occasion was marred by the death of Huskisson, the M.P. for Liverpool, who was run over by a train, the speed of which he had underestimated.

The story of the development of railways is told in numerous works dealing with the history of the coal trade and of railways themselves. In 1830 there appeared Henry Booth's *Account of the Liverpool and Manchester Railway*; he realized the importance of the occasion in transforming ideas of distance. In the same year came Joseph Priestley's *Navigable Rivers and Canals*, which include information on the railways, which were to displace the canals, giving a general description and details of costs of construction and of charges. The first edition was in quarto; a revised octavo edition appeared in 1831. The book was designed to accompany a map, but can stand on its own. In 1836 Sir George Head published his *Home Tour through the Manufacturing Districts*, which is a very readable and valuable account of the textile and mining industries, and contains accounts of trips by canal and railway. His brother Sir Francis Head published *Stokers and Pokers* in 1849; it is a popular and entertaining story of how the railways were run, including a description of the Clearing House in which accounts were adjusted for passengers, goods, and waggons which travelled over the lines of a number of different companies in one journey. Head describes the construction of railways, the rolling stock – including travelling post-offices – and the stations. The most famous passage is the description of Wolverton Tea Rooms; for before the era of restaurant and buffet cars, trains had meal-stops at stations, as they still do in some parts of the world. Head also reprinted, in the early editions, the rules and regulations of the London and North Western Railway; in later editions, this was replaced by expostulations against a hostile review of another book by Head, *Highways and Dryways*, describing the Conway and Britannia tubular bridges.

The new railways attracted topographical artists who produced volumes of picturesque *Views*. Railways in the coalfields are depicted in T. H. Hair's superb *Series of Views of the Collieries*, published in 1844. The watercolours

57 Coal-drops at Wallsend (Hair, T., *Series of Views of the Collieries*, 1844, pl opposite p 13)

58 Elephant drawing coal trucks in Austria (Hall, T. Y., *Treatise on various British and Foreign Coal and Iron Mines* [1854], 2nd treatise, pl facing p 32)

from which these etchings were made were done in the late 1830s; they show scenes down the pit, and a pleasant view of Telford's handsome iron bridge at Sunderland, as well as a locomotive and stationary steam engines. It is remarkable how many trains in the coal districts were hauled by stationary engines; locomotives were not yet powerful enough to manage more than very moderate gradients. A good standard history of the coal industry is Robert Galloway's *History of Coal Mining in Great Britain*; Galloway was himself a mining engineer, and describes the improvements in practice instituted by such 'viewers' as the great John Buddle, who managed the famous Wallsend Colliery, and was chiefly responsible for getting the safety-lamp rapidly adopted in collieries. Davy's little book on the safety-lamp was published in 1818. Other noteworthy mining books are Hodgson's *Newcastle upon Tyne*, and Hall's *Treatises*.

Further useful accounts of railways are to be found in J. M. Francis' *History of the English Railway* and in Dionysius Lardner's *Railway Economy*. The former, published in two volumes in 1851, is an excellent standard work, particularly strong on the commercial side of railway operations; written just after the railway boom, it is interesting on George Hudson the railway king and his financial operations. Like more modern railway kings, Hudson favoured trunk lines, and disliked branches. The book also describes the electric telegraph, which came into use with the railways, catching the public eye when it was employed in the arrest of a murderer. Lardner's book is more concerned with the economics of rail transportation; what emerges from his book is how heavily capitalized the British railways were compared to those abroad. The specifications set by British engineers in terms of avoidance of sharp curves and steep gradients, and the heavy use of track in this country compared to any other, entailed more solid and more expensive construction. To get around the corners and bumps, American locomotives, trucks, and carriages were mounted on bogies. Lardner was alarmed at the speeds attained by express trains – not unreasonably so when there was no continuous braking system – and included a chapter on how to avoid accidents if you were a passenger. He described railway arrangements, legislation, and costs in all the countries which had railways; and added a chapter on American steamboats for good measure. For a view of railways from the point of view of the contractor, Arthur Helps' biography of Thomas Brassey, who built railways all over the world, should be noted; as should Joseph Devey's life of Joseph Locke. Further standard histories are Frederick S. Williams' *Iron Roads*, 1852, and *Midland Railway*, 1876: and the construction of the London and Birmingham line was described by Peter Lecount in 1839. In the same year an illustrated account of it by J. A. Bourne was published. Wood's *Practical Treatise on Railroads* deserves to be mentioned; as does Weale's *Railway Making*.

We can conclude this chapter with mention of some books on the factory system, although it owes little directly to science. The standard work on this is Andrew Ure's *Philosophy of Manufactures*, Ure being a chemist who taught at the Andersonian Institution of Glasgow, and many of whose pupils went into industry in England and abroad. His book is an *apologia* for the factory-owners, and its relatively rosy picture of working conditions should be treated with caution; but he is strong on the technical side of manufacturing,

59 Locomotive engine (Wood, N., *Practical Treatise on Railroads*, 1831, 2nd ed, pl 7)

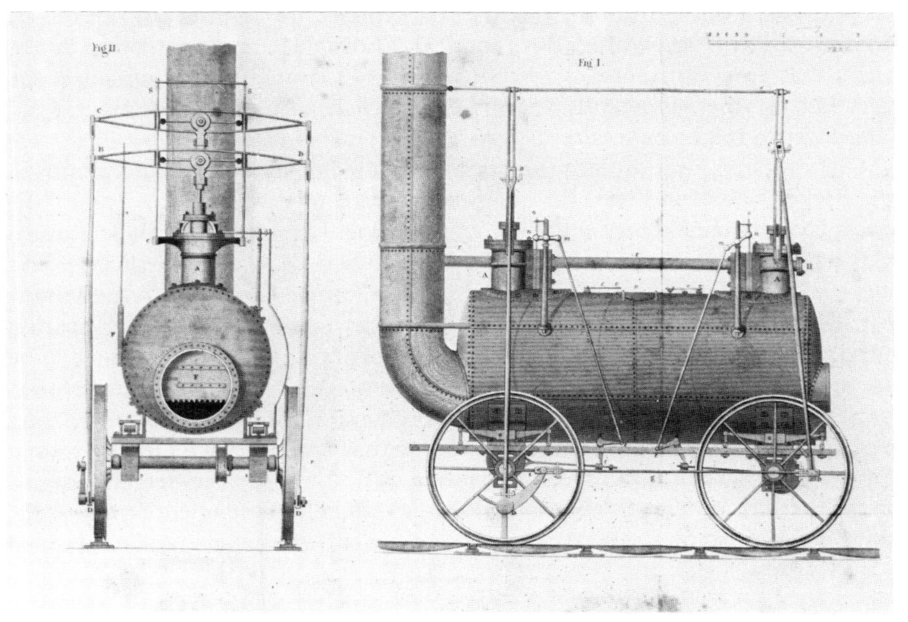

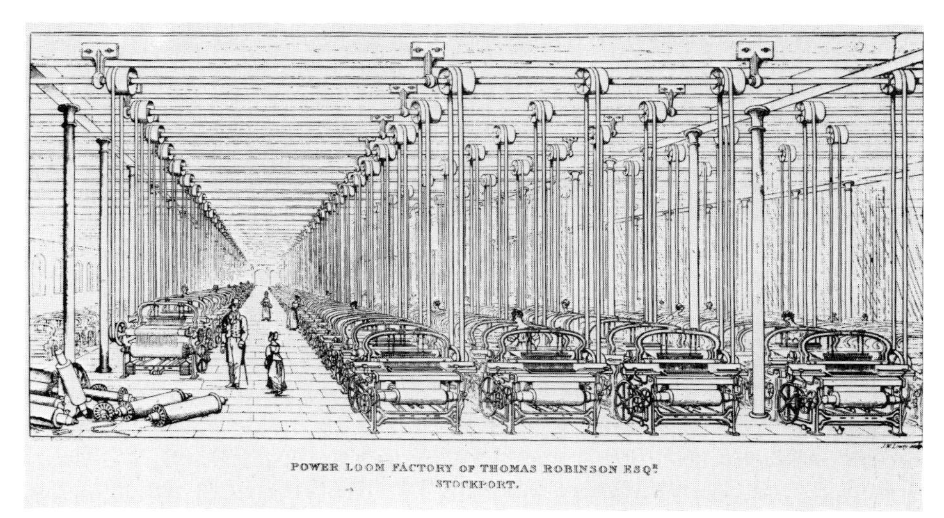

60 Power-loom factory (Ure, A., *Philosophy of Manufactures*, 1835, pl facing p 1)

and the book is deservedly a classic, as is Felkin's technical treatise on lace-manufacture. Charles Babbage, the cantankerous inventor of the 'calculating engine', an ancestor of the electronic computer, wrote *The Economy of Manufacturers* which can also still be read with great interest. Charles Thackrah, a physician of Leeds, made a pioneering study of *The Effects of Arts, Trades and Professions on Health and Longevity*, published in 1831. He drew attention to the shorter average life in the industrial areas of Yorkshire as compared to the rural parts of the county, estimating that 50,000 people died annually from the direct or indirect effects of manufactures, and suggesting how this figure could be reduced. Edwin Chadwick's famous *Report on the Sanitary condition of the Labouring Population*, compiled from accounts sent in by local informants, also stressed the high death rates to be attributed to overcrowding, dirt, and lack of drainage and ventilation in the cities that grew up so fast as the population of Great Britain doubled in the first half of the nineteenth century. Chadwick added that a longer-lived labour force might be less rebellious.

Chadwick was not directly concerned with factories, and for an account of the darker side of industrialization one can turn to such books as John Fielden's *Curse of the Factory System*, Charles Wing's *Evils of the Factory System*, Gaskell's *Artisans and Machinery*, Richard Fynes' *Miners of Northumberland and Durham*, and Alfred Williams' *Life in a Railway Factory*. Wing's book consists of evidence presented before various commissions investigating conditions in factories, of which it gives a horrifying and vivid picture to be set against Ure's account. Gaskell's book sets out the decline in happiness, health, and morals which followed the changes from handloom weaving as an activity for peasant farmers, to the factory system. Like Ure, he took it for granted that machinery would displace skilled labour, as indeed happened in the textile industry. Fyne's book is an account of the struggles of the miners, their various strikes, and the combinations of employers and employed against one another, with the background of the rapid expansion of the coal trade. It is worth remarking that coal and agriculture came under the control of the landed interest: it seemed to the advantage of manufacturers to point out that conditions in factories were better than on the land or in the mines, and to landowners to do the opposite. Williams' book was written in the early years of this century, and praised as a work of literature by Robert Bridges; it is also a superb source of social and technological history, showing the grim working conditions around 1900.

In this chapter we have not investigated the effects of technology upon science in terms of improvements in apparatus: nor have we dealt with the scientist-entrepreneurs so important in the gas, electrical, steel, and chemical industries; such men as Murdock, William Thomson (Lord Kelvin), Wheatstone, the Siemens brothers, Edison, Bessemer, and Perkin. In the following chapter, on chemistry in the Romantic period, we shall discuss chemical apparatus, some of which is attractively illustrated. And chapters after that, on the so-called decline of science in England about 1830 and on the dissemination of science, will give us the opportunity to look at those men who applied nineteenth-century science in the later phases of the Industrial Revolution, and to refer to the Great Exhibition of 1851; though we cannot properly cover nineteenth-century technology.

6 THE INDUSTRIAL REVOLUTION

Agriculture, Board of, Authors of *General Views of the Agriculture of Counties*:

Aiton, W., *Ayr*, Glasgow, 1811.
Anderson, J., *Aberdeen*, Edin., 1794.
Bailey, J., *Durham*, 1810.
  and Culley, G., *Cumberland*, 1794.
  Northumberland, 1794.
Baird, T., *Middlesex*, 1793.
Batchelor, T., *Bedford*, 1808.
Beatson, R., *Fife*, Edin. 1794.
Belsches, R., *Stirling*, Edin., 1796.
Billingsley, J., *Somerset*, 1794.
Bishton, J., *Salop*, Brentford, 1794.
Boys, J., *Kent*, Brentford, 1794.
Brown, T., *Derby*, 1794.
Claridge, J., *Dorset*, 1793.
Clark, J., *Brecknock*, 1794.
  Hereford, 1794.
  Radnor, 1794.
Crutchely, J., *Rutland*, 1794.
Davies, W., *N. Wales*, 1810.
  S. Wales, 2 vols., 1814.
Davis, R., *Oxford*, 1794.
Davis, T., *Wiltshire*, 1794.
Dickson, R. W., *Lancashire*, 1815.
Donaldson, J., *Banff*, Edin., 1794.
  Carse of Gowrie, 1794.
  Elgin or Moray, 1794.
  Nairn, 1794.
  Northampton, Edin., 1794.
  Kincardine, 1795.
Douglas, R., *Roxburgh*, Edin., 1798.
Driver, A. and W., *Hants*, 1794.
Duncomb, J., *Hereford*, 1805.
Erskine, J. F., *Clackmannan*, Edin., 1795.
Farey, J., *Derby*, 3 vols., 1811–17.
Foot, P., *Middlesex*, 1794.
Fox, J., *Monmouth*, Brentford, 1794.
  Glamorgan, 1796.
Fraser, R., *Cornwall*, Brentford, 1794.
  Devon, 1794.
Fullarton, W., *Ayr*, Edin., 1793.
Gooch, W., *Cambridge*, 1813.
Graham, P., *Stirling*, Edin., 1812.
Grainger, J., *Durham*, 1794.

Griggs, Messrs, *Essex*, 1794.
Hassall, C., *Carmarthen*, 1794.
Pembroke, 1794.
Monmouth, 1812.
Headrick, T., *Angus*, Edin., 1813.
Henderson, J., *Caithness*, 1812.
Sutherland, 1812.
Hepburn, G. B., *East-Lothian*, Edin., 1794.
Heron, R., *The Hebrides*, Edin., 1794.
Holland, H., *Cheshire*, 1808.
Holt, J., *Lancaster*, 1794.
Johnston, B., *Dumfries*, 1794.
Johnston, T., *Selkirk*, 1794.
Tweedale, 1794.
Kay, G., *N. Wales*, Edin., 1794.
Keith, G. S., *Aberdeen*, Aberdeen, 1811.
Kent, N., *Norfolk*, 1794.
Kerr, R., *Berwick*, 1813.
Leathan, I., *E. Riding*, 1794.
Leslie, W., *Nairn and Moray*, 1813.
Lloyd, W., and Turner, *Cardigan*, 1794.
Lowe, A., *Berwick*, 1794.
Nottingham, 1794.
Macdonald, J., *The Hebrides*, Edin., 1811.
Mackenzie, G. S., *Ross and Cromarty*, 1813.
Malcolm, W., *Buckingham*, 1794.
Surrey, 1794.
Marshall, W., *Highlands of Scotland*, 1794.
Martin, A., *Renfrew*, 1794.
Mavor, W., *Berkshire*, 1808.
Maxwell, G., *Huntingdon*, 1793.
Middleton, J., *Middlesex*, 1798.
Monk, J., *Leicester*, 1794.
Murray, A., *Warwick*, 1813.
Nasmyth, J., *Clydesdale*, Brentford, 1794.
Parkinson, R., *Huntingdon*, 1813.
Pearce, W., *Berkshire*, 1794.
Pitt, W., *Stafford*, 1794.
Leicester, 1809.
Northampton, 1809.
Worcester, 1813.
Plymley, J., *Shropshire*, 1803.
Pomeroy, W. T., *Worcester*, 1794.
Priest, St J., *Buckingham*, 1810.
Pringle, A., *Westmorland*, Edin., 1794.
Quayle, B., *Isle of Man*, 1794.
Rennie, G., *et al.*, *W. Riding*, 1794.
Robertson, G., *Mid Lothian*, Edin., 1793. *Kincairdshire*, 1813.
Robertson, J., *Perth*, 1794.
Inverness, 1808.
Robson, J., *Argyll*, 1794.
Roger, Rev., *Angus or Forfar*, Edin., 1794.
Rudge, T., *Gloucester*, 1807.
Shirreff, J., *Orkneys*, Edin., 1814.
Shetlands, Edin., 1814.
Sinclair, J., *Northern . . . Scotland*, 1795.
Scotland, 5 vols., Edin., 1815.
Singer, *Dumfries*, Edin., 1812.
Smith, J., *Argyll*, Edin., 1798.
Smith, S., *Galloway*, 1810.
Somerville, R., *East Lothian*, 1805.
Souter, D., *Banff*, Edin., 1812.
Stevenson, W., *Surrey*, 1809.
Dorset, 1812.
Stone, T., *Huntingdon*, 1793.
Bedford, 1794.
Lincoln, 1794.
Strickland, H. E., *E. Riding*, York, 1812.
Thomson, J., *Fife*, Edin., 1800.
Trotter, J., *W. Lothian*, Edin., 1794.
Tuke, J., *N. Riding*, 1794.
Turner, G., *Gloucester*, 1794.
Ure, D., *Dumbarton*, 1794.
Roxburgh, 1794.
Kinross, Edin., 1797.
Vancouver, C., *Cambridge*, 1794.
Essex, 1795.
Devon, 1808.
Hampshire, 1813.
Walker, D., *Hertford*, 1795.
Webster, J., *Galloway*, Edin., 1794.
Wedge, J., *Warwick*, 1794.
Wedge, T., *Chester*, 1794.
Whyte, A., and Macfarlan, D., *Dumbarton*, Glasgow, 1811.
Wilson, J., *Renfrewshire*, Paisley, 1812.
Worgan, G. B., *Cornwall*, 1812.
Young, A. *Suffolk*, 1794.
Lincoln, 1799
Hertfordshire, 1804.
Norfolk, 1804.
Essex, 2 vols., 1807.
Oxford, 1809.
Young, A. (the younger), *Sussex*, 1793.

Babbage, C., *The Economy of Manufactures and Machinery*, 1832.
Bewick, T., *A General History of Quadrupeds*, Newcastle, 1790.

A Memoir . . . written by Himself, 1862.
Blackmore, J. and Carmichael, J. W., *Views on the Newcastle and Carlisle Railway*, 1836–8.
Blith, W., *The English Improver*, 1649; rev. ed., *The English Improver improved*, 1652.
Booth, H., *An Account of the Liverpool and Manchester Railway*, Liverpool, 1830.
Bourne, J. C., *Drawings of the London and Birmingham Railway*, 1839.
The History and description of the Great Western Railway, 1846.
Brees, S. C., *Railway Practice*, 1837–47.
J. C., *The Compleat Collier . . .* , 1708; 2nd ed., Newcastle, 1845.
Caus, I. de, *New and Rare Inventions of Water Works . . .* , tr. J. L(eak), 1659; 2nd ed., incl. Savery's engine, 1704.
(Chadwick, E.), *Report . . . on . . . the sanitary condition of the labouring population of Great Britain . . .* , 1842.
Cresy, E., *Encyclopedia of Civil Engineering*, 2 vols., 1847.
(D'acres, R.), *The Elements of Water-Drawing*, 1659.
Daubeny, C., *Miscellanies*, 2 vols., 1867.
Davy, H., *Elements of Agricultural Chemistry*, 1813.
On the Safety Lamp for Coal Mines, with some researches on flame, 1818.
Devey, J., *The Life of Joseph Locke*, 1862.
Donaldson, J., *A Treatise on Manures*, 1842.
The Elements of Agriculture, botanical and zoological, 1847.
Improved Farm Buildings, 1851.
Agricultural Biography, 1854.
Doré, L. A. G. and Jerrold, W. B., *London; A Pilgrimage*, 1872.
Duhamel de Monceau, H. L., *A Practical Treatise of Husbandry* (tr. J. Mills), 1759.
The Elements of Agriculture, tr. P. Miller, 1764.
Evans, D. M., *Facts, Failures, and Frauds*, 1859.
Farey, J., *A Treatise on the Steam Engine . . .* , 1827.
Felkin, W., *A History of the Machine-wrought Hosiery and Lace Manufactures*, 1867.
Fielden, J., *The Curse of the Factory System . . .* (1836).
Francis, J. A., *A History of the English Railway . . .* , 2 vols., 1851.
Fynes, R., *The Miners of Northumberland and Durham . . .* , Blyth, 1873.
Galloway, R. L., *The Steam Engine and its inventors*, 1881.
A History of Coal Mining in Great Britain, 1882.
Annals of Coal Mining and the Coal Trade, 1898.
Gaskell, P., *Artisans and Machinery*, 1836.
Griffiths, S., *Guide to the Iron Trade of Great Britain*, 1873.
Grisenthwaite, W., *A New Theory of Agriculture*, Wells, 1819.
Hair, T. H., *A Series of Views of the Collieries . . . of Northumberland and Durham*, 1844; engraved title page dated 1839.
Hall, A. D., *The Book of the Rothamsted Experiments*, 1905.
Hall, T. Y., *Treatises on various British and Foreign Coal and Iron Mines and Mining*, Newcastle, n.d. (1850s).
(Head, F. B.), *Stokers and Pokers*, 1849.
Highways and Dryways, 1849.
Head, G., *A home tour through the manufacturing districts of England*, 1836.
Helps, A., *The Life and Labours of Mr Brassey*, 1872.
(Hodgson, J.), *The picture of Newcastle upon Tyne . . . And a detailed history of the coal trade*, Newcastle, 1807.
Home, F., *The Principles of Agriculture and Vegetation*, Edinburgh, 1756.
Home, H., Lord Kames, *The Gentleman Farmer*, Edinburgh, 1776.
Johnston, J. F. W., *Elements of Agricultural Chemistry and Geology*, Edinburgh, 1842.
Catechism of Agricultural Chemistry and Geology, Edinburgh, 1844.
Lectures on Agricultural Chemistry, Edinburgh, 1844.

Contributions to Scientific Agriculture, Edinburgh, 1849.
The Chemistry of Common Life, 2 vols., Edinburgh, 1855.
Lardner, D., *Railway Economy . . .*, 1850.
Lecount, P., *A Practical Treatise on Railways*, Edinburgh, 1839.
The History of the Railway connecting London and Birmingham, 1839.
Liebig, J. von, *Organic Chemistry in its application to Agriculture and Physiology*, ed. L. Playfair, 1840.
Marshall, W., *The Review and Abstract of the County Reports to the Board of Agriculture*, 5 vols., 1818.
Matthew, W., *An Historical Sketch of the Origin, Progress, and Present State of Gas-lighting*, 1827.
Morton, J., *On the Nature and Property of Soils*, 1838.
Murdock, W.; M(urdock), A., *Light without a wick; a century of gas lighting*, Glasgow, 1892.
Paine, E. M. S., *The two James's and the two Stephensons*, 1861.
Plat, H., *The Jewell House of Art and Nature*, 1594; 2nd ed., 1653.
Priestley, J., *Historical Account of the Navigable Rivers, Canals, and Railways, throughout Great Britain . . .*, 1831; 8vo ed., 1831.
Pyne, W., *Microcosm*, 1845.
Ransome, J. A., *The Implements of Agriculture*, 1843.
Rigg, A., *A Practical Treatise on the Steam Engine*, 1878; 2nd ed., 1894.
Royal Agricultural Society, *Journal*, *I* (1839–40)–.
Savery, T., *The Miner's Friend*, 1702.
Smiles, S., *Self Help*, 1859.
Industrial Biography, 1863.
Lives of the Engineers, 5 vols., 1874.
Snow, J., *On the Mode of Communication of Cholera*, 1849; 2nd ed., 1855, very greatly enlarged.
Somerset, E., marquis of Worcester, *A Century of the Names and Scantlings of such inventions as I can call to mind to have tried and perfected . . .*, 1663.
Telford, T., *Life . . . written by himself*, ed. J. Rickman, 1838; there was an accompanying atlas of copper plates.
Thackrah, C. T., *The Effects of the Principal Arts, Trades, and Professions . . . on Health and Longevity . . .*, 1831; 2nd ed., 1832.
(Tomlinson, C.), *Cyclopedia of Useful Arts*, 2 vols., 1852–4.
Tredgold, T., *A Practical Treatise on rail-roads and carriages*, 1825.
The Steam Engine . . ., 1827; 2nd ed., 2 vols., 1838 (vol. II, plates).
(Tull, J.), *The New horse-houghing husbandry . . .*, 1731; an expanded version of this is *The horse-hoing husbandry . . .*, 1733.
Ure, A., *The Philosophy of Manufactures . . .*, 1835.
Walker, J. S., *An Accurate Description of the Liverpool and Manchester Railway*, Liverpool, 1830.
Weale, J., *Ensamples of Railway Making*, 1843.
Williams, A., *Life in a Railway Factory*, 1915.
Williams, F. S., *Iron Roads*, 1852.
The Midland Railway, 1876.
Wood, N., *A Practical Treatise on Rail-roads*, 1825; 2nd ed., 1831.
Young, A., *A Six Weeks Tour through the Southern Counties . . .*, 1768.
Letters concerning the present state of the French Nation, 1769.
A Six Months Tour through the North of England, 4 vols., 1770.
A Tour in Ireland, 1780.
Autobiography, ed. M. Betham-Edwards, 1898.

Recent Publications

Atkinson, F., *The Great Northern Coalfield*, Barnard Castle, 1966.
Chambers, J. D., and Mingay, J. E., *The Agricultural Revolution*, 1966.
Coleman, T., *The Railway Navvies*, 1965.
Darby, H. C., *The Draining of the Fens*, Cambridge, 1940.
Dickinson, H. W., *John Wilkinson, Ironmaster*, Ulverston, 1914.
James Watt, Cambridge, 1935.
Matthew Boulton, Cambridge, 1937.
A Short History of the Steam Engine, Cambridge, 1939.
The Water Supply of Greater London, 1954.
and Titley, A., *Richard Trevithick*, Cambridge, 1934.
Fletcher, H. R., *The Story of the Royal Horticultural Society*, 1969.
Gillispie, C. C. (ed.), *A Diderot Pictorial Encyclopedia of Trades and Industry*, 2 vols., New York, 1959.
Goodison, N., *English Barometers*, 1969.
Henderson, W. O. (ed.), *Industrial Britain under the Regency*, 1968.
The Industrialization of Europe, 1780–1914, 1969.
Hudson, D., and Luckhurst, K. W., *The Royal Society of Arts, 1754–1954*, 1954.
Hughes, E., *North Country Life in the Eighteenth Century*, 2 vols., 1952–65.
Keller, A. G., *A Theatre of Machines*, 1964.
Kirby, R. S., Withington, S., Darling, A. B., and Kilgour, F. G., *Engineering in History*, New York, 1956.
Klingender, F. D., *Art and the Industrial Revolution*, ed. A. Elton, 1968.
McCloy, S. T., *French Inventions of the Eighteenth Century*, Lexington, Ky, 1951.
Ottley, G., *A Bibliography of British Railway History*, 1966.
Palissy, B., *Admirable Discourses*, tr. A. la Rocque, Urbana, Ill., 1957.
Perkins, W. F., *British and Irish Writers on Agriculture*, 3rd ed., Lymington, 1939.
Pike, E. R., *Human Documents of the Industrial Revolution in Britain*, 1966.
Human Documents of the Victorian Golden Age, 1967.
Raistrick, A., *Quakers in Science and Industry*, 1950.
Rolt, L. T. C., *Thomas Telford*, 1958.
Isambard Kingdom Brunel, 1959.
George and Robert Stephenson, 1960.
Russell, E. J., *A History of Agricultural Science in Great Britain*, 1966.
Singer, C., Holmyard, E. J., Hall, A. R., and Williams, T. I. (eds.), *A History of Technology*, 5 vols., Oxford, 1954–8.
Smeaton, J., *A Catalogue of Civil and Mechanical Engineering Designs*, ed. H. W. Dickinson and A. A. Gomme, 1950.
Steeds, W., *A History of Machine Tools, 1700–1910*, Oxford, 1969.
Sung Ying-Hsing, *T'ien-Kung K'ai-Wu*, tr. E-tu Zen Sun and Shiou-Chaun Sun, 1966.
Wright, L., *Clean and decent . . .*, 1960.
Home fires burning . . ., 1964.

Additional Reading

Baynes, K., & Pugh, F., *The Art of the Engineer*, 1981.
Bud, R. F., & Roberts, G. K., *Science Versus Practice*, Manchester, 1984.
Carnot, N. L. S., *Reflexions on the Motive Power of Fire*, ed. & tr. R. Fox, Manchester, 1986.
Knight, D. M., *The Nature of Science*, 1977.
Maurice, K. & Mayr, O., *The Clockwork Universe*, Washington, DC, 1980.
Mayr, O., *Authority, Liberty and Automatic Machinery in Early Modern Europe*, Baltimore, 1986.
Turner, G. L'E., *Nineteenth Century Scientific Instruments*, 1983.

7 *Chemistry in the Romantic Period*

It may seem surprising to link chemistry and Romanticism together, but the opening years of the nineteenth century were a period of great general interest in the science. Lecturers in London, Glasgow, and Paris attracted enormous audiences to their orations and demonstration experiments. The chemical philosopher was expected to develop and discuss a world-view; his science seemed the key to the nature of matter, and he was in a position to throw light upon such questions as the truth of materialism or the rôle of mechanistic explanations in psychology and biology. It seemed possible that chemists, using novel methods of analysis such as the electric battery of Alessandro Volta, might pin down the Proteus of matter and discover the one basic stuff which, in different arrangements or electrical states composed all the manifold substances which we find in the world. Coleridge compared the chemist to the poet as one searching through a multiplicity of forms for unity of substance.

This is of course only half the story. The chemistry books of about 1800 frequently ran into several volumes, mostly full of mere recipes and ill-organized experimental material relating to organic chemistry. The Chemical Revolution associated with Lavoisier had brought order into the chemistry of inorganic bodies, and crystallography began to make some progress. Though even in this region, naturally occurring minerals presented daunting problems: and there was dispute between devotees of the theory of René Haüy that crystals were built of integrant molecules having a certain geometrical form, and those such as William Hyde Wollaston who saw them as composed of spheroidal atoms. By about 1830, the sober fact-collecting side of chemistry had become more evident to the public mind than the spirited chemical philosophy of the previous generation, and those in search of excitement turned to geology. The Chemical Society of London, founded in the 1840s, would not publish in its journal papers of a purely theoretical nature. Nevertheless, it would be a mistake to suppose that contemporary philosophers of science were correct in describing chemistry as a science concerned only with collecting and classifying facts; not all chemistry books were by any means as dry as dust. Translations were made of works of Berzelius, Liebig, Dumas, Laurent and Mendeleev; and books first published in English often have, as well as a certain period charm, that concern with broad general issues which keeps a work of science alive when its period of immediate utility is past.

One who exemplifies the interaction between chemistry and the Romantic Movement is Thomas Beddoes. He was associated with the Lunar Society of Birmingham in its last years, and published papers with James Watt. The experiments they described, on the administration of gases, arose out of Beddoes' hope that the newly discovered gases would be valuable in curing diseases, particularly tuberculosis, which was at that period a terrible killer. Josiah Wedgwood subsidized Beddoes when he resigned his readership at Oxford – where he had attracted audiences larger than any lecturer since the thirteenth century – and set up the Pneumatic Institution at Clifton to test these views. Beddoes married an Edgeworth; his son, Thomas Lovell Beddoes, became a notable poet; and at Clifton he became the friend of Coleridge to whom he prescribed opium.

Beddoes had been responsible for publishing translations of the great Swedish chemists Tobern Bergman and Charles William Scheele. He had himself translated Bergman's *Dissertation on Elective Attractions*; he annotated the first volume of Edmund Cullen's translation of Bergman's *Physical and Chemical Essays*; and he put into good English the clumsy and literal translation of Scheele's *Chemical Essays* made by F. X. Schwediauer. Jeremy Bentham similarly polished Schwediauer's rendering of Bergman's *Usefulness of Chemistry*. Beddoes was soon converted to Lavoisier's new chemistry; he was also involved in the introduction to England of Kantian philosophy, to which he was not a convert but which he reviewed with a certain amount of sympathy in the *Monthly Review*. Beddoes appointed Humphry Davy, an apothecary-surgeon's apprentice from Penzance, his assistant at the Pneumatic Institution, and introduced him into the literary society of the region; and Davy became the friend of Coleridge and Southey, and later of Wordsworth and Walter Scott. Beddoes published Davy's first papers, full of speculations – to some extent anticipated by some Gentlemen of Exeter – on the rôle of light in chemistry and biology in the work called *Contributions to Medical and Physical Knowledge, principally from the West of England* in 1799, which also contained papers by Beddoes, notably one on relationships between the metals. The *Contributions* were quite well received; Davy's papers were praised as promising in some journals, but were laughed at in others. They are now best known for Davy's experiments on heat; which he endeavoured to prove a kinetic phenomenon depending upon the motion of particles rather than on a weightless caloric fluid.

The promise thus displayed by Davy in his teens was fulfilled in his first book, written at Clifton and published in 1800, on nitrous oxide and its respiration. Beddoes was interested both in the effects of opium and in the respiration of gases; and Davy's discovery of the anaesthetic and exhilarating effects of nitrous oxide – laughing gas – united these interests. It was brave of Davy to breathe the gas which was, according to the American Samuel Mitchill, the principle of contagion and must therefore prove instantly fatal; and when Davy repeated the experiment with carbon monoxide he almost put an early end to his career. Davy's own subjective accounts of nitrous oxide anaesthesia are among the best ever written; and the book also contains descriptions of their experiences with the gas by many people, including Coleridge, whose 'sensations were highly pleasurable', and whose heart beat violently; and the prosy Southey, who declared that, 'The sensation is not painful, neither is it in the slightest degree pleasurable'. Breathing the gas became highly popular; and there is a famous Gillray cartoon illustrating a session at the Royal Institution in London, with Davy looking like a mischievous schoolboy, holding some bellows. The book also contains competent volumetric analyses of the various oxides of nitrogen; and on the strength of it Davy was appointed in 1801 assistant lecturer at the Royal Institution, becoming professor in 1802 at the age of twenty-three and being elected F.R.S. in the following year.

Davy is perhaps the greatest of Romantic scientists and we shall notice his other books in due course. While he was at Clifton, chemical science was dramatically enriched by Volta's discovery that when two different metals are immersed in water, or a salt solution, and connected together, an electric current will flow. Volta's paper was published in French in the *Philosophical Transactions* of the Royal Society for 1800; a translation appeared in Alexander Tilloch's *Philosophical Magazine* for 1800, and in the same year William Nicholson and Anthony Carlisle showed that the electric current could be used to decompose water into hydrogen and oxygen. Their paper appeared in Nicholson's *Journal of Natural Philosophy, Chemistry and the Arts;* and indeed it is striking how many important chemical papers in the first decade or so of the nineteenth century appeared in these three journals. The first numbers of Nicholson's *Journal* appeared in quarto format, from 1797. In 1802 a new series in octavo began, which survived until 1813, when it was amalgamated with the *Philosophical Magazine*, which still appears, but during the nineteenth century became increasingly devoted to physics rather than the chemistry which had filled many of its pages in its early days. It had begun in 1798; and its first volume contained a translation of Haüy's views on crystal structures. Later volumes of both journals contained translations of papers by Ritter, and Berzelius; the *Philosophical Magazine* also translated important papers by Berthollet, Grotthus and Ampère; and later papers on thermochemistry by Dulong and Hess. These journals usually also reprinted important chemical papers from the *Philosophical Transactions* as well as printing original papers and book-reviews rapidly themselves; they are the first sources to turn to for anybody wanting to find out about science in English in the opening years of the nineteenth century.

In 1813 there appeared a new journal, *Annals of Philosophy*, edited by Thomas Thomson. In 1821 this began a new series, under Richard Phillips, becoming more geological; and after 1826 it too amalgamated with the *Philosophical Magazine*. It was important for publishing essays by Berzelius, including one in which he proposed the chemical notation which we still employ; for trenchant advocacy of Dalton's atomic theory by Thomson; for publishing Prout's hypothesis, and for useful annual reviews of the progress of chemistry, by Thomson. There were also chiefly chemical journals published at the Royal Institution, of which the *Quarterly Journal of Science* and the *Proceedings* should be noted; but the latter did not begin until 1851. In the first part of the century, the *Philosophical Transactions* contained many important chemical papers; and the *Proceedings of the Royal Society* developed through the century from an irregularly published series of abstracts into another important journal.

Chemistry not only dominated many scientific journals in the opening decades of the century, but accounts of chemical discoveries and reviews of chemical books and papers appeared in such publications as the *Monthly Review*, the *Edinburgh Review*, the *Westminster Review* and the *Quarterly Review*. While none of these essays count as original works of science, they do provide an indication of popular interest in the sciences; and we shall mention in later chapters such important publications as Playfair's review of Laplace in the *Edinburgh*, and Wilberforce's review of Darwin in the *Quarterly*. The *Athenaeum* can also be a useful source, particularly for reports of meetings of scientific societies; accounts of papers read to, but not later printed by, the British Association, can sometimes be found in these.

In the years immediately following the appearance of Lavoisier's *Elements of Chemistry*, standard chemical books were translated from the French; though curious assaults upon the French theories were made by Mrs Fulhame and by R. Harrington. Thus William Nicholson, editor of the *Journal* already described and compiler of a *Dictionary of Chemistry* published in 1795, translated the *Elements of Chemistry* of J. A. C. Chaptal, in three volumes; and the *System of Chemistry* of A. F. Fourcroy in eleven volumes. Robert Heron's *Elements of Chemistry*, based upon Fourcroy, appeared in 1800; the first textbook in English which could compete with the French compilations was Thomas Thomson's *System of Chemistry*. This passed through numerous editions, and received what was at this time the accolade of being translated into French, with a preface by Berthollet. In his third edition of 1807, Thomson published the first account of John Dalton's atomic theory; the book remained for many years a standard work, and was read and praised for example by John Stuart Mill.

In textbooks, and we shall examine some more elementary treatises in the chapter on the dissemination of science, the chief points to notice are the arrangement of the material, and the author's attitude to new theories. Fourcroy was an associate of Lavoisier, and one of those responsible for drawing up the recommendations for chemical nomenclature when the theory of phlogiston was overthrown. Thomas Thomson was a prominent atomist, and in his books, his journal and his papers a tireless popularizer of Dalton.

Later he also came to favour Prout's hypothesis that all the chemical elements have atomic weights which are integral multiples of that of hydrogen; that is, they are whole numbers expressed on the scale of which hydrogen is assigned an atomic weight of one unit. He published analyses demonstrating this, first in the *Annals of Philosophy* and then in a book: *An Attempt to Establish the First Principles of Chemistry upon Experiment*, of 1825. Unfortunately the book did not live up to its grandiloquent title; the atomic weights which concorded so beautifully with Prout's hypothesis were mostly the results of students' analyses, and seem to have been cooked, or selected, to fit. Berzelius, and then Edward Turner in England, utterly undermined the book; but Prout's hypothesis survived, because many atomic weights are close to whole numbers, to trouble chemists through the nineteenth century, and to find an explanation a hundred years after its appearance, in the doctrine of isotopes. Thomson also published a valuable *History of Chemistry* in 1830 which is particularly interesting for the events of the early nineteenth century, seen through the eyes of a very widely read participant. By the mid century, Germans were setting the pace in chemistry, and the famous *Handbook* of Gmelin is a monument to the industry and sagacity of its compiler.

The history of the introduction of the atomic theory into chemistry is an involved one. Dalton's *New System of Chemical Philosophy* appeared from 1808 onwards; the theory only takes up the last few pages of the first part, which is chiefly devoted to discussions of gases and of heat. The second part, of 1810, and the first part of the second volume, of 1827, are closer to ordinary textbooks and in their own day, and since, attracted little attention; and the second part of the second volume never appeared. Chemists before Dalton, including for example Thomson, had generally been atomists; but about 1800 there were two new developments which made Dalton's doctrine unacceptable. The first was the publication by Berthollet of his doctrine that chemical composition was not constant; that chemical reactions were an equilibrium, the outcome of which depended upon the masses of the reacting substances present. Berthollet's papers, read originally in Egypt whither he had accompanied Napoleon, were translated into English in the *Philosophical Magazine* in 1801, and a different translation by Dr Farrell appeared as a book in 1804, with an American edition at Baltimore in 1809. That in the *Philosophical Magazine* is a condensed version for the first part, but a full translation of the later sections. Berthollet expanded this work into a two-volume *Chemical Statics*, published in English in 1804.

Berthollet's views were inconsistent with Dalton's theory, according to which chemical composition was constant because chemical reaction was a matter of the juxtaposition of atoms in simple fixed ratios. This question was resolved by the work of Jean-Louis Proust, Dalton, Thomson, and Wollaston; and chemical compounds came to be defined in terms of constant composition. Alloys, solutions, and glasses were therefore regarded as mixtures, although some chemists throughout the century maintained Berthollet's interest in equilibria and maintained that alloys were a kind of compound. Subsequently Wollaston, in the *Philosophical Transactions*, and Berthollet in his preface to the French translation of Thomson's *System*, argued against Dalton that it was unnecessary to postulate atoms; it was enough to determine the relative equivalent weights in which elements combined, and this involved no hypothesis. Until the 1860s when Kekulé and his school began to explain in detail how chemical differences resulted from different arrangements in space of the atoms, it was not unreasonable to rest content with equivalents; and this is what most authors of textbooks did. They praised Dalton for the laws of chemical combination in definite and multiple proportions, describing these laws as 'the atomic theory divested of hypothesis.' Thomson thought this over-cautious; but we find it for example in the *Elements of Chemistry* of Edward Turner, and in Sir Robert Kane's *Elements*, a textbook widely read in the U.S.A. Kane was President of Queen's College, Cork, and an editor of the *Philosophical Magazine*; his book and Turner's are good standard works, and their attitude to atomism was shared by the authors of most of the catechisms, dialogues, and elementary works on chemistry written in this period.

One could also be suspicious of atomism on deeper philosophical grounds, if one believed with Schelling that stories of hypothetical particles and mechanisms could never constitute explanations. On this Heraclitean view, there were no such things as permanent atoms and elements. Force was the ground of matter, and all phenomena were to be seen in terms of the inter-

61 Black and white figures (Goethe, W., *Theory of Colours*, 1840, pl 1)

action of forces, usually polar forces like electricity and magnetism which exhibit a positive–negative and north–south duality of aspect; Goethe applied this view to colour. Apparent rest was simply equilibrium. Interest in mechanisms had led to a situation where men and animals were looked upon as machines; in reaction to this the followers of Schelling looked upon the inorganic realm as alive, and sought analogies from organisms in chemistry, physics and geology. In Germany, this movement led to the setting up of a great annual meeting of *Natürforscher*; which was the example followed when the British Association for the Advancement of Science, a much more sober body, was founded. Whereas Germans could not escape *Natürphilosophie*, the Kantian philosophy made slow progress in England and most books of the early nineteenth century reflect the Newtonian tradition, and, in chemistry, the scepticism towards the atomic hypotheses of Lavoisier. But in F. C. Gren's *Principles of Modern Chemistry*, which was translated in 1800, we do find the science presented by one who abandoned atomism for the dynamical system, and the continuum. It is difficult in England to distinguish adherents to a Newtonian dynamical chemistry, embracing particles but with primary interest in the forces between them, from those, if any, who accepted a Kantian position and denied the reality of atoms. It has been suggested that Davy and Faraday, though neither of them read German, absorbed elements of *Natürphilosophie* which guided them in their choice of theories and analogies.

In the field of electricity the disciples of Schelling made advances indisputable by their more sceptical contemporaries. J. W. Ritter's studies on the electric battery, and his discovery of ultra-violet radiation, were reported in *Nicholson's Journal* and the *Philosophical Magazine*. His friend H. C. Oersted published accounts of Ritter's work which appeared in the same journals; and a 'phlogistic' theory not unlike Ritter's was published by G. S. Gibbes. After Oersted discovered the magnetic effect of an electric current, a discovery which followed from his belief in the unity of all the forces of nature, he published an account of it in the *Edinburgh Encyclopaedia*. This article is one of the best sources in English for *Natürphilosophie*; but other essays by Oersted were also translated, and appeared in English in1851 under the title *The Soul in Nature*. This gives a fascinating picture of the thought of a very important scientist who completely fails to fit the paradigm of the sober and somewhat unimaginative seeker after facts. Oersted's acquaintances included the Schlegels and Schelling; of his essays in the book, that on the history of chemistry is perhaps the most rewarding, but others such as 'The Fountain' with its Heraclitean view of the world, and its statement that the laws of thought and the laws of nature are the same, are also exciting.

Oersted did not believe that the sciences were a rather dull enterprise, or that the scientist could separate his work and his world-view. A more extreme version of *a priori* chemistry is to be found in Lorenz Oken's *Elements of Physiophilosophy*, written under 'a kind of inspiration' in 1810 and translated into English in 1847. Oken began the annual gathering of German savants already mentioned, and he occupies an important place in the history of

Peacock butterfly on a peach tree (Wilkes, B., *120 Copper plates of English Moths . . .*, 1773, pl 106)

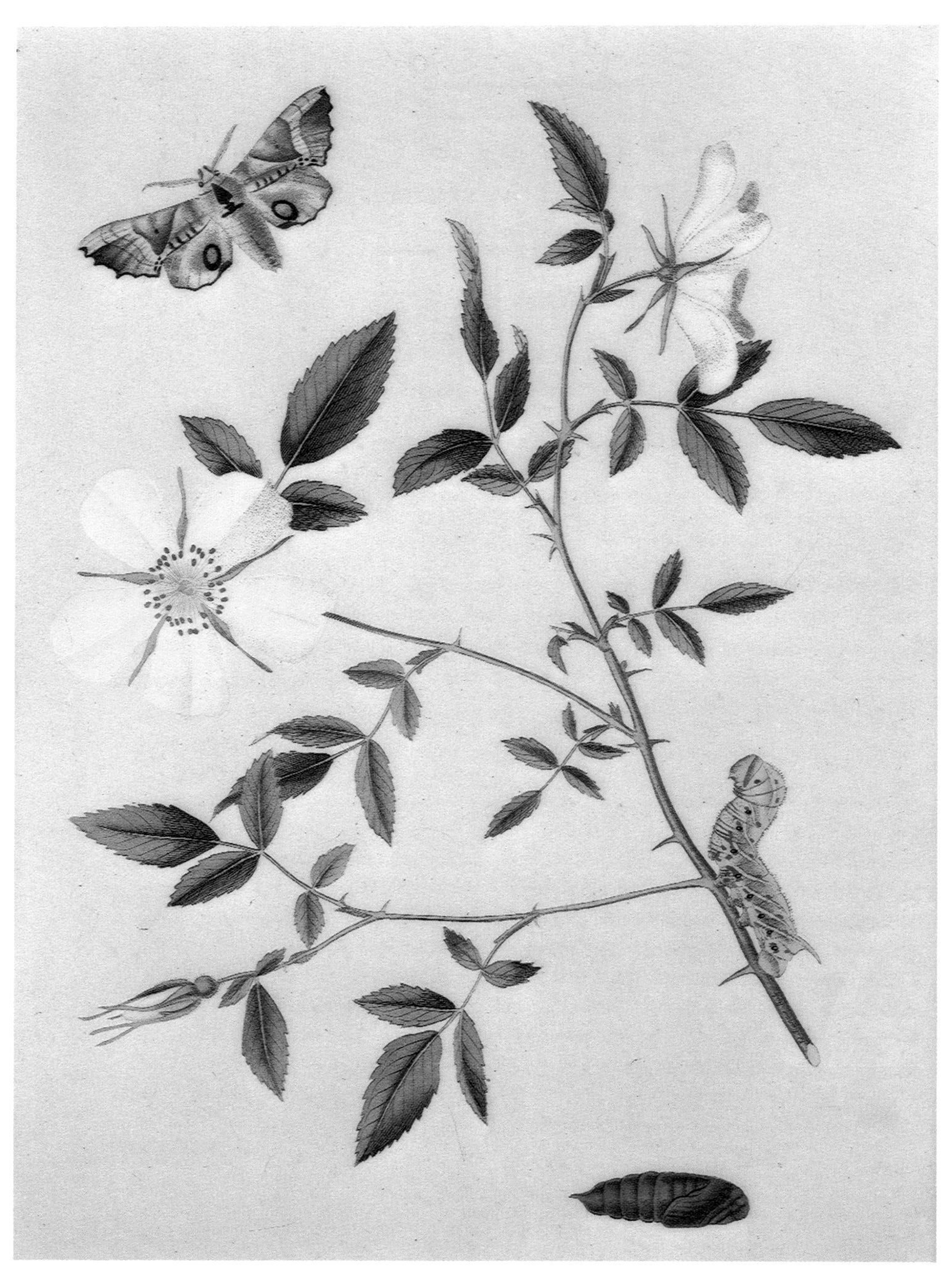

American eyed hawk moth and Carolina Rose (Smith, J. E., *The Rarer Lepidopterous Insects of Georgia*, 1797, vol 1, pl 25)

physiology. For Oken in this treatise, all bodies are composed of hydrogen (or caloric), oxygen (or light), and carbon (or matter). The world is full of force and polarity; an inert material substrate is nowhere to be found. A similar world, in which all is in flux and solid matter dissolves into evanescent gases, is to be found in Friedrich Schlegel's *Philosophy of Life*, translated into English in 1847, and in Oersted. Such a doctrine ran counter to a crude materialism, and received support from the discoveries of Davy and Berzelius as well as from the theorizing of *Natürphilosophie*.

We shall return to Davy shortly; the best account of *Natürphilosophie* in English was written by J. B. Stallo, a German immigrant to America, in 1848, with the title *The Philosophy of Nature*. All nature is an organism, and all movement results from vital forces; a physics of atoms and impacts is absurd. Life is 'a process of phenomenal variations, bodying forth an unvarying, permanent principle of existence'; matter only exists by virtue of its inner vitality, and everything which exists is alive. Explanations of chemical change in terms of the rearrangements of atoms of dead matter cannot be allowed; chemical change is a true synthesis, with the generation of new qualities not present in the so-called components of the compound. The latter part of the book describes the natural philosophy of various German philosophers, particularly Hegel; an English translation of Hegel's physics and chemistry is now available, and it is to be regretted that Stallo's book has not yet been reprinted. His views underwent considerable modification, and his later book, *The Concepts and Theories of Modern Physics* has appeared in a handsome modern edition; the doctrine there presented is a positivism close to that of Ernst Mach, with whom Stallo corresponded. Atomism is opposed as an intrusion of hypothesis and metaphysics into science, which should be concerned simply with laws connecting observables. It is one of the perplexing features of nineteenth-century science and philosophy of science that the rejection of 'hypotheses', even perhaps the same hypotheses, was a feature of a number of traditions, some of which are closely related while others are not.

One of Stallo's pupils was the geochemist Sterry Hunt, whose *Chemical and Geological Essays* were published in Boston in 1875. An essay of 1853 on chemical change and equivalent volumes expounds the doctrine of chemical synthesis as interpenetration and not juxtaposition, and denies the utility of any concept of chemical elements or atoms. No hypotheses as to *noumenon* or *substance*, which transcends all sensible knowledge, can be permitted in the sciences, which must be concerned with phenomena. Hunt elsewhere argues that solution is the typical chemical reaction; and following Laurent, that chemical formulae can be no more than shorthand descriptions of reactions – they can never express real structures. Another essay, on the chemistry of the primeval earth, sketches a theory of inorganic evolution. Hunt's later book, *A New Basis for Chemistry*, 1887, is a more systematic presentation of these ideas.

Such works may appear eccentric, speculative, sometimes incomprehensible, and in any case only remotely concerned with the history of chemistry in which the idea of the juxtaposition of particles proved so powerful compared to that of the generation of new qualities in chemical synthesis. But it is worth remembering that the principle of conservation of energy, which revolutionized physics and later chemistry in the nineteenth century, came as much out of *Natürphilosophie* as from the careful experiments of J. P. Joule; and that electromagnetism sprang from the speculations of Oersted. It was this speculative chemistry, set off by dramatic demonstrations, which attracted to lectures those great audiences composed of those in the main interested neither in chemical industry nor in the acquisition of numerous facts. Davy is famous for discovering the alkali metals sodium and potassium; for proving that chlorine was a chemical element and not an oxide; and for inventing a safety-lamp for coal-miners. The first two of these triumphs would have attracted little attention unless they had been believed to be connected with issues of some general interest; and it seems safe to say that they were. His work illuminated the nature of the imponderable agents – heat, light, and electricity; led to certain conclusions on the ultimate nature of matter; and overthrew the doctrine that chemical characteristics such as acidity or metallic properties depended upon the presence of material principles.

Davy's researches were published in the *Philosophical Transactions*; and in his *Elements of Chemical Philosophy* of 1812. Only part I of volume I of this book ever appeared; subsequent parts were to deal with the discoveries of others, but since Davy was unsystematic and not very good at reviewing the works of those in fields removed from his own, we may agree with contemporary critics that further volumes would not have increased his reputation. As it is, the book is spirited, in the style of his lectures at the Royal Institution; it has a section on the history of chemistry, and a speculative final chapter on the relationships between the chemical elements. Sandwiched between comes his own work on electrochemistry. If matter were simply bounded force, then if an electric current passed through a substance the products need not be components of the original substance. Boyle in the seventeenth century had demonstrated that the products of heating a body were by no means necessarily pre-existent in it. Again, if electricity were like ordinary matter, it might form compounds, 'galvanates', which some workers indeed detected. The critical case was water, which was known from synthesis to be composed of oxygen and hydrogen only. The experiment of Nicholson and Carlisle in 1800 showed that an electric current decomposed water; but more detailed investigations showed that acid and alkali formed around the poles where the current entered and left. Davy used apparatus of silver, gold, and agate, contained in an atmosphere of hydrogen, and found that he only obtained oxygen and hydrogen. The acid and alkali had come from dissolved nitrogen, or from the glass apparatus used by others. Electric charge modified the chemical properties of substances, but it did not generate new substances. It was a powerful tool for the chemical analyst, and Davy concluded that chemical affinity and electricity were manifestations of one power.

He used the battery to isolate sodium and potassium which, being extraordinarily light and reactive metals, caused a sensation and lent themselves to spectacular demonstrations. Ammonia forms a series of salts, the ammonium salts, very similar to those of potassium, and 'ammonium' was known to be a compound of nitrogen and hydrogen. Just as Lavoisier had supposed that all acids must contain oxygen, although none had been found in muriatic (hydrochloric) acid, Davy was tempted to revive the hypothesis of phlogiston in the form that all metals were compounds of hydrogen even though this had, so far, only been obtained from 'ammonium'. In his attempts to analyse the chemical elements and justify the common belief that the world must be composed of a few distinct elementary bodies. Davy invesigated muriatic acid; but electric sparks, metallic potassium, and other agents could not extract oxygen from it unless water, known to contain oxygen, was present. Davy therefore concluded that the acid was composed of hydrogen and another element, which he called chlorine, only; it was therefore not an oxide, and the doctrine that all acids contained oxygen was false. Oxygen had to share its throne with chlorine; and soon with bromine and iodine too. Davy also abandoned the idea that all metals had a common constituent; chemical properties were the result of peculiar arrangements of matter. Soon afterwards he proved that graphite and diamond were chemically identical, showing that physical properties, too, depended upon configurations.

Davy's next researches were on the safety-lamp, and on flames. He then became President of the Royal Society, and his *Discourses*, stressing the need for closer relations between government and scientists, and giving accounts of the labours of those awarded prizes and medals by the society, were published in 1829. He resigned through ill-health in 1827; and his two best-selling works, *Salmonia* and *Consolations in Travel* were written in retirement, the latter appearing posthumously. *Salmonia*, which won the approval of Scott, is a set of dialogues, modelled on those of Izaak Walton, concerned with fly-fishing, but containing remarks on other topics than fish; notably an attack on Mechanics' Institutes. In *Salmonia* too, is a lament for the loss of the power and freshness of youth; and the curious, and too-little-known, *Consolations* is an attempt to come to terms with this. One dialogue presents a vision of progress; another sets forth the dynamic balance between death and decay, and renovation; and perhaps the central tenet of the book is the pre-existence and transmigration of souls, a process which Davy believed, like Origen, would go on until all were purified. Another dialogue describes the ideal chemical philosopher, raised by his science above worldly ambition; and another is concerned with the possibility of a mechanistic physiology and psychology, which Davy could not accept. All these, and selections from his notebooks, were published by his brother John Davy in nine volumes of *Collected Works of Sir Humphry Davy* in 1839–40.

It is striking to find so important a chemist so impatient of the march of intellect, and so remote from cold philosophy. But our pursuit of Davy into these regions has directed us from the mainstream of chemical authors; of which the most important on galvanism were John Bostock: Peter Mark Roget, the author of the Bridgewater Treatise on physiology, but best-known for his *Thesaurus*; and Michael Donovan. These are not, like Davy's, works of originality but useful textbooks and invaluable surveys of the science of the time, although Donovan did not believe that galvanism was electricity. Donovan, an Irishman, also wrote a somewhat acid textbook of chemistry which is interesting in connection with atomism because he denied that the atomic theory differed in assumptions from that of equivalents; both depended simply upon the laws of chemical combination. Such an attenuated atomism, involving no assumptions about indivisible entities, became known as chemical atomism. This doctrine assumed only that bodies reacted chemically in definite lumps, which in fact differed from equivalents in the important respect that it is sensible to talk of different spatial arrangements of such lumps, but not of mere relative weights. This debate rumbled on, and the international Karlsruhe Conference of 1860 resolved that equivalents were more empirical than atomic weights; but then the acceptance of Avogadro's hypothesis, and the convergence of physical and chemical atomic models, made such discussions obsolete by the early years of the twentieth century.

The standard English work on chemical atomism in the generation after Dalton was Charles Daubeny's *Atomic Theory*, which appeared in 1831; a supplement was published in 1840, and a second edition, incorporating the material from the supplement, in 1851. Daubeny was Professor of Chemistry, and later also of Botany, at Oxford, in the difficult period of the first half of the nineteenth century when the university threw off its eighteenth-century lethargy and began to teach and examine the Classics seriously. This meant that students who had had plenty of time to go to the lectures of scientists such as Beddoes now tended to forgo such distractions. Daubeny, with such allies as William Buckland the geologist, fought for a museum and laboratories, and for honours degree courses in the sciences. His account of the atomic theory was a sound and sensible one, free from 'obscure and abstract speculations'. In an appendix Daubeny published correspondence with Prout on the hypothesis that atomic weights were whole numbers; and that the elements were all composed of hydrogen, or possibly some more-primitive first matter. Prout distinguished the second part as speculation; by then in fact the first part was being proved untrue, while the second lived on to torment or inspire through the century.

Prout himself turned more towards biology; but his Bridgewater Treatise on *Chemistry, Meteorology and the Function of Digestion* as exemplifying the power, wisdom and goodness of God deserves a mention as a work of chemistry. Not indeed for the depth of its theology; George Wilson criticized it, and George Fownes' *Chemistry as Exemplifying the Wisdom and Benificence of God*, for their facile view of the world. But in his treatise Prout gave an account of the rôle of hydrochloric acid in digestion, set out a theory that the elements were composed of simpler 'sub-molecules', and put forward an independent postulation of Avogadro's hypothesis. Wilson's *Religio Chemici*, too, is not confined to natural theology; it contains various historical articles, in one of which Wilson gave the now generally accepted view of how Dalton arrived at his atomic theory. Wilson's interpretation, later strengthened by Roscoe and Harden's study of Dalton's notebooks, was that it was his physical studies of gases rather than, as Thomson had supposed, his work on the laws of chemical combination, which put him on the track. Wilson was the first Professor of Technology at Edinburgh; he published a textbook of chemistry in 1830, and in 1859 a work on the *Electric Telegraph and the Chemistry of the Stars* – the two topics fit uneasily together – which includes some reflections on the abundance and distribution of the elements. Wilson noted that some elements, such as gold, were rare upon the Earth; and suggested that in other heavenly bodies they were probably common, there being no sufficient reason why the quantities of each element in the universe should not be different.

Wilson had in 1844 written a review article on a number of Proutian works which had recently appeared; notably one by Robert Rigg, urging that plants synthesized carbon, and another by D. Low, Professor of Agriculture at Edinburgh, who considered that the family relationships between the elements proved that they could not be irreducibly distinct bodies. It was absurd to suppose that different elements as similar as for example platinum and its congeners, or sodium and potassium, or chlorine, bromine and iodine, should have been created. Even more dramatic was the work of Samuel Brown, who believed that he had succeeded in preparing silicon from carbon, which would have been the first transmutation to be observed in modern chemistry had it been possible to confirm it. He was, on the strength partly of this claim, a candidate for the Chair of Chemistry at Edinburgh in 1843; and delivered a course of lectures there, which were published in 1858, after his death. Erudite and speculative, they provide entertaining reading, as ponderous rhetoric alternates with interesting remarks on chemical metaphysics, and descriptions of implausible experiments. Wilson was one of the referees who attempted to repeat Brown's transmutation, another being Robert Kane; both of them shared Brown's distaste for a chemistry of many irreducible elements, but neither could get anywhere with his experiments.

Davy's insight into the identity of chemical affinity and electricity was the basis of the system, 'dualism', erected by Berzelius. Accounts of this in English can be found in Berzelius' articles in *Nicholson's Journal* for 1812, and in his little treatise on mineralogy, published in 1814. The system failed to account for the facts of organic chemistry, where in a number of cases the electronegative element chlorine can replace the electropositive element hydrogen in a compound, and no startling change of properties results from the substitution. Dumas and Liebig proposed the radical theory, according to which certain groups of the elements carbon, hydrogen, oxygen and nitrogen played in organic chemistry the rôle of the elements in the inorganic realm, the electrical nature of the elements being played down. Liebig's chemistry is most accessible in his *Familiar Letters on Chemistry*, which first appeared in 1843 and grew into a fat volume in subsequent editions as fresh essays were added to it. It provides an excellent introduction to his formal treatises on animal chemistry and agricultural chemistry.

The next major work on theoretical chemistry to be translated was Auguste Laurent's *Chemical Method*, in 1855. The translator was William Odling, who later became professor at Oxford, and was a delegate at the Karlsruhe Conference. The story of Laurent and his associate Charles Gerhardt is a sad one; cut off from academic posts in Paris because of the opposition they aroused among the scientific pluralists there, they struggled in garrets and died young without due recognition. Dismayed at the controversies over atoms and equivalents, dualism and the radical theory, and other hypotheses, they proposed a chemistry based upon observables, in which formulae were no more than condensed recipes, since it was impossible without hypotheses to write down atomic structures. The formulae were arranged in formal series called types; and atomic weights were determined from equivalents according to a conventional rule, the unitary system. This became unnecessary when in 1860 Cannizzaro resurrected Avogadro's hypothesis, and it therefore became possible to determine atomic weights unambiguously. But in England, Gerhardt's attitude to formulae was taken by Benjamin Brodie in his lecture *Ideal Chemistry* of 1867, reprinted as a pamphlet in 1880; he argued for a theory-free operational calculus in place of Daltonian atomism and formulae. But his use of Boolean algebra alarmed contemporary chemists, and, although Brodie used his calculus to elucidate the nature of ozone, it proved a dead-end.

With Brodie, whom one must not confuse with his father, a physiologist of the same name, we have left the Romantic period far behind. To conclude the chapter, we should look at some British chemists younger than he. William Thomas Brande, who held a post at the Royal Institution, wrote a *Manual of Chemistry* and a *Dictionary of Science*, both of which were important in their day; and George Fownes wrote, as well as his theological chemistry, a very successful *Manual of Elementary Chemistry* which went through numerous editions. Like Kane and Turner, these authors were careful to separate the empirical and speculative parts of Dalton's atomic theory. J. F. Daniell, of King's College, London, also wrote a textbook of chemistry. He is famous for his electric cell; and his book is strong on electrochemistry. The most celebrated chemist of his day in England was the Scotsman Thomas Graham, whose researches on arsenates and phosphates in the 1830s led to an understanding of the polybasic acids, and whose work on the diffusion of gases was of great importance. His *Elements of Chemistry* was an important textbook; his collected papers were privately published in a limited edition by his friend and biographer R. A. Smith. They include a brief paper giving a sketch of a kinetic theory of matter. A less systematic account of chemistry than in any of these authors is to be found in J. M. Good's *Book of Nature*, which was based upon a series of lectures given at the Surrey Institution at Blackfriars in London. Good, like many chemists of the day, was a physician, and much of the book is devoted to the biological sciences; he is worth reading for his remarks on matter and the status of the imponderable substances. The

62 A chemical laboratory (Parkes, S., *Chemical Catechism*, 1822, 10th ed, frontispiece)

American J. P. Cooke was a Proutian and a pioneer in the classification of elements; he was a professor at Harvard, and his *New Chemistry* appeared in 1874.

Lavoisier's *Elements* contained a section on apparatus and manipulation, but most chemical textbooks used terms like 'trituration' without any clear description of how the process was performed. When he and Wordsworth contemplated setting up a chemical laboratory, Coleridge wrote to Davy for advice. He would not have needed to after 1827, when Michael Faraday's *Chemical Manipulation* appeared. This is Faraday's only monograph; it gives a splendid picture of his experimental genius, his ability to turn things to new uses, and also of the laboratory practice of the day. Chemists had to cut out their filter papers and make rubber tubing from sheets of 'caoutchouc', this being itself a great innovation. The section on 'tube chemistry', explaining how one could make apparatus for performing complex operations from glass tubing, was recognized as a masterpiece. Lavoisier's remarks had been illustrated by copperplates; Faraday's less-splendid illustrations were woodcuts, which had the advantage of appearing on the page to which they referred. A striking difference between Lavoisier and Faraday is that the former described calorimetry at some length; but reliable, repeatable measurements of heat changes were very difficult at this date, and had fallen out of fashion into disrepute, and Faraday has nothing on thermochemistry. The book went through three editions, with few changes of substance; parts of it at least can still be read with profit by a chemist wishing to improve his basic laboratory techniques.

7 CHEMISTRY IN THE ROMANTIC PERIOD

Aitkin, A. and C. R., *A Dictionary of Chemistry and Mineralogy . . .*, 2 vols., 1807.

Beddoes, T., *Contributions to Physical and Medical Knowledge . . .*, Bristol, 1799.
and Watt, J., *Considerations on the Medicinal Uses of the Factitious Airs*, Bristol, 1795.

Berthollet, C. L., *Researches into the Laws of Chemical Affinity*, tr. M. Farrell, 1804.
Essay on Chemical Statics, tr. B. Lambert, 2 vols., 1804.

Berzelius, J. J., *Animal Chemistry*, tr. G. Brunnmark, 1813.
An Attempt to Establish a Pure Scientific System of Mineralogy, 1814.
The Use of the Blowpipe in Chemical Analysis, tr. M. Fresnel and J. G. Children, 1822; tr. J. D. Whitney, Boston, Mass., 1845.

Bostock, J., *An account of the history and present state of galvanism*, 1818.

Brande, W. T., *A Manual of Chemistry*, 1819.

Brodie, B. C., *Ideal Chemistry*, 1880.

Brown, S., *Lectures on the Atomic Theory . . .*, 2 vols., Edinburgh, 1858.

Chaptal, J. A. C., *Elements of Chemistry*, tr. W. Nicholson, 3 vols., 1791.
Chemistry applied to Arts and Manufactures, 4 vols., 1807.

Cooke, J. P., *Religion and Chemistry . . .*, New York, 1864.
The New Chemistry, 1874.

Dalton, J., *Meteorological observations and essays*, 1793.
A New System of Chemical Philosophy, Manchester; vol. I, pt I, 1808, pt II, 1810; vol. II, pt I, 1827; no more published.

Daniell, J. F., *An Introduction to the Study of Chemical Philosophy*, 1839.

Daubeny, C., *Introduction to the Atomic Theory*, Oxford, 1831; a *Supplement* appeared in 1840, and was incorporated in the 2nd ed., 1850.

Davy, H., *Researches Chemical and Philosophical chiefly concerning Nitrous Oxide . . . and its Respiration*, 1800.
Elements of Chemical Philosophy, vol. I, pt I, 1812; no more published.
Salmonia, or days of fly-fishing, 1828.
Consolations in Travel . . ., 1830.

Donovan, M., *Essay on Galvanism*, Dublin, 1816.
Chemistry, 1832.

Dumas, J. B. A. and Boussingault, J. B., *The Chemical and Physiological Balance of Organic Nature*, 1844.

Essays by a Society of Gentlemen at Exeter, 1796.

Faraday, M., *Chemical Manipulation*, 1827.

Fourcroy, A. F., *Elements of Natural History and Chemistry* (tr. W. Nicholson), 1788.
Elements of Chemistry, tr. R. Heron, 4 vols., 1796.
Synoptic tables of chemistry, tr. W. Nicholson, 1801.
A General System of Chemical Knowledge, tr. W. Nicholson, 11 vols., 1804.

Fownes, G., *A Manual of Elementary Chemistry*, 1844.

Fulhame, Mrs, *An Essay on Combustion . . . wherein the phlogistic and antiphlogistic hypotheses are proved erroneous*, 1794.

Gibbes, G. S., *A phlogistic theory, ingrafted upon M. Fourcroy's Philosophy of Chemistry . . .*, Bath, 1809.

Gmelin, L., *Hand-book of Chemistry*, tr. H. Watts, 6 vols, 1848–52.

Goethe, W. *Theory of Colours*, tr. C. Eastlake, 1840.

Good, J. M., *The Book of Nature*, 3 vols., 1826.

Graham, T., *Elements of Chemistry*, 1842; 2nd ed., 2 vols., 1847.
(ed.), *Chemical Reports and Memoirs*, 1848 (Cavendish Society).
Chemical and Physical Researches, ed. R. A. Smith, Edinburgh, 1876.

Gren, F. C., *Principles of Modern Chemistry*, 2 vols., 1800.

(Harrington, R.), *A New System of Fire and Planetary Life; shewing that the sun and planets are inhabited . . .*, 1796.
The Death-warrant of the French theory of chemistry . . ., 1804.

An elucidation . . . of the Harringtonian System of Chemistry, 1819.
Heron, R., *Elements of Chemistry*, 1800.
Hunt, T. S., *Chemical and Geological Essays*, Boston, Mass., 1875.
A New Basis for Chemistry, Boston, Mass., 1887.
Kane, R., *Elements of Chemistry*, Dublin, 1846.
Laurent, A., *Chemical Method*, tr. W. Odling, 1855.
Liebig, J. von, *Animal Chemistry . . .*, tr. W. Gregory, 1842.
Familiar Letters on Chemistry, 1843.
Low, D., *An Inquiry into the Nature of the Simple Bodies of Chemistry*, 1844; 2nd ed., 1848; 3rd ed., 1856.
Newberry, W., *The Chymical Delectus: or companion to Newberry's cabinet laboratories . . .*, 1842.
Nicholson, W., *A Dictionary of Chemistry*, 1795.
Oersted, H. C., *The Soul in Nature*, tr. L. and J. B. Horner, 1852.
Oken, L., *Elements of Physiophilosophy*, tr. A. Tulk, 1847.
Rigg, R., *Experimental Researches, shewing Carbon to be a Compound Body, made by Plants*, 1844.
Roget, P. M., *Thesaurus of English Words and Phrases . . .*, 1852.
Treatises on electricity, galvanism, magnetism and electromagnetism, 1832.
Schlegel, F., *The Philosophy of Life . . .*, tr. A. J. W. Morrison, 1847.
(Southey, R.), *Letters from England: by Don Manuel Alvarez Espriella . . .*, 1807.
(and Coleridge S. T.,) *Omniana . . .*, 2 vols., 1812.
Stallo, J. B., *General Principles of the Philosophy of Nature*, 1848.
The Concepts and Theories of Modern Physics, New York, 1882.
Thomson, T., *A System of Chemistry*, 4 vols., Edinburgh, 1802; 3rd ed., 1807, contains first publication of Dalton's atomic theory.
An Attempt to Establish the First Principles of Chemistry upon Experiment, 2 vols., 1825.
The History of Chemistry, 2 vols., 1830–1.
Tobin, J. J., *Journal of a Tour . . . accompanying the late Sir Humphry Davy*, 1832.
Turner, E., *Elements of Chemistry*, Edinburgh, 1827.
Ure, A., *A Dictionary of Chemistry*, 1821.
Willich, A. F. M., *Elements of the Critical Philosophy*, 1798.
Wilson, G., *Chemistry*, Edinburgh, 1850.
Electricity and the Electric Telegraph, together with the Chemistry of the Stars, 1852.
in: *Edinburgh Essays*, Edinburgh, 1857.

Recent Publications

Cardwell, D. S. L., *John Dalton and the Progress of Science*, Manchester, 1968.
Crosland, M. P., *The Society of Arcueil*, 1967.
Dalton, J., *A Bibliography of Works by and about him*, by A. L. Smyth, Manchester, 1966.
Davy, H.: Fullmer, J. Z., *Sir Humphry Davy's Published Works*, Cambridge, Mass., 1969.
Grabo, C. H., *A Newton Among Poets*, Chapel Hill, 1930.
Greenaway, F., *John Dalton and the Atom*, 1966.
Hegel, G. W. F., *Logic*, tr. A. V. Miller, 1969.
Philosophy of Nature, tr. A. V. Miller, 1970.
Philosophy of Nature, tr. M. J. Petry, 3 vols., 1971.
Knight, D. M. (ed.), *Classical Scientific Papers – Chemistry*, 1st series, 1968; 2nd series, 1970.
Palmer, W. G., *A History of the Concept of Valency to 1930*, Cambridge, 1965.
Piper, H. W., *The Active Universe*, 1962.
Schelling, F. W. J., *On University Studies*, tr. E. S. Morgan, ed. N. Guterman, Athens, Ohio, 1966.
Smeaton, W. A., *Fourcroy*, 1962.
Szabadváry, F., *History of Analytical Chemistry*, tr. G. Svehla, Oxford, 1966.

8 *'The Decline of Science in England'*

In 1830 Charles Babbage published his *Reflections on the Decline of Science in England*. To the observer with advantage of hindsight, the charge that science was in decline seems ridiculous: Lyell's *Principles of Geology* had begun to appear, Darwin was setting off on the *Beagle*, and Faraday was beginning his revolutionary researches in electricity and magnetism. But the book, with its attacks on the Royal Society and the English universities, and its demand for more honours and sinecures for scientists, caused a sensation. Shortly afterwards, after a hard-fought campaign, the duke of Sussex was elected President of the Royal Society, defeating the astronomer John Herschel. And in 1831 the British Association for the Advancement of Science was set up and met for the first time at York; its origin can be traced on the one hand to dissatisfaction with the Royal Society, and on the other to the powerful provincial sentiments in England at the time. Babbage's book provoked an answer by a 'Foreigner', Gerrit Moll, a pamphlet of thirty-three pages which Faraday published; and the Royal Society's affairs were discussed anonymously by Augustus Bozzi Granville in his *Science without a Head* which showed how few Fellows of the Royal Society did any research. Although the reformers lost this battle in the Royal Society, they went on to win the war; gradually the amateurs, who had formed the majority down to this period, were replaced, as they died, by genuine, though as time passed increasingly elderly, men of science. This professionalization may not have been pure gain: with it was associated an increase of jargon and obscurity in the papers presented. The scientists of the late nineteenth century could not follow all one another's arguments as the natural philosophers of a century earlier had expected to do.

The occasion for lamenting the decline of science was death within a year of Davy, William Hyde Wollaston, and Thomas Young. Of their generation, Dalton remained; but his later papers were of little value, and the second volume of his *New System of Chemistry* remained on the Royal Society's shelves with its pages uncut. Dalton was idolized in Manchester, but would not have been reckoned elsewhere among the other three. Davy we have already discussed; he was a man to whom it was, and is, hard to be indifferent, and biographies of him are legion. Young is also well known; but Wollaston's reputation has unfairly declined, perhaps because he wrote no books and because of the dullness of his private life. His cool manner prevented any great success in his profession of physician; but he then discovered how to make platinum malleable, and on the revenue from this invention he was able to devote his time to research, which he published in the *Philosophical Transactions*. His papers include studies on fairy rings and sea-sickness, and one with Sir Thomas Lawrence on why the eyes in portraits seem to follow one around; but his reputation was made in chemistry, by his researches on the platinum metals, his analyses which supported Dalton's theory, his paper on chemical equivalents, and his crystallographic studies, including his invention of the goniometer for measuring crystal angles. He was a very cautious and painstaking worker, famous for his use of minute quantities by the standards of the day. His papers are lucid and readable, and often raise interesting questions, but they never appeared as a book.

Young lacked Wollaston's caution, and he is remembered chiefly for the successful introduction of the wave or undulatory theory of light, and for the first steps in the decipherment of the Rosetta stone. In both these fields, his work was completed by Frenchmen beginning independently; Fresnel and Champollion. Young also was a physician, and he studied the mechanism of vision, making experiments on himself, as well as theories of optics. He lectured at the Royal Institution before Davy, and his lectures were published. They are a mine for anyone seeking information on physical science at the beginning of the nineteenth century. As lectures they were not a success; he lacked the capacity displayed by Davy, and later Faraday, of making his subject come alive for a lay audience. Young's strategy in his optical papers had been to set out all the passages from Newton which favoured some kind of wave interpretation; to Newtonians who believed that their master had unequivocally embraced a corpuscular theory of light, this was heresy. As such it was stridently denounced by Henry Brougham in the first volume of the *Edinburgh Review* in 1803. Brougham was no scientist, and a mathematical paper he wrote had justly provoked Young's censure; in his attack on Young he employed the legal arts which later brought him both eminence and widespread hatred.

Later volumes of the *Edinburgh Review* carried attacks upon the teaching of mathematics in the English universities; Playfair's review of Laplace, remarking how few mathematicians in England could begin to understand it, is particularly noteworthy. In fact Laplace's popular *System of the World* was translated by the Astronomer Royal, J. Pond, in 1809; this book contains the nebular hypothesis, advanced independently by Kant, according to which the sun and planets had developed out of a mass of nebulous matter which had begun to rotate. Laplace's highly mathematical *Celestial Mechanics* was translated by an American sea captain, Nathaniel Bowditch, who amended the original by giving further proofs and drawing attention to the work of others – for Laplace, like Lavoisier, was less than scrupulous in regard to his intellectual debts. So abtruse did the work appear that Bowditch had to pay the costs of publication himself: the translation is one of the first great achievements of American science. In this work, Laplace had been able to show that the perturbations in the solar system would, in the long run, cancel one another out, and that there was no need for Newton to have invoked the hypothetical finger of God to maintain the stability of the system. Bowditch turned down the offer from Jefferson of a chair at the university of Virginia, because the teaching would have been too elementary and the stipend too small.

Laplace, like all the great Continental mathematicians, employed the notation and concepts of differential and integral calculus as devised by Leibniz. Mathematicians in Britain, on the other hand, adhered to the fluxional methods of Newton, and thereby cut themselves off and became hopelessly provincial. This was the burden of the charges made by Playfair and Babbage; but in the second decade of the nineteenth century this began to change. One of those chiefly responsible was George Peacock, who later became dean of Ely, but who is noteworthy for his biography of Thomas Young and for his part, particularly with Babbage who later held Newton's Chair, in introducing the Continental methods into the Cambridge syllabus. Peacock, Babbage, and John Herschel had been undergraduates at Cambridge together. Cambridge was throughout the eighteenth and nineteenth centuries the university where mathematics was taken most seriously; the form and context of the courses there can be found in Rouse Ball's *History of Mathematics at Cambridge*, and in the more general work by Christopher Wordsworth, *Scholae Academicae*. The latter gives details of the disputation subjects in other fields also; and lists the books in use at Cambridge in the mid-eighteenth century under various headings. It is clear that in about 1800 it was possible to be very idle at Cambridge; and that even if one were hard-working, the courses were much less modern than those available at, for example, the Ecole Polytechnique. For much of the nineteenth century, too, the course was distorted, according to some critics, by the system of putting the candidates for highest honours, the Wranglers, in order rather than in classes, because this encouraged the brightest students to cram rather than to follow their interests.

When Babbage referred to the decline of science, it was mathematical physics that he had in mind; and it was true that there had been a decline from the days of Newton, Halley, Cotes, and Maclaurin, and that new French giants, such as Fourier, Cauchy, Poisson, Leverrier, and Ampère were appearing to carry on the work of Laplace. Babbage, Herschel and Peacock collaborated on a translation of a book by Lacroix on the differential and integral calculus in 1816; and in 1820 published a textbook of examples for undergraduates. Thus, and by letting it be known when they were appointed examiners that they would set questions on the Continental methods, the reformers had succeeded by the early 1820s in modernizing the Cambridge course. Babbage did not remain at Cambridge very long but began to develop his mechanical calculating engines, which are ancestors of our electronic computers. This is a clear case of a technological frontier; Babbage worked out the principles, but his contraptions of gears and pulleys were too clumsy, government money for development was eventually cut off, and successful computers only appeared in the twentieth century. For this and other reasons, Babbage became somewhat embittered and his autobiography, *Passages from the life of a Philosopher*, contains some furious contributions to controversies long since dead. He had a particular dislike of Davy, whom he believed had prevented him from being made Secretary of the Royal Society, and from being awarded one of the first Royal Medals of the Society. He accused the long-dead Davy of diverting Royal Society funds into his own pocket. But other parts of the book are very light-hearted; there are some splendid reminiscences of research on vibrations of early railway trains, with the modern-sounding suggestion that all trains should carry automatic recording devices as 'incorruptible witnesses of the immediate antecedents of any catastrophe'. The book is very much alive, and so wide were Babbage's interests that nobody could find it dull.

When the Bridgewater Treatises appeared, there was none dealing with applied mathematics; and Whewell's, on astronomy, contained a sneer at the pretensions of purely deductive reasoners. Babbage's extraordinary and unofficial *Ninth Bridgewater Treatise*, anticipating some works of our century in its fragmentary form, is the liveliest of the series, and is of importance for its discussion of induction, laws of nature, and free will in terms of the programming of a computer; for its remarks on Mossotti's theory of matter of which we shall say more later; and for its reference to the possibility of tree-ring dating of fossil wood, to provide an absolute time-scale for geologists. Babbage also wrote on machinery, for the *Encyclopaedia Metropolitana*; and on *The Exposition of 1851*, giving his sometimes-pungent views on the industry, science, and government of England. Later mathematicians whose interests were almost as wide and writings equally amusing include, of course, Lewis Carroll, and also Augustus de Morgan, whose *Budget of Paradoxes* is splendid reading, and whose bibliography of early arithmetical works is invaluable; his gossip about Newton is entertaining.

Babbage's great friend and ally was John Herschel, son of the discoverer of Uranus, who inherited his father's interests in astronomy. His defeat in the election for the Presidency of the Royal Society was in a sense a blessing for science, because he left England for the Cape of Good Hope and drew up a catalogue of stars in the Southern Hemisphere. His most influential book was probably his *Preliminary Discourse on the Study of Natural Philosophy*, with a florid engraved title page showing a portrait of Francis Bacon and the date 1830. It formed the first volume of the 'Cabinet Cyclopedia' edited by Dionysius Lardner; which series later also included in 1833 Herschel's popular *Treatise on Astronomy*, expanded in 1849 into *Outlines of Astronomy*. The *Preliminary Discourse* influenced the philosophy of science of John

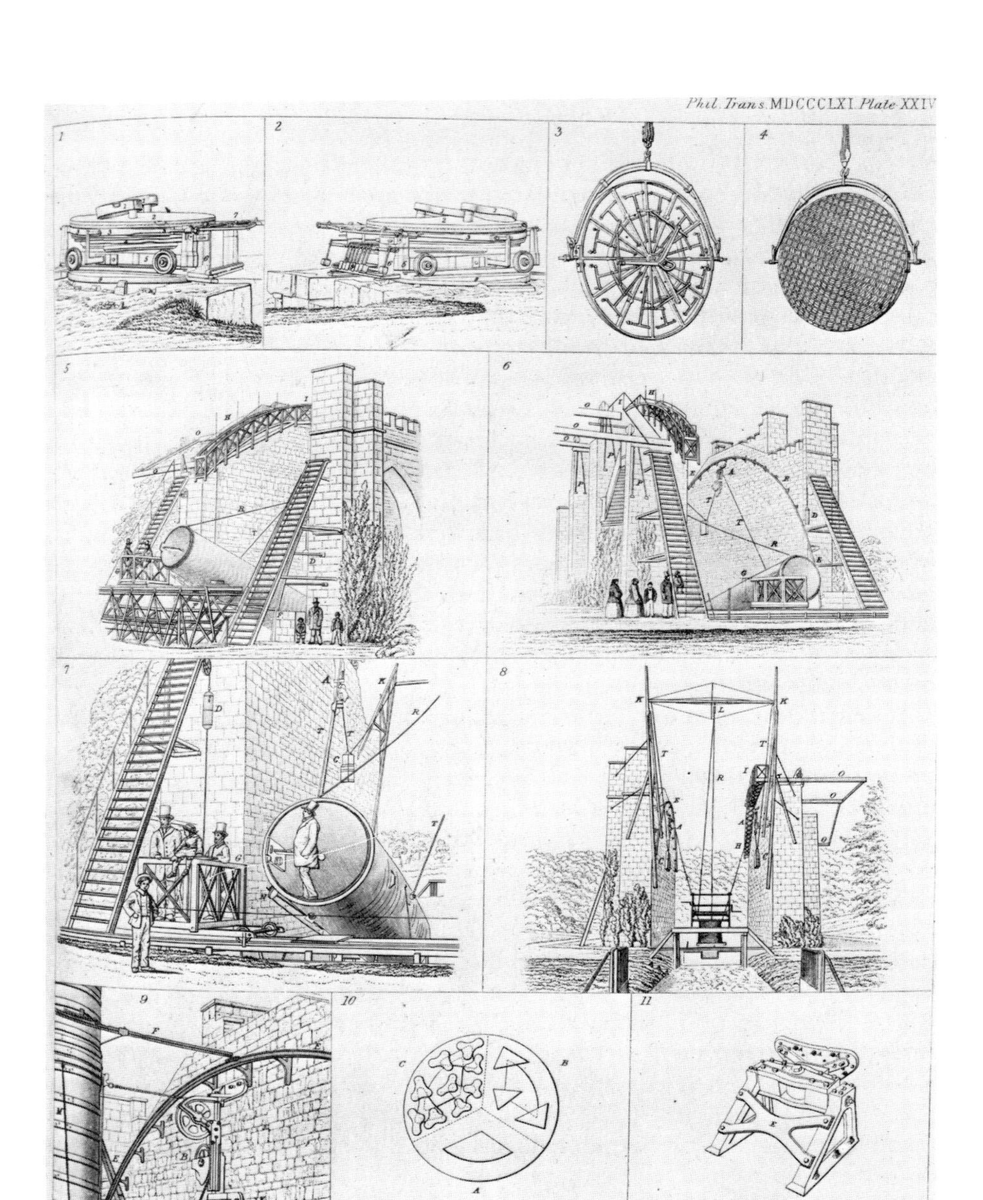

63 Lord Rosse's telescope (*Philosophical Transactions*, 1861, pl 24)

Stuart Mill and William Whewell; and was read by Darwin and indeed by most scientists of his generation. It is an informal, personal and discursive work, in which almost every page contains some insight invaluable to the student of the science of the period, and gets across the intellectual interest of the sciences, the relationships between them, and the excitement of scientific thought. The little book on astronomy is similarly a splendid piece of popularization, devoid of mathematics, very clear and never condescending, although it is made clear that mathematics is the only gateway of a real understanding of science. Herschel's *Essays for the Edinburgh and Quarterly Reviews*, published in 1857, includes reviews of works by Whewell, Humboldt, Gauss, Laplace, and Quetelet, displaying the judgement and lucidity which made him in his day such a pundit.

John Herschel continued the investigations of the nebulae begun by his father; and the most important subsequent steps in this direction were taken by the earl of Rosse who built, at Parsonstown in Ireland, telescopes first with a reflector of three feet diameter, and then one of six feet. Although he died in 1867, his papers were not published in book form until 1926, when their editor was his son Charles Parsons, the builder of the first marine turbines. Rosse's papers give practical details of the building of giant reflecting telescopes, and have plates illustrating various beautiful nebulae.

His work stimulated William Nichol, a professor at Glasgow, who wrote a handsome and edifying work of a pantheistic tendency, *The Architecture of the Heavens*, from which much about contemporary astronomy can be learnt. Nichol gave William Thomson his first impulse towards physics, introduced Wordsworth to the poetry of Longfellow, and impressed George Eliot. More straightforward accounts of nineteenth-century astronomy in English include Robert Grant's *History of Physical Astronomy* of 1852 which won him the Gold Medal of the Royal Astronomical Society; this volume began to come out in parts, but only a few parts so appeared. The volume is particularly useful for the light it casts upon astronomy in the first half of the nineteenth century, when the two most dramatic developments were the first measure of stellar parallax, yielding for the first time the distance of a fixed star from the Earth; and the discovery of the planet Neptune, the position of which was predicted independently by John Couch Adams in England, and by Urbain Leverrier in France. To explain the irregularities in the behaviour of the planet Uranus, they postulated a hitherto unobserved planet beyond it perturbing its orbit; and computed where it would be

64 Spiral nebula (Nichol, J. P., *Architecture of the Heavens*, 1850, 2nd ed, pl 12)

65 Lunar crater (Ball, R. S., *Story of the Heavens*, 1885, pl 8)

found. At first no observatory would disturb its programme to look for the new planet; but when Galle in Berlin observed the planet, the prediction was immediately recognized as a triumph of Newtonian physics, and became an occasion of Anglo-French recrimination. On this controversy, and generally, Grant is judicious; and it would be hard to better his *History* for scope, detail, and judgement.

Agnes Clerke's *History of Astronomy during the Nineteenth Century*, of 1885, is less of a monument than Grant's; it is divided into two parts, the first dealing with the progress in sidereal and planetary astronomy in the first half of the century, and discussing advances in instruments, and the second describing at greater lengths the new science of astrophysics, through which spectroscopists such as Bunsen, Kirchhoff, Huggins, and Lockyer were transforming astronomy and linking it to terrestrial physics and chemistry. A handsome and popular essay of much the same date was Robert Ball's *Story of the Heavens*, which appeared in 1886; Ball became Astronomer Royal for Ireland. His book contains excellent illustrations, in colour, of the planets and solar prominences; those of lunar craters are very striking. The book is reliable and clearly written, and went through numerous editions. Other works on the history of astronomy which deserve to be mentioned are Robert Small's *Account of the Astronomical Discoveries of Kepler* which

Calxfluor (Sowerby, J., *British Mineralogy*, 1804, vol 1, pl 73)

Yellow butterflies (Lewin, W., *Papilios of Great Britain*, 1795, pl 31)

appeared in 1804 and remains a standard work for its painstaking analysis of the *Astronomia Nova*; and the standard *History of the Royal Astronomical Society*, which reprints manuscript material relating to the pursuit of the science in nineteenth-century England.

In Babbage's *Bridgewater Treatise* he included an appendix discussing the theory of matter of Fabrizio Mossotti, who was Professor of Physics at the University of the Ionian Islands, which were for much of the nineteenth century under British rule. Mossotti's idea was that the matter was composed of two kinds of particles, which repelled similar particles and attracted dissimilar ones according to an inverse square law. The theory grew out of eighteenth-century particulate theories of electricity; an example in English is Cavendish's paper of 1771, which was reprinted in his *Electrical Researches*. On Mossotti's theory, the electrical particles form an atmosphere around the particles of ponderable matter; by different arrangements and associations, different positions of equilibrium would result, accounting for the various states of matter: solid, liquid, and gaseous. The atoms, Babbage remarked, might be no more than centres of force; it seems to have been in this form that the theory recommended itself to Faraday, who communicated the paper to Richard Taylor for inclusion in his journal *Scientific Memoirs*, which printed only translations of papers originally published in foreign languages.

This kind of theory involved more different kinds of entity than Boscovich's theory, but did not require the transitions from attractive to repulsive force which seem inexplicable and *ad hoc* in Boscovich's account. A somewhat similar theory to Mossotti's was developed in England, apparently independently, by the encyclopedist Thomas Exley, whose *Principles of Natural Philosophy* appeared in 1829 and was followed by a series of papers in the *Philosophical Magazine* and the *Reports* of the British Association. Exley in the *Principles* urged that attraction was the thread to follow in the sciences; and that attraction and repulsion should, if we apply Newton's Rules of Reasoning, be regarded as primary properties of matter. Both probably follow the same inverse square law. We know nothing of matter, he declared, except 'the forces which it exerts, and which doubtless constitute its nature'.

The problem for Exley, as for anybody proposing a theory of matter, was to apply these ideas in some detail to the phenomena of physics and chemistry; and this he tried to do in his papers. He soon found it necessary to increase the number of different particles; by 1844 his theory involved four different kinds, which were arranged in various shells. He defended his hypothetico-deductive approach by an appeal to Newton; but his theory had by then lost the simplicity which gave it such attractiveness as it had ever had, and nobody tried to develop it further. It was not until late in the nineteenth century that any theory of matter became sufficiently articulated to generate predictions or explanations testable in detail; before that time, choice between theories was a matter of metaphysics, depending upon one's views of simplicity, harmony, and the nature of explanation.

The first theory which did become testable in any strict sense was the dynamical, or more properly kinetic, theory of gases, according to which gases are composed of particles, or molecules, in rapid motion. In various forms, the kinetic theory had a very long history; one clue to its development in the nineteenth century is, as Lord Kelvin pointed out, Davy's account in 1799 of repulsive power as the outcome of motion, and hence even perhaps of attraction. But the particles of gases in Davy's theory, as in those of most earlier scientists, were only in rapid vibrational or rotational motion about their mean position. They were not, as in the modern theory, rushing hither and thither all over the vessel, and colliding with each other and the walls. This idea is to be found in the theory of John Herapath, which appeared in a paper in the *Annals of Philosophy* for 1821, having been turned down by the Royal Society. Herapath later gave an account of the theory in his *Mathematical Physics* of 1847; by then he had become a publisher of railway guides.

The difficulty of a theory involving collisions was that the particles, if they were the atoms of Newton, must be hard. An elastic body is one in which the parts can be relatively displaced by a force, but will spring back into their old places; hence a body not composed of parts could not be yielding or elastic, but must be absolutely hard. Seventeenth-century mathematicians, notably Wren, had investigated the collisions of hard bodies; and Herapath was forced to pursue this recondite and paradoxical line of enquiry.

His success was not complete, although his theory did account for some properties of gases; but when the principle of Conservation of Energy became generally accepted, it was apparent that a gas composed of hard particles would be impossible; that instead the molecules must be perfectly elastic, and hence lose no energy in collisions. J. J. Waterston, probably the most unfortunate Scottish man of science, took up the problem in the 1840s; his papers on this and other topics did not get published in full in his lifetime, but may be found in his *Collected Scientific Papers* which appeared in 1928, or in the *Philosophical Transactions* for 1893. Waterston's aims were lower than those of Herapath, who intended an account of the physical constitution of the universe; Waterston described an idealized model which might fit the real world.

Neither Herapath nor Waterston had any impact in their own day, and the kinetic theory became established only when J. P. Joule, Rudolph Clausius, and James Clerk Maxwell took it up. Joule's paper appears in his *Scientific Papers*, a collection which he supervised in his old age; and Maxwell's discussions in the superb edition of his *Scientific Papers* which was published after his early death. Joule's most important papers related to conservation of energy, and will come to our notice later; Maxwell's contributions to kinetic theory were of extreme importance, and he had a rare capacity for making clear to the layman what were in fact the results of complex mathematical reasonings. For in the generation following Babbage's an important school of mathematical physicists had begun to appear in Britain, who were, like Newton, to unite physical insights with mathematical ability. They have been censured, notably by Pierre Duhem, whose *Aim and Structure of*

Physical Theory appeared early in the twentieth century, for their insistence upon physical models rather than simply upon equations which fit the facts. But the kinetic theory of gases is the classic case where a model, that of a kind of three-dimensional game of billiards, was used by Maxwell to predict an unexpected phenomenon, namely that the viscosity of gases would be found to be independent of density.

In the same pages we find the revolutionary introduction by Maxwell, a few weeks before *The Origin of Species* transformed biology, of statistical methods into physical science. Statistics had been used by astronomers in finding a mean value from a number of observations, and by social scientists such as Quetelet; but when Maxwell applied it to gases, the particles of which must, as they collide, be moving with a wide range of velocities, he opened up new territory and paved the way for the abandonment of simple ideas of causality.

One of the chief problems connected with the study of gases was that of their diffusion and mixture. Dalton had been perplexed that the atmosphere was not a kind of sandwich, with the densest gas, carbon dioxide, at the bottom, then oxygen, then nitrogen, and then water vapour, the lightest; and had invoked repulsive forces between like particles to account for the uniform mixing of these gases. Contrarily, kinetic theorists had to explain why, if the molecules were moving at the velocities that the theory required, gases did not mix in an instant. Studies on diffusion were made by Thomas Graham, and are to be found in his *Collected Papers*, which also include a speculative paper of 1863 where Graham suggests that the different kinds of matter really differ only in the degree of motion given to the identical particles in a primordial impulse. This view is incompatible with the collisions required by the kinetic theory. Of more importance than this speculation are the investigations of Joule and William Thomson on the cooling of gases upon sudden expansion. If there were no repulsive forces between particles, but only attractive ones, then the gas would do work on expansion and be cooled, as Joule and Thomson found that it was. The paper is reprinted in the second volume of Joule's *Scientific Papers*; the discovery, which holds for all gases at low temperatures, was of great importance for the liquefaction of gases which had previously refused to condense.

Indeed at the beginning of the nineteenth century a distinction was made between permanent gases, like oxygen and carbon dioxide, which no cooling would condense; and vapours like steam. Davy and Faraday evolved a technique for liquefying gases by generating them in a sealed tube; chlorine being the first gas to be so liquefied. These researches are described in Faraday's *Experimental Researches in Chemistry and Physics*; the experimental arrangements and necessary precautions may be found in his *Chemical Manipulation*. The pressure attained was high and, with the then available apparatus, explosions were very frequent. While these experiments indicated how the vapour and liquid states merged into one another, the detailed researches on the condensation of vapour and the continuity of the liquid and vapour states were performed by Andrews, whose studies of carbon dioxide have become a classic. The original lengthy articles are reprinted in his *Collected Papers*. By combinations of cooling, pressure, and rapid expansion, all gases except helium had been liquefied by the end of the nineteenth century.

These investigations into changes of state and the nature of gases bring us to the related question of the nature of heat. By about 1800 the chemists' view that heat was a substance was generally received; but the *Mathematical and Philosophical Dictionary* of Charles Hutton, a splendid monument of late eighteenth-century Newtonianism and a most useful source, also sets out the kinetic theory of heat, that it is the effect of the motion of particles. George Gregory, in his *Economy of Nature*, set out lucidly the reasons for preferring the view that heat was an imponderable substance, caloric fluid. This theory had been developed by Black and Lavoisier, and accounted happily for the phenomena of chemistry and of changes of state. The first assaults upon it were made by Rumford and Davy in the last years of the eighteenth century; but Rumford's observation that an indefinite quantity of heat could be produced by friction in cannon-boring was felt to be an anomaly which could not be allowed to overturn a generally useful theory. Rumford applied his studies on heat to the construction of cooking stoves and fireplaces, initially for the workhouses in Bavaria where he was employed in government after fleeing America as a Tory. While many people had their stoves and fireplaces Rumfordized, his revival of the Newtonian theory of heat gained few adherents.

The experiments which did finally prove convincing were those of Joule, who heated water by churning it in an insulated vessel with paddle-wheels driven by falling weights. He proved that a fixed quantity of mechanical work was used up in the generation of a given quantity of heat. It therefore became implausible to think of a caloric fluid being somehow created, and sensible to think rather of indestructible energy or force, of which heat, electricity, and mechanical work were different manifestations, which could be transformed one into another. Joule drew this great conclusion in a lecture 'On Matter, Living Force, and Heat', of 1847. Joule was not alone in announcing the principle of conservation of energy; indeed, it was in the air, and the problem was to enunciate it clearly and distinctly, and to perceive the difference that it made to science. The palm here is generally given to Hermann von Helmholtz, whose essay, translated by John Tyndall, appears in Helmholtz's *Popular Scientific Lectures*. Helmholtz's was one of the most wide-ranging intellects of the nineteenth century; his works on the perception of colours and tones are classics, his mathematical treatment of vortices set off a string of exciting work, and his popular lectures include a lucid discussion of non-Euclidean geometries, and a sympathetic, but devastating, account of Goethe's scientific researches. He excelled in bringing together into a synthesis, which he expounded with clarity, the fragmentary discoveries of others in disparate fields. As the doctrine of conservation of energy became generally

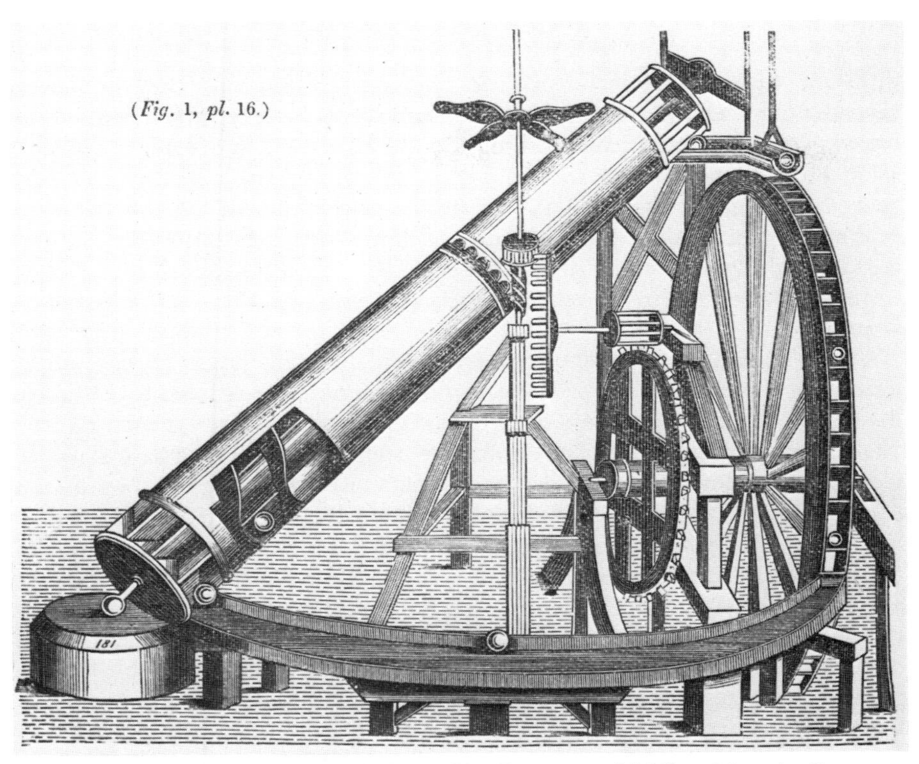

66 Perpetual-motion machine (Dircks, H., *Perpetuum Mobile*, 1861, pl 16)

accepted, the impossibility of perpetual-motion machines became even more evident; the standard account of such devices, which is often very entertaining, is Henry Dircks' *Perpetuum Mobile* of 1861.

In England the work of Davy and others on the Voltaic cell indicated that chemical and electrical energy had a common source; and the terms 'energy' and 'power' were used in something like their modern scientific sense by Wollaston in his Bakerian Lecture to the Royal Society in 1805, and by Young two years later. The lawyer William Grove published *On the Correlations of Physical Forces* in 1846; it was an expansion of lectures delivered at the London Institution, and is based upon the idea that force cannot be lost, but only transformed. It went through various editions as the implications of the doctrine were further developed by Joule, Helmholtz and others. Faraday used the principle *en passant* in his electrical researches; and in 1857 published a paper on the subject, based upon an address to the British Association. It is to be found in his *Experimental Researches in Chemistry and Physics*, with a modification following correspondence with Maxwell.

It was the young William Thomson who recognized the merits of Joule's paper; and the textbook of physics which Thomson wrote with Peter Guthrie Tait, *A Treatise on Natural Philosophy*, was the first to approach the subject from the new standpoint. Parts of physics which had seemed quite distinct now converged; instead of being concerned with different imponderable fluids or with mechanical work, all physicists found that they had been investigating the different manifestations of force, or energy as it came to be called. Thomson, later Lord Kelvin, was the dominant figure in the physical sciences in Britain in the second half of the nineteenth century. For fifty years he was professor at Glasgow; at the celebration of his Jubilee, with scientists from all over the world to pay homage, he declared that the one word which characterized his efforts was 'failure'. To suppose that Kelvin can stand for the alleged intellectual arrogance of later Victorian scientists is absurd.

One who better fits that picture is John Tyndall, whose *Fragments of Science for Unscientific People* went through numerous editions. It contained his notorious Belfast Address of 1874, a high-flown apology for materialism, some entertaining stories of psychical research (Tyndall being a determined sceptic), and some interesting essays on many aspects of physics. His *Heat a Mode of Motion*, is a classic textbook of the new doctrines of Joule and Helmholtz; and his *Floating Matter in the Air* popularized and extended the discoveries of Pasteur in bacteriology. Tyndall and T. H. Huxley were the most influential agnostic scientists of their day; with a number of like-minded friends, including J. D. Hooker, Edward Frankland the chemist, and Spottiswoode the mathematician, they founded the X Club which played a mildly sinister rôle behind the scenes in the scientific societies of late Victorian London.

William Thomson was responsible, at the same time as Rudolph Clausius, for resurrecting the second law of thermodynamics, which Sadi Carnot had derived in his *Reflections on the Motive Power of Fire*. Carnot had used the caloric fluid theory of heat, with which he subsequently began to be disillusioned. Thomson and Clausius showed that his conclusion, that to take heat from a cool body to a warmer one requires work, and his elegant method of considering reversible, cyclic processes, could be made compatible with the kinetic theory of heat. Thomson's researches ranged over wide areas of physics; there are famous papers on the sizes of atoms, on the age of the Sun and the Earth (Thomson's estimates proved wildly wrong when an unexpected source of energy, radioactivity, was discovered), on the vortex atom, and on electricity. His *Papers* make tough reading; his *Popular Lectures and Addresses* serve as a useful introduction to his thought. He delivered a course of *Lectures on Molecular Dynamics and the Nature of Heat* in Baltimore, and a small edition in facsimile of notes taken at the lectures was produced in Baltimore in 1884. The book appeared in expanded form, and in ordinary print, in 1904. Among his important practical contributions was his work on the Atlantic cable; he stands as the last great classical physicist before quantum theory and relativity revolutionized the science.

Thomson was associated from his Cambridge days with George Gabriel Stokes, who is best-known for his studies on viscous fluids, and on fluorescence – a word he coined. For thirty-one years from 1854 he was Secretary of the Royal Society and devoted much time to the improvement of papers for its journals; he was also very interested in natural theology. Stokes' own *Mathematical and Physical Papers* were published in five volumes between 1880 and 1905; his Gifford Lectures on natural theology appeared in 1891 and 1893.

Another whose interests were almost as diverse as Thomson's was William Rankine, who was an important figure in early thermodynamics; and who wrote on a whole range of engineering problems. *His Miscellaneous and Scientific Papers* of 1881 establish him as a great engineer and physicist. The *Miscellaneous Papers* of the engineer Joseph Whitworth appeared in 1858; he was an advocate of standardization and established a uniform system of screw-threads. Translations of thermodynamical papers by Clausius and Ludwig Boltzmann appeared in the *Philosophical Magazine*; and the next dominant figure in this field was Josiah Willard Gibbs, the mandarin mathematician from Yale. His papers appeared in the *Transactions of the Connecticut Academy of Science*, few of the members of which can have made much of them; but he circulated offprints and Maxwell, and then Wilhelm Ostwald, made his work well known. He is particularly famous for the introduction of the 'phase rule' which inaugurated chemical thermodynamics. His *Collected Works* did not appear until the twentieth century. Thermodynamics, an elegant deductive science requiring no assumptions about the nature of matter, made a considerable appeal to some chemists, notably Ostwald, who hoped to establish a mathematical chemistry from which hypothetical entities were excluded.

While the rise of thermodynamics was one of the most dramatic episodes in nineteenth-century science, of equal importance was the rapid development of the science of electricity. At the end of the eighteenth century electricity was seen as either one or two imponderable fluids; the science provided valuable material for the itinerant lecturer and the quack, and enthralled the *Natürphilosophen.* Davy proved the importance of electricity in chemistry; and Ampère, Joseph Henry, Faraday, and Maxwell converted electricity from a curious into a fundamental science. It was the electric telegraph, in the development of which the inventor W. F. Cooke and the physicist Charles Wheatstone played a crucial rôle, which first made evident the utility of the science. Cooke published a book furiously urging his claims against those of Wheatstone, with whom he had been in partnership. Wheatstone's *Scientific Papers* contains articles on telegraphy – he determined the velocity of an electric wave in a conductor; on binocular vision; on sound, especially the music of wind instruments; and on his best-known invention, the Wheatstone bridge for determining the electrical resistance of a conductor.

Joseph Henry was one of the few outstanding American scientists of his period; he was involved in the development of the telegraph, and used it at the Smithsonian Museum, of which he was first Director, in weather forecasting. He discovered the phenomenon of self-induction at the same time as Faraday, whose account of it was published first. Henry's *Scientific Writings* were published at the Smithsonian in 1886. He was drawn into the co-ordination and administration of science in the U.S.A.; and with Alexander Dallas Bache had a hand in the organization of much of the scientific life of the country. Faraday, by contrast, turned down civic honours and consultancies, and withdrew for a long time from active participation in the affairs of the Royal Society, of which he later declined to be President. His *Experimental Researches in Electricity* do not need attention called to them; they came out in three volumes in 1839, 1844, and 1855. First editions of the first two volumes are rare; they were reprinted in facsimile by Quaritch in the nineteenth century, and again recently by Dover Books. Most of the researches first appeared in the *Philosophical Transactions*; the paragraphs are numbered consecutively, and the work is cross-referenced. Faraday's *Diary*, really his laboratory notebook, was published in a splendid edition of seven volumes and an index, with Faraday's drawings reproduced on the pages, between 1932 and 1936, and casts valuable light on the *Researches*.

The second and third volumes of *Researches* include some articles not in the series of numbered paragraphs, of which the 'Speculation touching electric conduction and the nature of matter' in volume II, and the 'Thoughts on Ray Vibrations' in volume III are particularly interesting as showing Faraday's mind at work. Faraday seems to have been influenced by the electrical theory of George Atwood, whose *Description of Experiments* appeared in 1776, and *Treatise on the Rectilinear Motion of Bodies* in 1784; by the dynamical Newtonianism of corpuscles and powers of John Rowning's *Compendious System of Natural Philosophy*, and of Gowin Knight, whose attempt to demonstrate that all the phenomena in nature may be explained by two simple principles, attraction and repulsion, appeared in 1748; and by the Boscovichean dynamical point atomism of Priestley. It has been suggested that his work shows traces of Kantian dynamics, and that he rejected orthodox atomism in favour of fields of force from very early on. Be that as it may, there is a transition in the *Researches* from the language of contiguous particles of the first volume to the lines and centres of force of the later work.

Faraday was an experimental and not a mathematical physicist, and his writings are therefore more accessible to the layman than are the formal papers of many of his contemporaries; in his own day his results were appreciated, but his theorizing was felt to be incoherent compared to that of the French. The first major contribution to the mathematical science of electricity in nineteenth-century England was George Green's *Essay on the Application of Mathematical Analysis to the Theories of Electricity and Magnetism*, published by subscription at Nottingham in 1828; it introduced the term 'potential function'. Then in 1845 William Thomson showed that Faraday's idea of lines of force could be given a mathematical form; and from 1855 Maxwell went beyond this, and in articles, reprinted in his *Scientific Papers*, and finally in his *Treatise on Electricity and Magnetism* of 1873, he presented the theory of the electromagnetic field, in which light became an electromagnetic disturbance in the ether, like electronic and magnetic phenomena. Heinrich Hertz in an experimental test of the theory, generated and detected wireless waves for the first time; this is described in his book *Electric Waves*, to the English translation of which William Thomson contributed an introduction.

It is clear that any general decline of science was in fact being reversed by 1830; as things turned out the French hegemony in mathematics and physical science did not last long, and the nineteenth century was a period of great flowering of physics in Britain, although chemistry did decline from the great days of Cavendish, Priestley, Dalton, and Davy. Indeed this science in Britain became provincial, as science generally was in the U.S.A. until into the twentieth century, with a few individuals enjoying an international reputation, but no strong national school. In pure mathematics, the work of, among others, William Rowan Hamilton, George Boole, Arthur Cayley, and J. J. Sylvester does not accord with any theory of decline. In geology and biology, the nineteenth century was a period of great achievement, and the history of these sciences can be satisfactorily, though not of course completely, traced from works available in English. It is to these sciences that we now turn, in the period in which the natural history of Gilbert White gradually evolved into the science of Charles Darwin and T. H. Huxley.

8 'THE DECLINE OF SCIENCE IN ENGLAND'

(Abbott, E. A.), *Flatland, A Romance of many Dimensions . . .* by A SQUARE, 1884.
Andrews, T., *The Scientific Papers*, ed. P. G. Tait and A. C. Brown, 1889.
Atwood, G., Half-title: *Description of the Experiments intended to illustrate a Course of Lectures . . .*, 1776.
Treatise on the Rectilinear Motion . . . of Bodies, Cambridge, 1784.
Babbage, C., *Reflections on the Decline of Science in England*, 1830.
The Exposition of 1851, or Views of the Industry, the Science, and the Government of England, 1851.
Passages from the Life of a Philosopher, 1864.
Ball, R. S., *The Story of the Heavens*, 1886.
Ball, W. W. R., *A History of the Study of Mathematics at Cambridge*, Cambridge, 1889.
Carnot, N. L. S., *Reflections on the Motive Power of Heat*, tr. R. H. Thurston, 1890.
Cavendish, H., *The Electrical Researches . . .*, ed. J. C. Maxwell, Cambridge, 1879; rev. J. Larmor as vol. I of *The Scientific Papers*, of which vol. II was *The Chemical and Dynamical Researches*, ed. E. Thorpe; 2 vols., Cambridge, 1921.
Cayley, A., *The Collected Mathematical Papers*, 14 vols., Cambridge, 1889–98.
Clausius, R. J. E., *The Mechanical Theory of Heat* (tr. J. Tyndall), 1867.
Clerke, A., *A Popular History of Astronomy during the Nineteenth Century*, Edinburgh, 1885.
Cooke, W. F., *The Electric Telegraph: was it invented by Professor Wheatstone?*, 1854; Wheatstone's *Reply*, 1855; Cooke's *Reply to Mr Wheatstone's Answer*, 1856; the whole exchange published in 2 pts, 1856–7.
De Morgan, A., *An Essay on Probabilities . . .*, 1838.
Formal Logic . . ., 1847.

Dircks, H., *Perpetuum Mobile*, 1861.
Exley, T., *Principles of Natural Philosophy*, 1829.
and Johnston, W. M., *The Imperial Encyclopedia . . .*, 4 vols., 1809–14.
Faraday, M. (ed.), *On the alleged decline of science in England . . .* (by G. Moll), 1831.
Experimental Researches in Electricity, 3 vols., 1839–55.
Experimental Researches in Chemistry and Physics, 1859.
Fourier, J. B. J., *The Analytical Theory of Heat*, tr. A. Freeman, Cambridge, 1878.
Gauss, C. F., *Theory of the Motion of the Heavenly Bodies . . .*, Boston, Mass., 1857.
Gibbs, J. W., *Elementary Principles in Statistical Mechanics*, New Haven, 1902.
The Scientific Papers, ed. H. A. Bumstead and R. G. Van Name, 2 vols., 1906.
The Collected Works, New York, 1928, includes both items.
Grant, R., *History of Physical Astronomy . . .* (1852).
(Granville, A. B.), *Science without a head; or the Royal Society dissected*, 1830.
Green, G., *An Essay on the Application of Mathematical Analysis to the Theories of Electricity and Magnetism*, Nottingham, 1828.
Mathematical Papers, ed. N. M. Ferrers, 1871.
Gregory, G., *The Economy of Nature . . .*, 3 vols., 1796.
A Dictionary of Arts and Sciences, 2 vols., 1806–7.
Grove, W., *On the Correlation of Physical Forces*, 1846.
Hamilton, W. R., *Elements of Quaternions . . .*, ed. W. E. Hamilton, 1866.
The Mathematical Papers, ed. A. W. Conway and J. L. Synge, Cambridge, 1931–.
Henry, J., *Meteorology in the connection with agriculture*, Washington, 1858.
Herapath, J., *Mathematical Physics*, 2 vols., 1847.
Herschel, J. F. W., *A preliminary discourse on the study of natural philosophy*, 1830.
A Treatise on Astronomy, 1833.
Outlines of Astronomy, 1849.
Essays from the Edinburgh and Quarterly Reviews, 1857.
Familiar Lectures on Scientific Subjects, 1866.
Hertz, H. R., *Electric Waves*, tr. D. E. Jones, 1893.
Miscellaneous Papers, tr. D. E. Jones and G. A. Schott, 1896.
The Principles of Mechanics, tr. D. E. Jones and J. T. Whalley, 1899.
Hutton, C., *The Principles of Bridges*, Newcastle, 1772.
Miscellanea Mathematica . . ., 1775.
Tracts on mathematical and philosophical subjects, 3 vols., 1812.
Joule, J. P., *Scientific Papers*, 2 vols., 1884–7.
Knight, G., *An attempt to demonstrate, that all the phenomena in nature may be explained by . . . attraction and repulsion*, 1748.
Lacroix, S. F., *An Elementary Treatise on the Differential and Integral Calculus*, tr. C. Babbage, G. Peacock, J. F. W. Herschel, Cambridge, 1816.
Laplace, P. S. de, *The System of the World*, tr. J. Pond, 2 vols., 1809.
Mécanique céleste . . ., tr. N. Bowditch, 4 vols., Boston, 1829–39.
Maxwell, J. C., *On the Stability of the Motion of Saturn's Rings*, Cambridge, 1859.
A Treatise on Electricity and Magnetism, 2 vols., Oxford, 1873.
Nichol, J. P., *Views of the Architecture of the Heavens*, Edinburgh, 1837; 2nd ed., greatly enlarged, 1850.
The Planet Neptune, Edinburgh, 1848.
A cyclopedia of physical sciences, 1857.
Peacock, G., *Life of Thomas Young*, 1855.
Quetelet, L. A. J., *Popular Lectures on the Calculation of Probabilities*, ed. R. Beamish, 1839.
A Treatise on Man and the Development of his Faculties, Edinburgh, 1842.
Letters addressed to the Grand Duke of Saxe-Coburg and Gotha on the Theory of Probabilities, tr. O. G. Downes, 1849.
Rankine, W. J. M., *A Manual of the Steam Engine*, 1859.

A Manual of Civil Engineering, 1862.
Miscellaneous Scientific Papers, ed. W. J. Miller and P. G. Tait, 1881.
Rosse, W. Parsons, 3rd earl of, *The Scientific Papers*, ed. C. Parsons, 1926.
Rowning, J., *A Compendious System of Natural Philosophy*, 4 pts, 2nd ed., Cambridge, 1735–42; 1st ed. not seen.
Rumford, Benjamin Thompson, Count, *Essays, political, economical, and philosophical*, 4 vols., 1796–(1812).
Philosophical Papers, vol. I, 1802; no more published.
The Complete Works . . . , 5 vols., Boston, Mass., 1870–5.
Small, R., *An Account of the Astronomical Discoveries of Kepler*, 1804.
Stokes, G. G., *Burnett Lectures. On Light*, 3 pts, 1884–7.
Natural Theology: The Gifford Lectures, 2 vols., 1891–3.
in: *Science Lectures at South Kensington*, 2 vols., 1878.
Sylvester, J. J., *Collected Mathematical Papers* (ed. H. F. Baker), 4 vols., Cambridge, 1904–12.
Unpublished Letters . . . (ed. R. C. Archibald), Bruges, 1936.
Tait, P. G., *An Elementary Treatise on Quaternions*, Oxford, 1867.
Lectures on some recent advances in physical science, 1876.
Properties of Matter, Edinburgh, 1885.
Scientific Papers, 3 vols., Cambridge, 1898–1911.
Thomson, W. (Lord Kelvin), *Baltimore Lectures on Molecular Dynamics and the Wave Theory of Light*, Baltimore, 1884 (cyclostyled ed.); London, 1904.
and Tait, P. G., *Treatise on Natural Philosophy*, 2 pts, Cambridge, 1879.
Turner, H. H. (ed.), *History of the Royal Astronomical Society*, 1923.
Tyndall, J., *Contributions to Molecular Physics in the domain of radiant heat*, 1872.
(Waterston, J. J.), *Thoughts on the Mental Functions; being an attempt to treat metaphysics as a branch of the physiology of the nervous system*, pt I, Edinburgh, 1843; no more published.
The Collected Scientific Papers, ed. J. B. S. Haldane, 1928.
Wheatstone, C., *The Scientific Papers*, 1879.
Whitworth, J., *Miscellaneous Papers on Mechanical Subjects*, 1858.
and Wallis, G., *The Industry of the United States*, 1854.
Wordsworth, C., *Scholae Academicae*, 1877.
Young, T., *A Course of Lectures on Natural Philosophy and the Mechanical Arts*, 2 vols., 1807.

Recent Publications

Boscovich, R. J., *A Theory of Natural Philosophy*, tr. J. M. Child, Chicago, 1922.
Brush, S. G., *Kinetic Theory*, 2 vols., Oxford, 1965–6.
Duhem, P., *The Aim and Structure of Physical Theory*, tr. P. Wiener, Princeton, 1954.
Faraday, M., *Faraday's Diary* . . . , ed. T. Martin, 8 vols., 1932–6.
A List of his Lectures and Published Writings, by A. E. Jeffreys, 1960.
Grosser, M., *The Discovery of Neptune*, Cambridge, Mass., 1962.
Haworth, O. J. R., *The British Association* . . . *1831–1931*, 1931.
Herschel, J. F. W., *Herschel at the Cape*, ed. D. S. Evans *et al.*, Austin, Texas, 1969.
Knight, D. M., *Atoms and Elements*, 1967; corrected reprint 1970.
Lodge, O. T., *Advancing Science: being personal reminiscences of the British Association in the Nineteenth Century*, 1931.
Morison, P. and E., *Charles Babbage and his Calculating Engines*, New York, 1961.
North, J., *The Measure of the Universe*, Oxford, 1965.
Scott, W. L., *The Conflict between Atomism and Conservation Theory 1644 to 1860*, 1970.
Williams, L. P., *Michael Faraday*, 1965.

9 *Biology and Geology in the Nineteenth Century*

This vast territory is beset with snares for the unwary; the number of workers begins to become very great, and some of the authors were very prolific. By the beginning of the century there was a professional readership for works of physiology, as by the end there was for biological works generally. As the other chapters have been, this survey will be concerned chiefly with books of a general nature. But it must be remembered that, for example, Lamarck and Darwin established their credentials with solid works on the classification of invertebrates, giving themselves the right to be heard in their more general and speculative works. This is in accord with what we found in looking at the kinetic theory of gases, which was only taken seriously when it was proposed by men with a reputation already made.

The main theme must be the development of these sciences to the point at which *The Origin of Species* was published, and the reception of the evolutionary doctrine in the following decades. For it was this which captured the public attention, and it would be perverse, though no doubt possible, to follow as most important a different line. This is particularly true of the science in Britain but we should remember that natural historians continued their activities through the nineteenth century in a manner not very different from that of their predecessors; and indeed that Darwin, pottering in his garden, and making little experiments in botany there, seems closer to this image than did Cuvier in his museum laboratory in Paris in the opening years of the century. Nevertheless, as in the seventeenth century natural history displaced the lore of bestiaries and herbals, so in the nineteenth natural history was eclipsed by biology.

In the field of natural history, the nineteenth century was the great epoch of the bird book. A magnificent bibliography, *Fine Bird Books 1700–1900*, by Sitwell, Buchanan, and Fisher, is available as a guide to these, and it is therefore not necessary to try to be exhaustive. The techniques of reproduction, from wood engraving to lithography were developed just as explorers in North and South America, Asia, and Australasia described new species and brought them home dead or alive. They were usually dead, and stuffed. Edward Lear, the nonsense poet, was rare among painters of exotic birds in that his *Parrots* of 1832 were drawn from living specimens in the zoo. Lear was one of the greatest bird-painters; the most famous of all was J. J. Audubon, whose elephant folios of plates, *The Birds of America*, need no recommendation, and whose accompanying *Ornithological Biography* makes fascinating reading for he, like William Bartram, described the eastern United States just before the wilderness was destroyed in the march of civilization.

Audubon's paintings were dramatic and living compared with anything that had been seen before; as far as possible he drew his birds life-size, a procedure which entailed the elephant folio format. The plates were engraved and hand coloured in Edinburgh. Audubon himself caused a sensation when he came to Britain with his wolfskin coat and long hair. In the backgrounds of his pictures there appear plants and insects, and he painted mammals, reproduced by lithography, in his *Viviparous Quadrupeds*; as did Lear, whose *Gleanings from the Menagerie and Aviary at Knowsley Hall* was published in 1846. Among compilers of bird books, John Gould stands out for the sheer quantity of his output of giant volumes, covering most of the world except for Africa. Gould was the organizer of the project; the drawings, and the lithography were done by collaborators who included Lear, Joseph Wolf and Mrs Gould. Another splendid book was W. L. Buller's *History of the Birds of New Zealand* which had more text than many bird books, with descriptions of Buller's adventures in search of specimens. This had lithographs by J. G. Keulemans; the stone was wiped after the first edition, and the second edition contained fresh plates by the same artist. Both editions are greatly prized; the first has thirty-five plates against the second's forty-eight. Other works illustrated by Keulemans include R. B. Sharpe's *Kingfishers* and G. E. Shelley's *Sunbirds*; both are superb books, which appeared in very small editions.

These great books were never cheap, and for the bird watcher the wood engraving provided another medium for illustration. Thomas Bewick's famous *History of British Birds* appeared in 1797 (the land birds) and 1804 (the water birds); the illustrations are lively, natural and delightful. William Yarrell's *History of British Birds* of 1843, with numerous later editions, also has very agreeable woodcuts and a text of greater ornithological interest than Bewick's. As well as pictures of the birds, these volumes and their companions on fishes have little vignettes after the manner of Bewick scattered about them. The great period of flower-paintings had come a little earlier; for an excellent account one should see Wilfrid Blunt's *Art of Botanical Illustration*, and the enormous bibliography, *Great Flower Books*, by Sitwell, Blunt, and Synge. The Sowerby family of illustrators have already been mentioned; R. Thornton's Romantic and magnificent *Temple of Flora* cannot be passed by, although it is not botanically important, and John Sibthorp's *Flora Graeca*, illustrated by Ferdinand Bauer, is a splendid example of botanical illustration. The largest, and one of the greatest, of all flower books was James Bateman's *Orchidaceae of Mexico and Guatemala*, which appeared between 1837 and

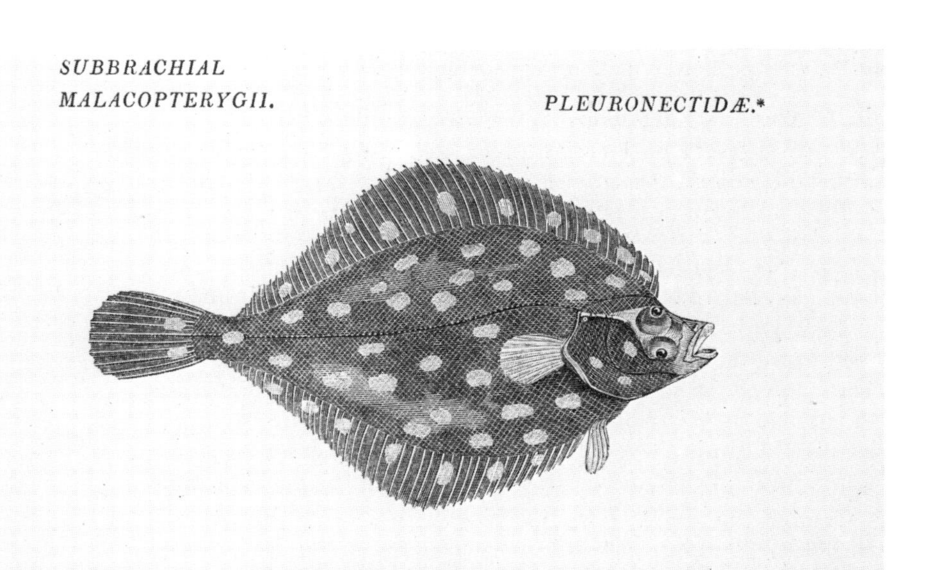

67 Plaice (Yarrell, W., *British Fishes*, 1836, vol II, p 209)

68 (*below left*) Night-blowing cereus; an accompanying colour-plate showed the flower blooming at midnight (Thornton, R. J., *New Illustrations of The Sexual System of Linnaeus*, 1807, pl 27 of the genera and species)

69 (*below right*) American Woodbine (Hooker, W. J., *Exotic Flora*, 1823–7, vol I, pl 27)

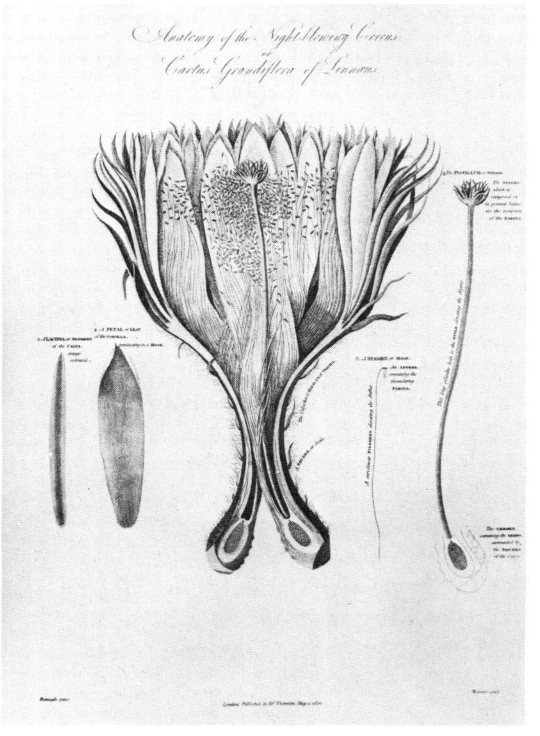

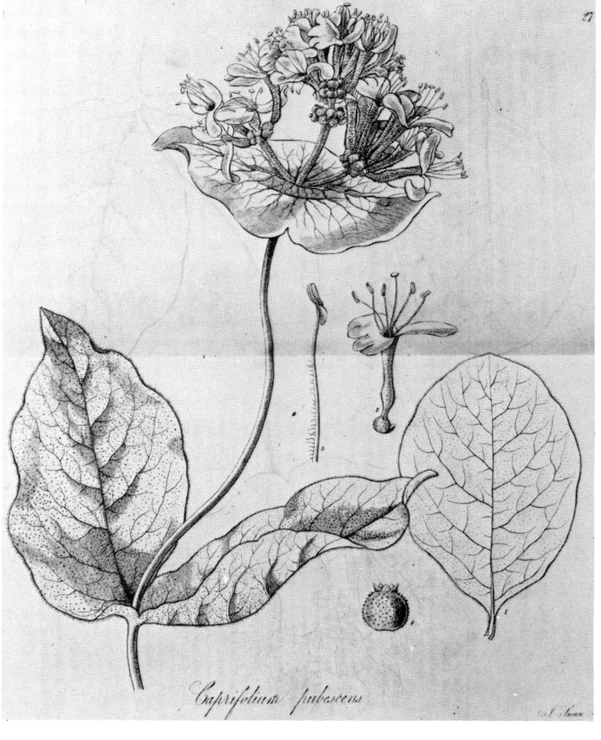

1841. In the first half of the nineteenth century the botanist Sir William Hooker, who became Director of Kew Gardens, illustrated a number of botanical works himself; and after 1826 was responsible for the plates of the *Botanical Magazine* which Blunt describes as 'a national institution of which Englishmen may justly be proud'. William Hooker's son, Sir Joseph Hooker, was one of the greatest of all botanists; he was Darwin's friend and first convert, and a pioneer in the study of plant distribution.

Of illustrated works devoted to insects, Benjamin Wilkes' *One Hundred and Twenty Copper-plates of English Moths and Butterflies*, published between 1737 and 1741, is magnificent; and in 1766 there appeared an even more handsome work, *The Aurelian* by Moses Harris, with a frontispiece of butterfly collectors with their apparatus. Each plate of the volume bears the coat of arms of the person to whom it was dedicated. Entomologists were fancifully called aurelians, apparently from the gold colour of the chrysalids of the silkworm. Harris also illustrated other insects, in Drury's *Illustrations of Natural History (1770–82)* and in his own *Exposition of English Insects*. Sir James Smith's *Lepidoptera of Georgia*, with its exquisite plates of butterflies and

70 Frontispiece, illustrating the fecundity of nature in insects and reptiles, of Harris, M., *Exposition of English Insects*, 1782

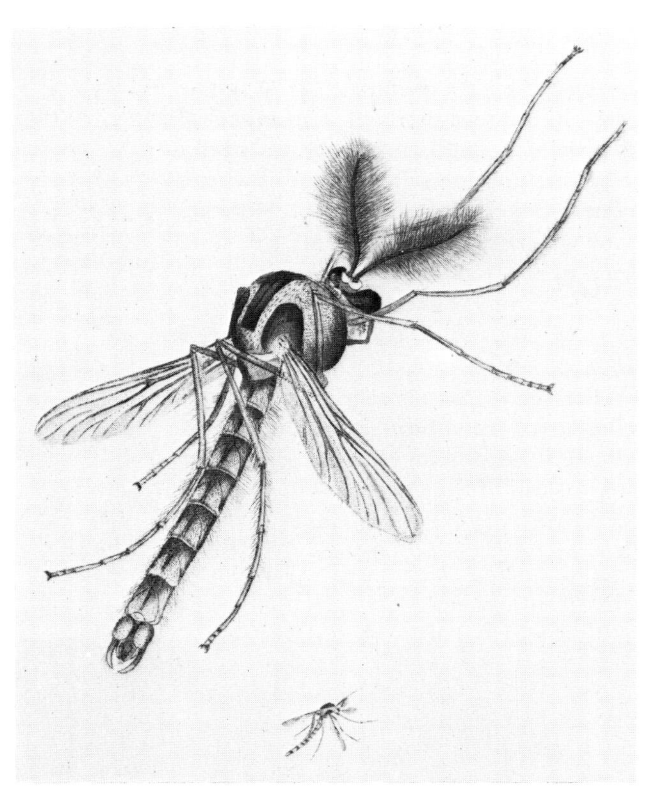

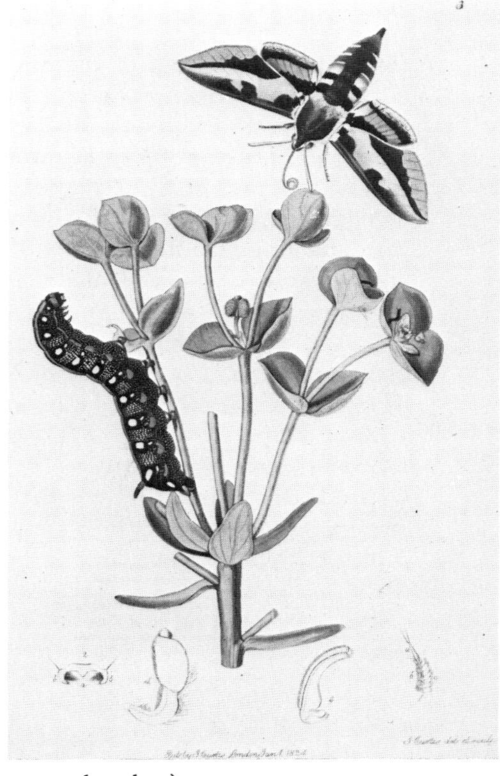

71 (*left*) Gnat (Donovan, E., *British Insects*, 1793, vol 1, pl 22)
72 (*right*) Spotted elephant moth (Curtis, J., *British Entomology*, 1824, pl 3)

moths on the plants they favour, is deservedly a collector's piece. Edward Donovan's *Natural History of British Insects*, the publication of which spanned two decades, is an important work in which certain species are first described; and John Curtis' great *British Entomology* has unusually careful and exact illustrations. J. F. Stephens, a civil servant seconded for a time to the British Museum to help arrange the entomology collection, published his important *Systematic Catalogue of British Insects* in 1829, and *Illustrations of British Entomology* from 1828 to 1835. In 1795 there had appeared William Lewin's exquisitely illustrated book on butterflies *Papilios of Great Britain*; this valuable work, the text of which is in English and French, added several species to the list of British butterflies. A. H. Haworth's *Lepidoptera Britannica*, which appeared from 1803, has been called 'the earliest work on British butterflies produced in a scientific manner'; this tradition was carried on by Stephens, and by Kirby and Spence whose *Introduction to Entomology* is a classic. Later butterfly books include Humphreys and Westwood's *British Butterflies*, which illustrates also various species not found in these islands; the first edition is apparently more accurate than the second. Edward Newman's *British Butterflies* of 1868 is another well-known insect book, useful and enjoyable both for its plates and its text, though not a great work of art. Drawings and paintings have the advantage over photographs, that the artist can tactfully draw attention to the most important features, and is not limited by the peculiar features of one individual specimen; and therefore they remain valuable in natural history.

The exact descriptions and accurate drawings which were gradually evolved up to the mid-nineteenth century were a necessary preliminary to the classification of animals and plants: and hence to biological theories which were, within limits, testable. In the second half of the nineteenth century, the

73 Clouded yellow butterflies on thistles (Humphreys, H., and Westwood, J. O., *British Butterflies*, 1841, pl 2)

interest of scientists shifted towards more detailed studies, usually concerned with evolution, and the period when beautifully illustrated books were simultaneously first-rate contributions to science drew to a close, though valuable and fine *Flora* and bird books do still continue to appear. In anatomy and physiology this stage had been reached much earlier, and the Renaissance was the great period of medical illustration. The early nineteenth century was nevertheless a period of rapid advance in physiology. Sir Everard Home published as his own what it seems were the researches of John Hunter, in the *Philosophical Transactions*; Sir Charles Bell's studies on the nerves were beautifully illustrated; and from abroad came the researches of M. F. X. Bichat, F. Magendie, and T. Schwann, and later of C. Bernard, Louis Pasteur, and Herman Koch. These works, concerned with the rôle of tissues and cells, the elucidation of the organic processes, and the cause of disease, belong perhaps more to the history of medicine; but Pasteur's experimental disproof of spontaneous generation was made generally known by Huxley and Tyndall, and was assailed in a series of works by H. C. Bastian. This controversy serves to show once again how uncertain in implication experiments, which later seem decisive, can appear to contemporaries. To those who favoured the Darwinian thesis of development by infinitesimally small steps, spontaneous generation was anathema; which it was not to those who accepted discontinuous goal-directed evolution proceeding constantly from inanimate matter, or who believed in the fixity of species. In plant physiology, a most important author was M. J. Schleiden whose *Principles of Scientific Botany* was translated in 1849; his more popular book, *The Plant; a biography*, appeared earlier.

Physiological ideas played a part in Romantic thought, for the new physiology in which John and William Hunter were pioneers was a revolt against the mechanistic science of the Cartesians, of whom Offray de la Mettrie, author of *L'Homme Machine*, translated 1749, was an extreme representative. The Hunters emphasized the individuality of specimens and the dynamic equilibrium in animals; the muscles, bones, and vessels in two members of the same species are never arranged exactly alike, and living creatures maintain their individuality despite the endless flux of their material particles, which constantly wear away and are replaced. Goethe and Oken stressed the homologies linking diverse animals and vegetables, and the transformations which organs have undergone to enable creatures to thrive in new circumstances. Indeed the idea of evolution or development in response to new conditions became widely held in the early decades of the nineteenth century; Darwin's achievement was to make it acceptable to the hard-headed.

At the close of the eighteenth century, we find physiological ideas derived chiefly from Hunter and from Boerhaave's most distinguished pupil, Albrecht von Haller, in J. G. von Herder's *Outlines of a Philosophy of the History of Man*, which came out in English in 1800. Haller's *First Lines of Physiology*, of which the best English translation of 1786 was supervised by William Cullen, taught the science to a generation of physicians and amateurs. Herder used the ideas of individuality and flux which we have noted to support his thesis that all the races of man belonged to one species, which had become adapted to the different environments of different countries. Each nation had therefore acquired its individuality and had its own destiny, however unpleasantly this might strike the 'idle cosmopolite'. Nationalism proved powerful outside science, and Herder's insistence that the various races and nations of man were different but equal is pleasanter than the idea, prevalent in the nineteenth century, that some races are more akin to apes than others. The analogy between the powers that vivify the inorganic components of an organism, and heat, light, and electricity which shape brute matter into chemical substances, has already been mentioned in our discussion of chemistry. In modern science, analogies from the biological to the inorganic realm have been somewhat infrequent; another example came in the later nineteenth century, when evolutionary explanations were applied to the various chemical elements.

We can now return to our main theme, the development of geological and biological ideas down to Darwin. It was in the later eighteenth century that geology emerged from natural theology and began its triumphal course which made it the most popular science of the mid-nineteenth century. In 1750 de Maillet's *Telliamed*, a rationalistic geology spiced with tall travellers' tales, was translated, and became widely known; but one of the first professional geologists to achieve renown was Abraham Werner, who attracted students to Freiberg from all over the world. Like John Woodward in his celebrated *Essay Toward a Natural History of the Earth* of 1695, Werner believed that all rocks had been laid down by deposition from water; those who followed him in this were called Neptunists, while those who urged the primacy of volcanic action were described as Vulcanists. Of Werner's works, the *Treatise on the external character of fossils* appeared in English in 1805, the *New Theory of the Formation of Veins* in 1809, and the *Nomenclature of Colours*, a subject of great importance in mineralogy, in 1814. The Wernerian Natural History Society, whose chief pillar was Professor Jameson of Edinburgh, was founded to propagate his doctrines.

In opposition to Werner, James Hutton published in the *Transactions of the Royal Society of Edinburgh* for 1788 a long paper he had read in 1785 on the theory of the Earth. Hutton was a Vulcanist, as one who studies granite must be, but more important was his insistence that past changes were to be explained not in terms of catastrophic events but by means of processes now observable, acting over a vast period of time. He ended his paper with the remark that geology discloses no vestige of a beginning and no prospect of an end. In 1795 the paper was expanded with proofs and illustrations into a book, which has become one of the classics of geology; but its arguments were not well marshalled and in its own day, and since, its readers seem to have been few.

Hutton was attacked, notably by Kirwan, whose Wernerian *Elements of Mineralogy* appeared in 1784, and *Geological Essays* in 1799; indeed Kirwan's remarks in a review, republished in the *Essays*, induced Hutton to publish his Theory as a book. Hutton's Vulcanist geology and his remark about beginnings and ends aroused alarm, but in 1802, John Playfair, Professor of Mathematics at Edinburgh, produced a lucid and well-organized book, *Illustrations of the Huttonian Theory of the Earth*, which made the uniformitarian thesis more popular. Playfair had been a pupil of Hutton, and supposed like him that even the coal measures had been consolidated by intense heat; but his general perspicuity has aroused the admiration of later geologists. He even argued that the Earth need never have been fluid, that it would still have taken up its flattened spheroidal form under the action of long-continued erosion. Another pupil of Hutton's, Sir James Hall, was able to produce by heat and pressure in the laboratory crystalline and glassy basaltic rocks, while all attempts to prepare such rocks from aqueous solution failed. To ascribe to water in the past powers which it lacks at present, would be, Playfair remarked, not merely gratuitous but absurd and impossible. Hall's papers appeared in the *Philosophical Transactions*.

The uniformitarians were opposed by J. A. Deluc, an upholder of deluges and inventor of the word 'Geology', who attacked Hutton in the *Monthly Review*, and in his *Elementary Treatise on Geology*; and by the notable authority of Georges Cuvier, whose *Essay on the Theory of the Earth* was translated by Robert Kerr, on the urging of Jameson. Cuvier's reputation was made by his anatomical studies; his *Lessons on Comparative Anatomy* had appeared in two volumes in 1802. In the Paris basin numerous fossils had been turned up in the quarries from which the city was being built in imperial splendour: the bones were jumbled together, but Cuvier succeeded in assigning them to their proper species. They were all large mammals, of species no longer found, and some of them belonging to genera only now represented in America. Cuvier's researches established that it was impossible to think in terms of a great chain or ladder passing up from crystals through plants, zoophytes and animals to man. There were a number of great groupings in the animal kingdom between which direct comparison was impossible; they belonged to different ladders. Cuvier therefore forcibly opposed the evolutionary doctrines of Lamarck and Geoffroy St Hilaire, which seemed to entail one ladder with everything working its way upwards. In his *Essay*, which was a popular work, he set out his arguments. He believed that there had been a series of great catastrophes in nature which accounted for the changes of fauna; for the mere passage of time was not sufficient, mummified cats from Egypt being indistinguishable from modern varieties. On the chronology defended at length by Cuvier in his *Essay* and held by most of his contemporaries, the world was only a few thousand years old, and this last argument therefore potent. Similarly, the entire mammoths found congealed in the Siberian ice were evidence that a region with sufficient vegetation to support these elephants had in an instant changed into a frozen desert. This last argument formed the basis of an attack on uniformitarian dogma in geology by H. H. Howarth, M.P. in his *The Mammoth and the Flood* of 1887. Cuvier's work, and his emphasis on the unity

74 Philosophers on Etna (Spallanzani, L., *Travels in the Two Sicilies*, 1798, vol 1, pl 2)

of an organism, were vitally important; and his admirers included Darwin. His studies on fossils inaugurated a truly scientific palaeontology, of which science the first exponent in English was J. Parkinson.

In England, meanwhile, interest in volcanoes continued. Sir William Hamilton had described in his *Travels* an eruption of Vesuvius; and R. E. Raspe had published in 1776 an illustrated *Account of some German Volcanoes*, in Hesse and Kassel. Spallanzani's *Travels in the Two Sicilies* vividly described volcanoes. Humphry Davy visited Italy and ascended Vesuvius; he also applied chlorine to papyri found in Herculaneum, in an effort to unroll and decipher them, which had the support of the Prince Regent. In his *Consolations*, he declared himself an adherent of the 'refined Huttonian system' of Playfair and Hall. Poulett Scrope, Secretary of the newly founded Geological Society, published in 1825 his *Considerations on Volcanos* which would lead, he hoped, to the establishment of a new theory of the Earth. This incorporated both Neptunist and Vulcanist elements; and the crucial part of his book is the insistence upon the analogy of the past to the present, and the avoidance of any element of the marvellous or the supernatural. Nothing could happen that was not caused by a law of nature; these laws do not vary; and 'similar results always are, have been, and will be produced, by similar preceding circumstances.'

Charles Daubeny, whom we have already met as agriculturalist, botanist and chemist, produced in 1826 *A Description of Active and Extinct Volcanos* based upon his own travels in Western and Central Europe and reports from further afield. He was particularly interested in the chemical reactions that might account for eruptions; and stressed the relative nearness of volcanoes to the sea or a large lake. His book has a long bibliography of books and papers on volcanoes; and a 'geological thermometer' upon which he graded his predecessors, with extreme Vulcanists in the highest temperatures and Neptunists near freezing point. He placed himself on the cool side. Also on the watery side should appear William McLure, whose *Geology of the United States*, with the first geological map of the country, appeared in 1817; he allowed water, fire, and electricity as important agents in geology, but urged the avoidance of speculative systems. His book abounds in confidence in the future of his country, especially when it began to lose its European habits.

The Neptunist–Vulcanist debate finally lapsed when William Smith in 1816 published his *Strata identified by Organised Fossils* and in the following year his *Stratigraphical System of Organised Fossils*, which suggested the dating of rocks from the fossil species found in them. Smith was a surveyor who worked for canal companies; he had worked out his ideas by about 1800, and they were first published in 1813 at Bath, by a friend, the Rev. J. Townsend, in *The Character of Moses established for Veracity as an Historian*. The subsequent progress of the science would have disappointed Townsend. Smith, as geologists often did, disclaimed interest in systems, and declared that his books were intended simply to support his geological map or *Delineation of the Strata of England and Wales* of 1815; but there occur phrases which indicate that he believed, like Cuvier, in catastrophes followed by new creations. It was now clear that comparative anatomy was of more importance than chemistry as a key to geology; and this became obvious in the next generation of geologists.

Charles Lyell was trained as a lawyer, and it would seem to be as much for his forceful presentation of his case as for any great novelty of views that his *Principles of Geology* became the great classic that it undoubtedly is. The book passed through numerous editions, of which perhaps the most interesting are the first, and the tenth, of 1867, when he accepted Darwin's theory. The book is very readable and persuasive; and made it difficult for anybody to go on believing in sudden catastrophes rather than in the uniform operation of laws over an immensely long period of time. Lyell included an account of the history of geology: his predecessors had been led into being prodigal of violence because they had been parsimonious of time, and they had been too preoccupied with the land, when in fact deposition was proceeding under the sea. Lyell's account of Lamarck's views, which was hostile and somewhat distorted, made these evolutionary doctrines known in the English-speaking world: but perpetrated the false idea that for Lamarck plants and animals developed into higher species because of their desire for self-improvement. The first edition of Lyell was in three volumes, which appeared sufficiently

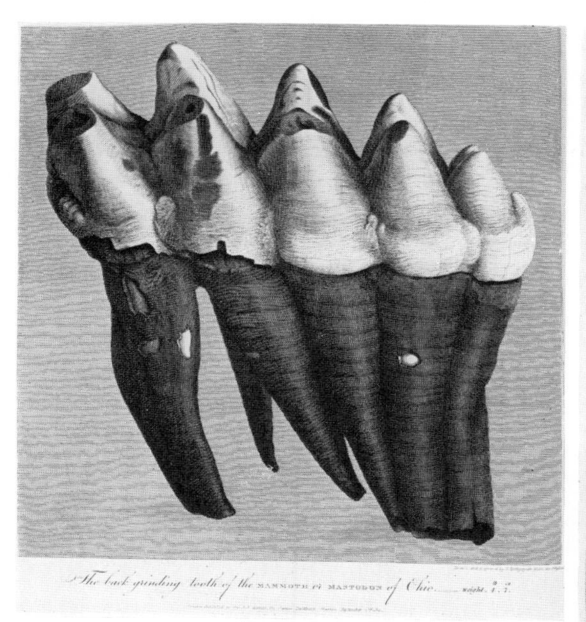

75 Extinct elephant's grinders (Parkinson, J., *Organic Remains of a Former World*, 1804–11, vol III, frontispiece)

76 (*right*) Temple at Puozzoli, showing how it has moved up and down relative to the sea (Lyell, C., *Principles of Geology*, 1830, vol I, frontispiece)

slowly for criticisms to be answered in the later volumes; the second edition of the first two volumes in fact appeared before the first edition of the third. Later editions give the dates of the earlier ones, and list the changes in content. As the quantity of knowledge grew, the book was divided into two, the practical part being called *Elements of Geology*, the first edition of which was published in 1838. In 1863 he brought out his *Geological Evidences of the Antiquity of Man*, which represents a move towards the Darwinian position; Lyell proved that man's history was far longer than had been believed. Three editions of this work came out in 1863.

In 1831 H. Witham published *Observations on Fossil Vegetables*, with handsome illustrations; his *Internal Structure of Fossil Vegetables* of 1833 is an expansion of the earlier work, which includes the earlier plates together with fresh ones. In 1824 Gideon Mantell, a surgeon of Lewes in Sussex, discovered fossil teeth resembling those of the iguana but much larger; this was the first fossil of a dinosaur to be recognized, and was named iguanadon. Mantell was a supporter of Lyell's uniformitarianism, as can be seen in his *Geology of the South-East of England* of 1833; in his popular works, *Wonders of Geology* of 1838 and *Medals of Creation* of 1844; and in *Petrifactions and their*

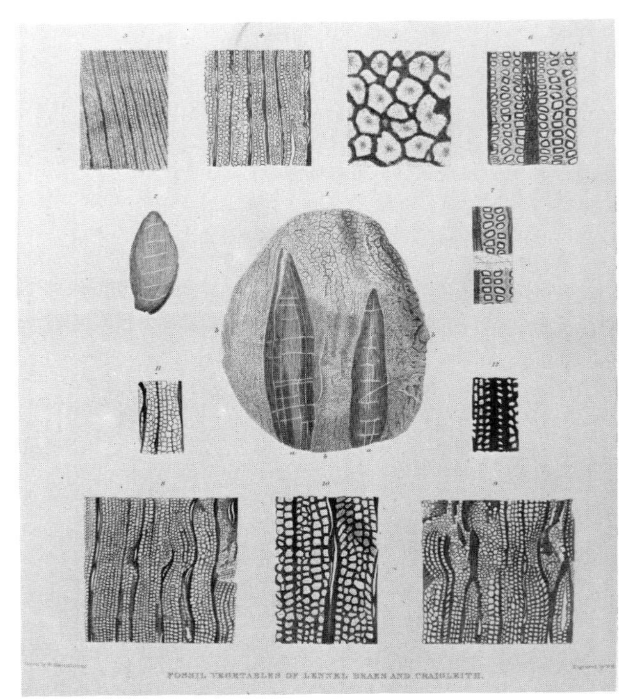

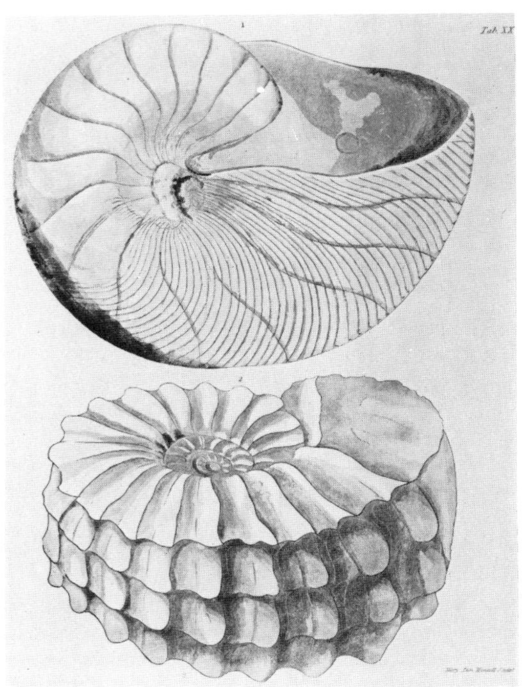

77 (*left*) Coal and lias fossils (Witham, H. M. W., *Fossil Vegetables*, 1833, pl 4)
78 (*right*) Nautilus and ammonite (Mantell, G., *Fossils of the South Downs*, 1822, pl 20)

Teachings, a solid work which appeared in 1851. Apart from his great discovery, Mantell was not among the greatest geologists; but his *Journal*, published in this century, is of great interest, revealing an ambitious country surgeon making his way through scientific research into the ranks of the Royal Society and the Athenaeum, while his wife left him when his house became, essentially, a geological museum. Mantell's works are a useful and readable guide to the geology of the period.

Humphry Davy claimed to have given the first public lectures on geology in London; and William Thomas Brande, his successor at the Royal Institution, published *Outlines of Geology* in 1817, which was a valuable introduction to the subject. Brande, in his chemical and geological writings was sound and thorough rather than original or exciting. Another competent introductory work was the land-surveyor Robert Bakewell's *Introduction to Geology* of 1813; this had an American edition, in 1829, edited by Benjamin Silliman, who added a Wernerian and catastrophist appendix. Wernerian works written in America included Parker Cleaveland's *Elementary Treatise on Mineralogy and Geology* which appeared in 1816, and Amos Eaton's *Index to the Geology of the Northern States* of 1818. Eaton, in his *Geological Text-book* of 1830 allowed some merit to Hutton and Playfair, as did van Rensselaer in his *Lectures on Geology* of 1825.

In England the standard introduction to geology in the 1820s was by William Conybeare and William Phillips, *Outlines of the Geology of England and Wales*, which first appeared in 1822 and went through a number of editions. They allowed for the Flood and various other catastrophes and Divine interventions, but the book represents generally a sound compromise. Conybeare opposed Lyell's uniformitarianism with determination, became a dean, and ceased to play an important part in the development of geology. Phillips had, in 1818, published *A Selection of Facts from the Best Authorities, Arranged so as to Form an Outline of the Geology of England and Wales*; like Brande and Bakewell, and like the Geological Society, he claimed an interest in facts rather than systems, and urged the utility of the science. A manual similar in tone was G. B. Greenough's *Critical Examination of the First Principles of Geology* of 1819; Greenough, first president of the

79 Regional geology; frontispiece to Young, G., and Bird, J., *A Geological Survey of the Yorkshire Coast*, Whitby, 1822

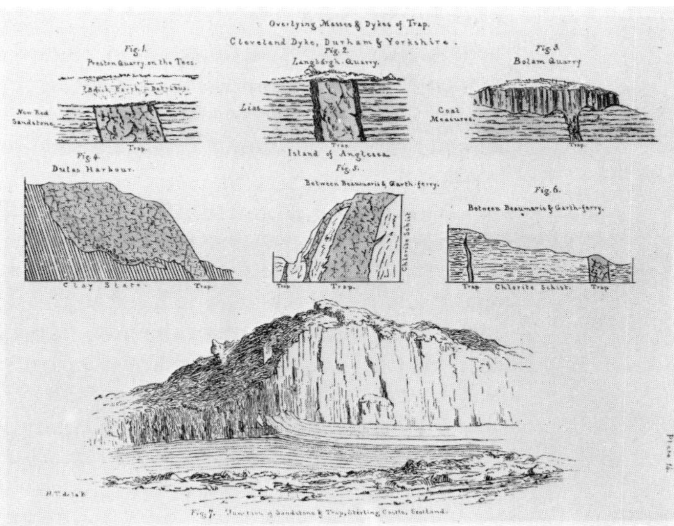

80 Geological sections (De La Beche, H., *Sections of Views*, 1830, pl 14)

Geological Society, and Phillips, could both be described as professional geologists, and Greenough in 1820 published a geological map of England based in the main upon mineralogy rather than palaeontology. This practical emphasis in geology culminated in the setting up of the Geological Survey in 1835 under Henry De La Beche. His *Report on the Geology of Cornwall, Devon and Somerset* is a classic; and his more popular works, *Sections of Views illustrative of Geological Phenomena* (1830); *Manual of Geology* (1831); *Researches in Theoretical Geology* (1834); and the most famous, *The Geological Observer* (1851), are all excellent examples of introductory and general books on the subject. His little book in the series *How to Observe*, a splendid example of popular science, appeared in 1835.

In contrast to this empirical geology were the writings of William Buckland, Professor of Geology at Oxford and later dean of Westminster, who sought constantly to reconcile the discoveries of the geologists with the text of *Genesis*, particularly the description of the Flood. His reputation was made when in 1821 the Kirkdale Cavern in Yorkshire was discovered, and found to be full of bones, which Buckland identified as mostly those of hyaenas. In 1823 he published an account of his researches, *Reliquae Diluvianae*; or *Observations on the Organic Remains Contained in Caves, Fissures, and Diluvial Gravel, and on other Geological Phenomena, Attesting the Action of an Universal Deluge*. The book was an enormous success. Buckland considered that he had proved the existence of the Flood, and the late creation of man; animals such as hyaenas had only existed in Europe in the pre-Diluvial period. The bones from Kirkdale were taken to York where they formed the nucleus of the museum of the Yorkshire Philosophical Society, whose first president, William Vernon Harcourt, son of the archbishop of York, became the chief architect of the British Association for the Advancement of Science, which first met at York in 1831; and in the deliberations of which the Geological Section was prominent. Buckland later wrote the Bridgewater Treatise on geology; by 1836, when this appeared, he had dropped the Universal Flood in deference to the uniformitarians, but remained a catastrophist, believing in 'perpetual destruction, followed by continual renovation', which tended to increase animal enjoyment. Buckland's son, Frank, is well known for his *Curiosities of Natural History*; he was a particular authority on fish, and an excellent naturalist generally and the *Curiosities* is a wonderful bedside book.

The professor at Cambridge was Adam Sedgwick, who began to learn geology when he was elected to his chair in 1818. He soon began making important field investigations, on one of which he was accompanied by the young Charles Darwin, which led to the founding of the Cambrian System, based upon the structure of the mountains of North Wales. His collaborator, until they fell out in a quarrel over priorities was Roderick Murchison, who had been a country squire fond of field sports until Davy persuaded him to take up scientific study. He studied the rocks of South Wales, and in 1839 published his classic work *The Silurian System*. Also in 1839, he and Sedgwick

81 The Urals (Murchison, R. I., *Geology of Russia in Europe*, 1845, vol 1, frontispiece)

founded the Devonian System; and in 1841 he visited Russia and coined the term Permian for another system of rocks. His researches were published in his book *Siluria* of 1854; the terms devised by Sedgwick and Murchison are still used to describe the periods in the geological time scale. Murchison held numerous posts of honour and responsibility; he followed De La Beche as Director-General of the Geological Survey, and was for many years President of the Royal Geographical Society; which is why his name is attached to rivers and waterfalls around the world discovered by British explorers in the last century.

Scientists had travelled in the eighteenth century, and of course earlier, but an immense impetus was given to scientific expeditions by Alexander von Humboldt's *Personal Narrative* of his travels in Latin America, which had been previously kept *terra incognita* by the Spanish government. Humboldt was an extraordinary man whose knowledge covered numerous fields: but his emphasis was chiefly geographical. His *magnum opus, Cosmos*, was based upon lectures delivered in Berlin and represents an attempt to see the world whole. An abridged translation by E. Sabine was published in four volumes between 1846 and 1858; the full translation in five volumes came out in Bohn's Library from 1849 to 1858, by E. C. Otté, B. H. Paul, and W. S. Dallas. Humboldt's *Views of Nature* with a coloured view of Chimborazo, which when Humboldt climbed it was believed to be the highest mountain in the world, appeared in the same series. Although dauntingly voluminous for twentieth-century tastes, Humboldt's writings are well worth reading if one wants to recapture the spirit of the science of the early nineteenth century; and they were very influential in their day. Darwin, J. D. Hooker, T. H.

Huxley, and Wallace all travelled; and John Herschel drew up *The Admiralty Manual of Scientific Enquiry*, published in 1849, for the guidance of naval officers and explorers generally.

The arguments of Cuvier, and later of Lyell, were generally felt to have refuted vague and Romantic evolutionary beliefs; and such works as John Fleming's *Philosophy of Zoology* of 1822 insisted upon the fixity of species. Then in 1844 there was published anonymously *Vestiges of the Natural History of Creation*; various putative fathers were suggested for this alarming infant, which turned out to have been written by Robert Chambers, a writer of popular works and the publisher of an encyclopedia. But this remained a secret until after Chambers' death; all was told in the preface to the twelfth edition, of 1884. Chambers traced the development of the solar system according to the nebular hypothesis of Kant and Laplace, from primeval 'Fire Mist' to its present state; and then the development of living forms up to man. The mechanism in *Vestiges* of biological development owed something to the railways; Chambers believed that the embryos of all animals began on the same main line, but took a turning off it at some point. Monstrous births occur when a turning is taken too soon: thus mammalian monsters sometimes have a heart appropriate to a reptile. Evolution happened when *per contra* an embryo stayed longer on the main line. Australia has a strange fauna because it is a younger continent; and the duck-billed platypus represents a half-way stage between a bird and a rat.

To account for monsters is one thing, but to account for the development of viable higher forms is another; and Chambers' critics were not impressed by his mechanism and his hypothetical genealogies. They also hated the determinism and materialism of his theory, which was assailed by almost all the experts. The longest, and perhaps windiest, attack was made by Adam Sedgwick. He had earlier published a book, *A Discourse on the Studies of the University*, which occupies a niche in the history of educational theory; to the fifth edition of this, which appeared in 1850, he added a prefatory essay which dwarfs the book and is concerned with refuting *Vestiges*. Chambers' mechanism did not stand up to examination; he was uncritical in the credence he gave to certain experiments in which living creatures seemed to have been generated from inorganic matter by an electric current, and his knowledge was mostly at second hand. His book therefore made an easy target: but that his general thesis became in a generation the orthodox one may encourage authors of syntheses which are condemned by authorities.

Two eminent scientists who gave at least a qualified welcome to *Vestiges* were Richard Owen and Baden Powell. Owen studied for a time under Cuvier, and became the leading comparative anatomist of his day: his greatest triumph being the recognition that a single bone brought back from New Zealand came from a hitherto unknown enormous flightless bird, the moa. Owen was interested in analogies and homologies, and in 1848 published the *Archetypes and Homologies of the Vertebrate Skeleton*; another work, on the *Nature of Limbs*, appeared in 1849; and Owen also wrote a great number of

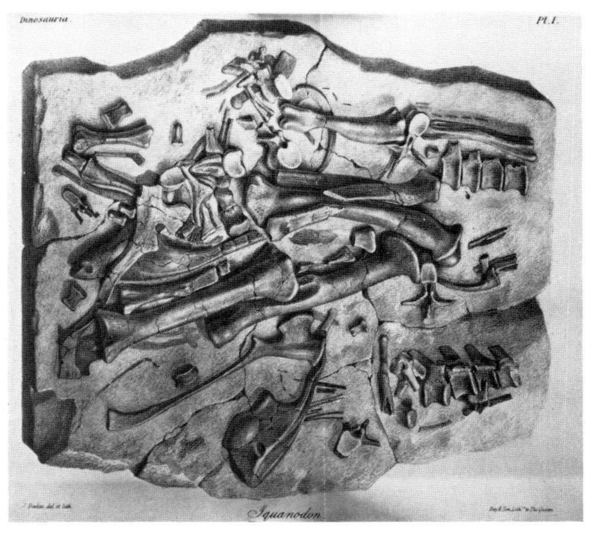

82 (*above*) Iguanadon as found (Owen, R., *British Fossil Reptiles*, 1849–55, pl 1 of dinosaurs)

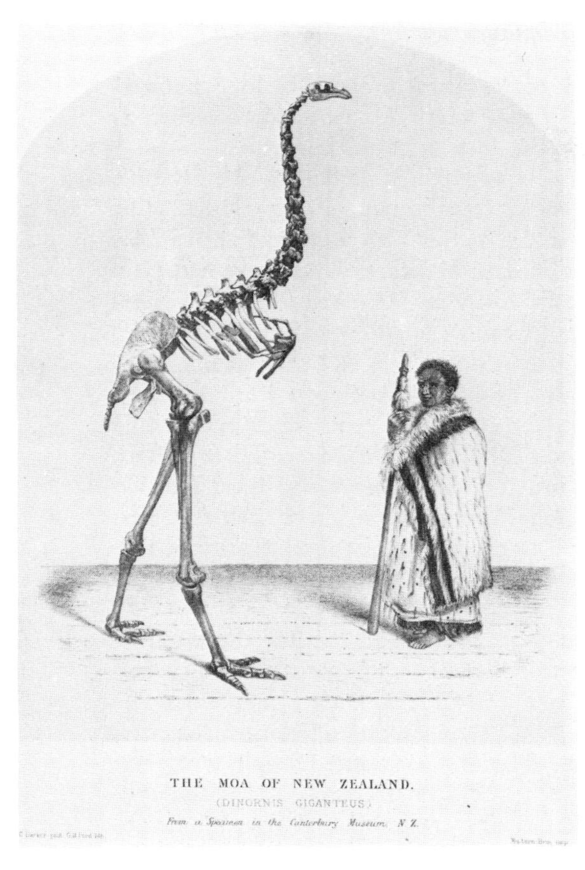

83 (*left*) Moa and Maori (Buller, W. L., *Birds of New Zealand*, 1873, frontispiece)

detailed monographs and papers, including an illustrated work on British fossil reptiles. It was Owen who briefed Bishop Wilberforce for his attack on *The Origin of Species* at Oxford in 1860: but as Darwin himself realized, Owen would not have objected to a theory of discontinuous evolution. It was the manner in which, for Darwin, species gradually became transformed that upset the comparative anatomist and believer in archetypes, not the mere suggestion that later species had in some sense sprung from earlier ones. Baden Powell on the other hand, welcomed both *Vestiges* and *The Origin of Species*; he was Savilian Professor of Geometry at Oxford and a liberal divine who contributed an essay against miracles to *Essays and Reviews* in 1860, thus diverting some *odium theologicum* from Darwin. Baden Powell believed that the best evidence for the existence of God was a world of law, harmony, and symmetry, in which the development of the species followed law rather than caprice or arbitrariness. In his *Essays on the Spirit of the Inductive Philosophy, the Unity of Worlds, and the Philosophy of Creation* of 1855, he urged these points; and therefore when *The Origin of Species* appeared he swallowed its arguments with a relish that surprised Darwin himself.

Vestiges sold well, and in 1845 there appeared *Explanations*, by the author of *Vestiges*, which contained answers to the critics, and was incorporated into later editions of *Vestiges*. Professional scientists were not convinced; but laymen had begun to get used to the idea of evolution which was also being disseminated in Herbert Spencer's writings. Parts of Tennyson's *In Memoriam* are derived ultimately from Lyell and from *Vestiges*. In 1831 in Patrick Matthew's *On Naval Timber and Arboriculture* there had appeared a summary of the doctrine of evolution by natural selection, as there had even earlier in an appendix to William Wells' *Essay on Dew*; but neither of these passages were remarked on until after Darwin's book appeared. The phrase 'the struggle for existence' had been used by Thomas Malthus in his *Essay on Population* of 1798, in reference to the contest between similar tribes. Malthus' doctrine that population tends to expand in geometrical ratio while food supply can only increase in arithmetical proportion was widely accepted, particularly in radical circles in the nineteenth century; and both Darwin and Wallace read Malthus at critical stages in the formation of their theories. Malthus believed that there was a limit to the amount of modification which might be produced in species like cows or sheep; he is not a precursor of the evolutionary doctrine.

When Darwin set out on the *Beagle*, his friend and mentor the botanist Henslow advised him to read but not to believe Lyell's *Principles* which had just begun to appear. Before he had seen a coral reef, Darwin explained the formation of reefs and atolls in terms of the slow sinking of islands, the coral going on building at the old shore line. When he got home Darwin wrote the *Naturalist's Voyage*, and opened his notebooks on transmutation; sketches of his theory that he wrote in 1842 and 1844 were published after his death. He wrote a monograph for the Ray Society on limpets, and then in 1859 produced *The Origin of Species* with its doctrine of evolution, or rather 'development', by natural selection. The book requires no commendation: both to the layman and the biologist the facts and arguments are of enormous interest, and the style is clear and unpretentious. The case was taken further in his *Descent of Man*, with its long section on sexual selection; and in *Animals and Plants under Domestication*. Shorter works were his *Expression of Emotions*, a reply to Charles Bell's view that we had been provided with facial muscles in order to help us communicate with one another; and *Earthworms and Vegetable Mould*, pointing out the importance of these lowly creatures.

The *Origin* was finally published because A. R. Wallace had hit upon the same idea and written to Darwin about it. Wallace was originally a surveyor, who had travelled with H. W. Bates, the author of the splendid travel book *The Naturalist on the River Amazons*, who described how tasty butterflies 'mimic' unpalatable species in colour and form. Wallace left Bates and went to Indonesia, where he noted the change from Asian to Australian fauna and flora between Bali and Lombok, and described orang-utangs, establishing that they were not wild men, and birds of paradise, confirming that they had legs. His book *The Malay Archipelago* is excellent reading, and pleasantly free from any condescension towards the Malays who were his sole compa-

84 (*left*) Naturalist under attack by toucans (Bates, H. W., *Naturalist on the Amazons*, 1863, frontispiece)

85 (*right*) Dyaks fighting an orang-utang (Wallace, A. R., *Malay Archipelago*, 1869, frontispiece)

nions. Wallace's autobiography, *My Life*, appeared in 1905, with an abridged version in 1908; he became increasingly interested in socialism and spiritualism as he got older, and also wrote against the plurality of worlds. In his *Darwinism* and his *Island Life* are to be found his great contributions to biology; he was one of the most modest and generous of men and refused to elbow his way on to a pedestal beside Darwin. In *Darwinism* there appears his divergence from Darwin himself, in that he rejected sexual selection, and did not accept natural selection as sufficient to account for the emergence of man. The mathematical faculty, for example, would have been useless to the savage, but when civilization appeared then mathematics sprang up rapidly with it. In short, there is a breach between man and animals as there is between animals and plants.

Another friend and correspondent of Darwin who refused to accept the whole thesis was the American botanist Asa Gray, whose *Darwiniana* contains his essays on the subject. He wanted to allow design; and suggested that the variation upon which natural selection worked might be 'planned'. This attempt to plug the Darwinians' most important gap with God became hopeless when Mendel's work on heredity was rediscovered. While Gray and Wallace accepted Darwinism with qualification, Louis Agassiz, the Swiss-born geologist who added the ice ages to the geological calendar, remained like Owen an adamant opponent. In 1857 he began to publish his magnificently illustrated *Contributions to the Natural History of the United States*, where he was then living; and in 1859 the major part of the first volume was printed separately as an *Essay on Classification*. It represents the last appearance of the idealism of Cuvier's school; and is extraordinarily theological, rhetorical, and old-fashioned beside Darwin's work. Even odder is Philip Gosse's *Omphalos* of 1857, urging that just as Adam would have been created with a navel, so the world was a recent creation filled with fossils of animals which had never lived. This Gosse was the father in Edmund Gosse's *Father and Son*, and a well-known naturalist.

An influential attack on Darwinian theory was the 8th duke of Argyll's *Reign of Law*; its arguments for design and contrivance were not new, but were forcibly put, and the book appeared in very many editions – far more than the *Origin* – after its first appearance in 1867. The only serious Darwinian apostate was St G. J. Mivart, son of the founder of what became Claridge's Hotel in London. Mivart was converted to Roman Catholicism, chiefly it seems by Pugin's writings on architecture; in London he came under the influence first of Owen and then of T. H. Huxley and began studies on the anatomy of apes. In 1871 he published his *Genesis of Species* which had a mildly evolutionary tone, but he was sceptical of many genealogies which evolutionists were constructing. He reacted strongly to Darwin's *Descent of Man*, to which he wrote a hostile review; and his later writings, which include a number of textbooks, were theological and anti-Darwinian in tone. More serious from Darwin's point of view were the opposition of Lord Kelvin, who apparently demonstrated from thermodynamics that the solar system could not be as old as the geology of Lyell and Darwin required; and of Fleeming Jenkin, who showed statistically that any variation would soon be swamped, since offspring were supposed to be a blend of the characters of both parents. To get around this, in later editions of the *Origin* Darwin allowed for the effects of the environment as well as natural selection. Finally, in 1879, Samuel Butler published his *Evolution Old and New* seeking quite entertainingly to discredit Darwin by calling attention to precursors, some genuine and some not; more recently a similar attempt to show that Darwin was neither original nor a gentleman has been made by C. D. Darlington.

So much then for the opponents of evolution, some endeavouring to disprove it and others to show that it was not new. Darwin's chief supporters were J. D. Hooker and T. H. Huxley. Hooker, like his father, was a very distinguished botanist who, like Asa Gray, had become convinced that the innumerable species described by some botanists could not represent genuine and natural groups. He sought to redescribe a great number of species as mere varieties, and was rapidly convinced by Darwin, who corresponded with him before the *Origin* was written, of the reality of evolution. Hooker argued for it first in his *Essay on the Tasmanian Flora*, which actually appeared before the *Origin*. Huxley on reading the *Origin* was convinced of its general thesis. He wrote a review of it for *The Times* and fought on its behalf against Owen and Bishop Wilberforce at Oxford in 1860; though this last confrontation, at the British Association meeting, probably assumes more importance to those with hindsight than it had at the time, or intrinsically. Huxley became Darwin's champion; he was combative while Darwin was retiring; but it is arguable that the theory would have prevailed anyway in the hands of milder persons like Gray and Wallace without provoking all the excitement and *odium theologicum* in which Huxley revelled. But Huxley's writings are splendid to read; his learning was wide and his polemical skill great. In *Darwiniana* are to be found most of the essays relating to Darwin; and in *Lay Sermons* his attempt to rout Lord Kelvin's incursion into geology. In *Man's Place in Nature*, published in 1863, he sought to show that man was not in an entirely distinct natural group, separated from the apes; and established indeed that man differed no more from the anthropoid apes in structure than they differed among themselves. Earlier work on apes, monkeys and man can be found in an agreeable little anonymous work of 1848, *A Sketch of the History of Monkeys*. Darwin had remarked in the *Origin* that his theory would cast light on the origin of man and his history; Huxley took this further, as Darwin was himself later to do. Himself an important biologist, Huxley

86 Chimpanzee (Anon, *History of Monkeys*, 1848, pl 10)

occupies a place in history chiefly on account of his abilities as a controversialist and popularizer, and in the next chapter we shall look at those who, like him, sought to make theories and discoveries of science, or what they believed to be the scientific method or the scientific attitude, known to a wider audience.

9 BIOLOGY AND GEOLOGY IN THE NINETEENTH CENTURY

Agassiz, L., *Bibliographia Zoologiae et Geologiae*, 4 vols., 1848–54.
Contributions to the Natural History of the United States, 4 vols., Boston, 1857–62.
Essay on Classification, 1859.

Anon., *A Sketch of the History of Monkeys*, 1848.

Argyll, Campbell, G. D., 8th duke of, *The Reign of Law*, 1867.

Audubon, J. J., *The Birds of America* . . . , 4 vols., 1827–38.
Ornithological Biography . . . , 5 vols., Edinburgh, 1831–9.
and Bachman, J., *The Viviparous Quadrupeds of North America*, 3 vols., New York, 1846–54; plates, 2 vols., 1845–6.

Bakewell, R., *Introduction to Geology*, 1813; ed. B. Silliman, New Haven, 1829.

Barton, W. P. C., *A Flora of North America*, 3 vols., Philadelphia, 1821–3.

Bartram, W., *Travels through North and South Carolina* . . . , Philadelphia, 1791; London, 1792.

Bastian, H. C., *The Mode of Origin of Lowest Organisms* . . . , 1871.
The Beginnings of Life . . . , 2 vols., 1872.
The Nature and Origin of Living Matter, 1905.

Bateman, J., *The Orchidaceae of Mexico and Guatemala*, 1837–44.

Bates, H. W., *The Naturalist on the River Amazons*, 2 vols., 1863; one vol., abridged ed., 1864.

Bernard, C., *Notes of M. Bernard's Lectures on the Blood*, by W. F. Atlee, Philadelphia, 1854.
An Introduction to the Study of Experimental Medicine, tr. H. C. Greene, New York, 1927.

Bewick, T., *History of British Birds*, 2 vols., Newcastle, 1797–1804.

Bichat, M. F. X., *Physiological Researches on Life and Death*, tr. F. Gold (1815).
General Anatomy . . . , tr. C. Coffyn, 2 vols., 1824.
Pathological Anatomy . . . , tr. J. Togno, Philadelphia, 1827.

Blackwell, J., *A History of the Spiders of Great Britain and Ireland*, 2 vols., 1859–64.

Blumenbach, J. F., *The Institutions of Physiology*, tr. J. Elliotson, 2nd ed., 1817; 1st ed. not seen.
The Anthropological Treatises, tr. T. Blandyshe, 1865.

Botanical Magazine, *1* (1787)–*42* (1815); new series, *1* (1815)–*11* (1826); 2nd new series, *1* (1827)–*17* (1844); 3rd series, *1* (1845)–*60* (1905); still continues.

Brande, W. T., *Outlines of Geology*, 1817.

Brown, R., *Miscellaneous Botanical Works*, ed., J. J. Bennett, 2 vols., 1866–7.

Buckland, F. T., *Curiosities of Natural History*, 1858; 2nd series, 1860; a new series, 2 vols., 1866.

Buckland, W., *Reliquae Diluvianae* . . . , 1823.

Buller, W. L., *A History of the Birds of New Zealand*, 1873; 2nd ed., 2 vols., 1887.

Butler, S., *Evolution, old and new* . . . , 1879.
Luck, or Cunning, as the main means of organic modification?, 1887.

Cambridge, O. P., *The Spiders of Dorset*, Sherborne, 1879–81.

(Chambers, R.), *Vestiges of the Natural History of Creation*, 1844; a sequel, *Explanations . . .*, appeared in 1845 and was incorporated into later editions of *Vestiges*.
Cleaveland, P., *An Elementary Treatise on Mineralogy and Geology*, Boston, Mass., 1816.
Cockburn, W., *A New System of Geology*, 1849.
Conybeare, W. D., and Phillips, W., *Outlines of the Geology of England and Wales*, 1822.
Curtis, J., *British Entomology . . .*, 2 vols., 1824.
Cuvier, G., *Lectures on Comparative Anatomy*, tr. W. Ross, 2 vols., 1802.
Essay on the Theory of the Earth, tr. R. Kerr, Edinburgh, 1813.
The Animal Kingdom . . ., 16 vols. 1827–35.
Dalyell, J. G., *Rare and remarkable Animals of Scotland*, 2 vols., 1847–8.
Darwin, C. R., *Journal of Researches . . .*, 1839.
Monograph on the sub-class Cirripedia, 2 vols., 1851–4.
On the Origin of Species, 1859.
On the various contrivances by which British and Foreign Orchids are fertilized by insects . . ., 1862.
The Variation of Animals and Plants under Domestication, 1868.
The Descent of Man . . ., 2 vols., 1871.
The Expression of the Emotions in Man and Animals, 1872.
The Formation of Vegetable Mould, through the Action of Worms, 1881.
Daubeny, C., *A Description of Active and Extinct Volcanos . . .*, 1826.
De La Beche, H. T., *Sections of Views, Illustrative of Geological Phenomena*, 1830.
A Geological Manual, 1831.
Researches in Theoretical Geology, 1834.
How to Observe: Geology, 1835.
Report on the Geology of Cornwall, Devon, and West Somerset, 1839.
The Geological Observer, 1851.
Deluc, J. A., *An Elementary Treatise on Geology*, 1809.
Letters on the Physical History of the Earth, 1830.
Eaton, A., *Index to the Geology of the Northern States*, Leicester, 1818.
Geological Textbook, Albany, N.Y., 1830.
Fleming, J., *The Philosophy of Geology*, 2 vols., Edinburgh, 1822.
Forbes, E., and Hanley, S., *A History of British Mollusca*, 4 vols., 1848–53.
(Gosse, E. W.), *Father and Son . . .*, 1907.
Gosse, P. H., *Omphalos: An Attempt to Untie the Geological Knot*, 1857.
Gould, J., *The Birds of Europe*, 5 vols., 1837.
The Birds of Asia, 7 vols., 1850–83.
The Birds of Australia, 7 vols., 1848 and Supplement, 1869.
The Birds of New Guinea . . ., 5 vols., 1875–88.
Gray, A., *Darwiniana*, Cambridge, Mass., 1876.
Gray, J. E., *List of Books, Memoirs, and Miscellaneous Papers by . . . J. E. G.*, ed. J. Saunders, 1872.
Greenough, G. B., *A Critical Examination of the First Principles of Geology*, 1819.
Memoir of a Geological Map of England, 1820.
Haller, A. von, *First Lines of Physiology*, 2 vols., Edinburgh, 1786.
Hamilton, W., *Observations on Mount Vesuvius . . .*, 1772.
Campi Phlegraei: Observations on the Volcanos of the Two Sicilies . . ., 2 vols., Naples, 1776–9.
Harris, M., *The Aurelian, or natural history of English moths and butterflies . . .*, 1778; Engl./French; engraved title page, 1766.
Exposition of English Insects . . ., 1776; Eng./French.
Natural System of Colours, ed. T. Martyn, 1811.
Haworth, A. H., *Lepidoptera Britannica*, 2 vols., 1803–28.

Herder, J. G. von, *Outlines of a Philosophy of the History of Man*, tr. T. O. Churchill, 1800.
Herschel, J. F. W. (ed.), *A Manual of Scientific Enquiry . . .*, 1849.
Hooker, J. D., *The Botany of the Antarctic Voyage of H.M. Discovery-Ships Erebus and Terror . . .* I, *Flora Antarctica*, 2 pts; II, *Flora Novae Zelandicae*, 2 pts; III, *Flora Tasmaniae*, 2 pts; 1844–60.
The Rhododendrons of Sikkim-Himalaya . . ., 1849.
Handbook of the New Zealand Flora, 1864.
The Flora of British India, 2 vols., 1872–97.
Hooker, W. J., *Journal of a Tour in Iceland . . .*, Yarmouth, 1811.
Exotic Flora . . ., 3 vols., Edinburgh, 1823–7.
Kew Gardens, 1847.
Filices Exoticae . . ., 1857–9.
Horsfield, T., *Zoological Researches in Java . . .*, 1824.
Howorth, H. H., *The Mammoth and the Flood . . .*, 1887.
Humboldt, A. von, *Personal Narrative of Travels . . .*, tr. H. M. Williams, 7 vols., 1814–29; tr. and ed. T. Ross, 3 vols., 1847.
Cosmos, tr. E. and Mrs Sabine, 4 vols., 1846–58; tr. E. C. Otté *et al.*, 5 vols., 1849–58.
Aspects of Nature . . ., tr. Mrs Sabine, 2 vols., 1849; tr. as *Views of Nature*, by E. C. Otté and H. G. Bohn, 1850.
Humphreys, H. N., and Westwood, J. O., *British Butterflies and their transformations . . .*, 1841.
British Moths and their transformations . . ., 2 vols., 1843–5.
Hunter, J., *The Works*, ed. J. F. Palmer, 5 vols, 1835–7.
Essays and observations on Natural History, Anatomy, Physiology, Psychology, and Geology, ed. R. Owen, 2 vols., 1861.
Hutton, J., *A Dissertation upon the philosophy of light, heat, and fire*, 7 pts, Edinburgh, 1794.
Dissertations on different subjects of Natural Philosophy, Edinburgh, 1794.
An investigation of the Principles of Knowledge . . ., Edinburgh, 1794.
The Theory of the Earth, 2 vols., Edinburgh, 1795; vol. III, ed. A. Geikie, Edinburgh, 1899.
Huxley, T. H., *Evidence as to Man's Place in Nature*, 1863.
Lay Sermons, Addresses, and Reviews, 1870.
The Scientific Memoirs, ed. M. Foster and E. R. Lankester, 5 vols., 1898–1903.
Jameson, R., *An outline of the mineralogy of the Shetland Islands . . .*, Edinburgh, 1798.
System of Mineralogy . . ., 3 vols., Edinburgh, 1804–8.
Manual of Mineralogy . . ., Edinburgh, 1821.
Jenkin, F., *Papers Literary, Scientific, &c*, ed. S. Colvin and J. A. Ewing, memoir by R. L. Stevenson, 2 vols., 1887.
Kirwan, R., *Elements of Mineralogy*, 1784.
Geological Essays, 1799.
Koch, R., *Investigations into the etiology of Traumatic Infective Diseases . . .*, tr. W. W. Cheyne, 1880.
Lamarck, J. B., *An Epitome of Lamarck's arrangement of Testacea . . .*, by C. Dubois, 1824.
An illustrated introduction to Lamarck's Conchology . . ., plates by E. Crouch, 1826.
Lamarck's Genera of Shells, with a catalogue of species, tr. A. A. Gould, Boston, Mass., 1833. All three from *Animaux sans Vertèbres*.
Lear, E., *Illustrations of the Family of Psittacidae, or Parrots*, 1832.
Gleanings from the Menagerie and Aviary at Knowsley Hall, 1846.
Lewin, W., *Papilios of Great Britain*, 1795; English/French.
Lindley, J., *Introduction to the Natural System of Classification*, 1830.
Lyell, C., *Principles of Geology . . .*, 3 vols., 1830–3; 10th ed., 2 vols., 1867.
Elements of Geology, 1838.

The Geological Evidences of the Antiquity of Man . . ., 1863.
McLure, W., *Observations on the Geology of the United States*, Philadelphia, 1817.
Magendie, F., *An Elementary Compendium of Physiology*, tr. E. Milligan, 2nd ed., Edinburgh, 1826; 1st ed. not seen.
Maillet, B. de, *Telliamed: or discourses between an Indian philosopher and a French missionary . . .*, 1750.
(Malthus, T.), *An Essay on the principle of population . . .*, 1798.
Mammatt, E., *A Collection of Geological Facts . . .*, Ashby-de-la-Zouch, 1836.
Mantell, G., *The Fossils of the South Downs*, 1822.
The Geology of the South East of England, 1833.
The Wonders of Geology, 1838.
The Medals of Creation, 1844.
Petrifications and their Teachings, 1851.
Matthew, P., *On Naval Timber and Arboriculture*, 1831.
Miller, J. S., *A Natural History of the Crinoida . . .*, Bristol, 1821.
Mivart, St. G. J., *On the Genesis of Species*, 1871.
Men and Apes, 1873.
The Origin of Human Reason, 1889.
Montagu, G., *Testacea Britannica . . .*, 1803.
Moore, T., *The Ferns of Great Britain and Ireland*, ed. J. Lindley, 1845; 8^{0} ed. 2 vols., 1859–60.
Morris, J., *A Catalogue of British Fossils*, 1843; 2nd ed., 1845.
Murchison, R. I., *The Silurian System*, 1839.
Siluria, 1854.
(ed.) *The Geology of Russia in Europe . . .*, 2 vols., 1845; vol. II is in French.
[Murray, J.], *A Comparative View of the Huttonian and Neptunian Systems of Geology*, Edinburgh, 1802.
Newman, E., *The Grammar of Entomology*, 1835; 2nd ed. (*A Familiar Introduction to the History of Insects*), 1841.
All the British Butterflies . . . (1868).
An illustrated natural history of British Moths, 1869.
Offray de la Mettrie, J. J., *Man a Machine . . .*, 1749.
Owen, R., *On the Archetype and Homologies of the Vertebrate Skeleton*, 1848.
History of British Fossil Reptiles, 1849–55.
On the Nature of Limbs. 1849.
Palaeontology, Edinburgh, 1860.
Parkinson, J., *Organic Remains of a Former World . . .*, 3 vols., 1804–11.
Outlines of Oryctology. An introduction to the study of fossil organic remains . . ., 1822.
Phillips, W., *An outline of Mineralogy and Geology*, 1815.
A Selection of Facts from the Best Authorities, arranged so as to form an outline of the geology of England and Wales, 1818.
Playfair, J., *Illustrations of the Huttonian Theory of the Earth*, Edinburgh, 1802.
The Works, 3 vols., Edinburgh, 1822.
Powell, Baden, *Essays on the Spirit of the Inductive Philosophy, the Unity of Worlds, and the Philosophy of Creation*, 1855.
Raspe, R. E., *An Account of Some German Volcanoes . . .*, 1776.
Rensselaer, J. van, *Lectures on Geology*, New York, 1825.
Schleiden, M. J., *Principles of Scientific Botany*, tr. E. Lankester, 1849.
The Plant: a biography . . ., tr. A. Henfey, 2nd ed., 1848–53; 1st ed. not seen.
Schwann, T., *Microscopical Researches, into the accordance of the structure and growth of animals and plants*, tr. H. Smith, 1847.
Scrope, G. P., *Considerations on Volcanos . . .*, 1825.
Sedgwick, A., *A Discourse on the Studies of the University*, 1833; 5th ed., 1850, attacks *Vestiges*.

Sharpe, R. B., *A Monograph of the Alcedinidae, or . . . kingfishers*, 1868–71.
Shelley, G. E., *A Monograph of the Cinnyridae, or . . . sun birds* (1876–7).
Sibthorp, J., *Voyages in the Grecian Seas*, in R. Walpole (ed.), *Travels in Various Countries of the East . . .*, 1820.
Flora Graeca, 10 vols., 1806–40.
Smith, W., *Observations on . . . water meadows*, Norwich, 1806.
A Memoir to the Map and delineation of the strata of England and Wales, with part of Scotland, 1815.
Strata identified by organised fossils, 1816.
Stratigraphical System of Organised Fossils, 1817.
Sowerby, G. B., *Popular British Conchology*, 1854.
Sowerby, J. E., *The Ferns of Great Britain*, 1855.
Spallanzani, L., *Travels in the Two Sicilies . . .*, 4 vols., 1798.
Stephens, J. F., *Illustrations of British Entomology . . .*, 11 vols., 1828–35; supplement, 1867.
A Systematic Catalogue of British Insects, 1829.
Strickland, H. E., and Melville, A. G., *The Dodo and its Kindred*, 1848.
Townsend, J., *The character of Moses established for veracity as an historian*, Bath, 1813.
Thornton, R. J., *New Illustration of the Sexual System of Carolus Linnaeus . . .*, 1807; pt III of this is *The Temple of Flora*; for bibliography see the ed. of this by G. Grigson and H. Buchanan, 1951.
Tyndall, J., *Essays on the Floating Matter of the Air*, 1881.
Wallace, A. R., *A Narrative of Travels on the Amazon and Rio Negro . . .*, 1853.
The Malay Archipelago, 2 vols., 1869.
Contributions to the Theory of Natural Selection, 1870.
The Geographical Distribution of Animals, 2 vols., 1876.
Tropical Nature and other essays, 1878.
Island Life, 1880.
Darwinism, 1889.
Werner, A. G., *A treatise on the external character of fossils*, tr. T. Weaver, Dublin, 1805.
A New Theory of the Formation of Veins, tr. C. Anderson, Edinburgh, 1809.
Nomenclature of Colours, tr. P. Syme, 1814.
Wilkes, B., *One hundred and twenty copper plates of English Moths and Butterflies, coloured . . .*, 1773.
Willdenow, C., *The Principles of botany, and of vegetable physiology*, Edinburgh, 1805.
Wilson, A., *American Ornithology*, 9 vols., Philadelphia, 1808–14.
Wilson, J., *A History of Mountains . . .*, 3 vols., 1807–10.
Witham, H. M. W., *Observations on Fossil Vegetables*, Edinburgh, 1831.
Internal Structure of Fossil Vegetables, Edinburgh, 1833.
Yarrell, W., *A History of British Fishes*, 2 vols., 1836.
A History of British Birds, 3 vols., 1843.
Young, G., and Bird, J., *A Geological Survey of the Yorkshire Coast*, Whitby, 1822.

Recent Publications

Archer, M., *Natural History Drawings in the India Office Library*, 1962.
Blunt, W., *The Art of Botanical Illustration*, 1950.
Burgess, G. H. O., *The Curious World of Frank Buckland*, 1968.
Butler, S., *A Bibliography*, by S. B. Harkness, 1955.
Coleman, W., *Georges Cuvier zoologist*, Cambridge, Mass., 1964.
(ed.) *The Interpretation of Animal Form*, 1967.
Darlington, C. D., *Darwin's Place in History*, Oxford, 1959.
Darwin, C., *The Works . . . An Annotated Bibliographical Handlist*, by R. B. Freeman, 1965.
Dupree, A. H., *Asa Gray: 1800–1888*, Cambridge, Mass., 1959.

Eiseley, L., *Darwin's Century*, 1959.
Ford, E. B., *Butterflies*, 1945.
Gillispie, C. C., *Genesis and Geology*, Cambridge, Mass., 1951.
Glass, B., Temkin, O., and Straus, W. L., *Forerunners of Darwin*, Baltimore, 1959.
Gosse, P. H., *A Bibliographical Check-list*, by R. Lister, Cambridge, 1952.
Gould, J., *An Analytic index to the works of the late John Gould . . .*, by R. B. Sharpe, 1893.
Gruber, J. W., *A Conscience in Conflict: the life of . . . Mivart*, New York, 1960.
Hugo, T., *The Bewick Collector . . .*, 1866; supplement, 1868.
Lurie, E., *Louis Agassiz: a life in science*, Chicago, 1960.
Mellersh, H. E. L., *Fitzroy of the Beagle*, 1968.
Millhauser, M., *Just before Darwin: Robert Chambers and Vestiges*, Middletown, Conn., 1959.
Sitwell, S., Blunt, W. and Synge, P. M., *Great Flower Books*, 1956.
Sitwell, S., Buchanan, H. and Fisher, J., *Fine Bird Books 1700–1900*, 1953.
Williams-Ellis, A., *Darwin's Moon. A biography of Alfred Russel Wallace*, 1966.

Additional Reading

Bowler, P. J., *The Eclipse of Darwinism*, Baltimore, 1983.
Browne, J., *The Secular Ark*, New Haven, 1983.
Burchfield, J. D., *Lord Kelvin and the Age of the Earth*, New York, 1975.
Burckhardt, F., & Smith, S. (eds.), *The Darwin Correspondence*, Cambridge, 1985–.
Coldeway, J., *Feathers to Brush*, Melbourne, 1982.
Gage, A. T. & Stearn, W. T., *A Bicentenary History of the Linnean Society of London*, 1988.
Grammiccia, G., *The Life of Charles Ledger 1808–1905*, 1988.
Geison, G. L., *Michael Foster and the Cambridge School of Physiology*, Princeton, 1978.
Greene, J., *Science, Ideology and World View*, Berkeley, 1981.
Gunther, A., *A Century of Zoology at the British Museum*, 1975.
Home, R. W. (ed.), *Australian Science in the Making*, Melbourne, 1988.
Hyman, S., *Edward Lear's Birds*, 1980.
Jordanova, L. J., *Lamarck*, Oxford, 1984.
Knight, D. M., *Ordering the World*, 1981.
Kohler, R.E., *From Medical Chemistry to Biochemistry*, Cambridge, 1982.
Lenoir, T., *The Strategy of Life*, Dordrecht, 1982.
Mabberly, D.J., *Jupiter Botanicus*, 1985.
MacLeod, R., *The Commonwealth of Science*, Melbourne, 1988.
Oldroyd, D., *Darwinian Impacts*, Milton Keynes, 1980.
Outram, D., *Georges Cuvier*, Manchester, 1984.
Rudwick, M., *The Great Devonian Controversy*, Chicago, 1985.
Rupke, N. A., *The Great Chain of History*, Oxford, 1983.
Ruse, M., *Taking Darwin Seriously*, Oxford, 1985.
Sauer, G. C., *John Gould*, Melbourne, 1982.
Secord, J., *Controversy in Victorian Geology*, Princeton, 1986.

10 *The Diffusion of Science in the Nineteenth Century*

It is difficult to draw a line between popular science and original expositions of science during the nineteenth century. In the geological and biological sciences, such works as Lyell's *Principles of Geology*, Darwin's *The Origin of Species*, and Huxley's *Man's Place in Nature* were books of great scientific importance but nevertheless written for the general reader. Eminent physicists, from Laplace to Helmholtz, Maxwell, and Lord Kelvin, popularized their own discoveries which they had written up in mathematical form for a smaller circulation. Articles written for encyclopedias, or books forming a part of a popular series, were sometimes at a very high level indeed, and made no concessions to the reader; they were often written by very eminent authorities, without too much editorial intervention. Beside them there is a whole spectrum of authors; some of whom were original in their interpretations, others up to date, interesting and clear, while others again cannot have greatly enlightened their readers. The productions of this last group may be allowed to slumber on undisturbed, unless they are in some way curious. A rather different class of books, which contributed to the diffusion of science, were works on the history of science, on philosophy, especially philosophy of science, and on logic; for it was during the nineteenth century that the term 'scientific' became prestigious and those pursuing such disciplines as political economy and theology sought to understand and follow the method, or methods, of the sciences. A spectrum of sciences was often described, from the pure sciences of mathematics and logic, which involved no observations of the world, through the mixed sciences which, on the way from physics through biology to the philosophy of manufactures, involved an increasing descriptive component.

At the beginning of the nineteenth century, the Newtonian tradition continued in the writings of such men as John Robison, whose *Natural Philosophy* is worthy of note, and Charles Hutton, whose *Mathematical and Philosophical Dictionary* has already been mentioned. The Scottish 'common sense' philosophers, notably Reid and Dugald Stewart, tried to disarm Hume's scepticism about causation and induction, and their voluminous writings were read by such notable scientists as Dalton and Young. Perhaps more important than any writings on induction by philosophers was the splendid example of inductive reasoning by William Charles Wells. Like Rumford, he was an American Tory, who came to England after the Revolution; and his *Essay on Dew* was rapidly recognized as a classic. It is odd that the edition of 1818 contains discussion of human pigmentation which entitles Wells to a place among precursors of Darwin, but the passage remained unremarked until after *The Origin of Species* had appeared. Wells' *Essay* was greatly praised by John Herschel, in his *Preliminary Discourse*; and it is indeed one of the few examples of the use of strict inductive logic in science. Herschel's book was very widely read by scientists, and was used by John Stuart Mill; while it is not an inductivist work, like Mill's *Logic*, it is written within the framework of empiricism. On the other hand, William Whewell, in his *Philosophy of the Inductive Sciences*, set out to combat 'the fallacies of the ultra-Lockean school', and like Kant stressed the way we impose categories on nature, and are not passive recording instruments. Unless one had the right basic idea, so that one knew what to look for, the piling up of facts was vain. Only when our conceptions are clear, our facts are certain and numerous, and the conceptions accord with the facts, have we science. Induction is not mere fact collection; a conception is 'super-induced' upon the facts.

Whewell's *Philosophy* followed upon his *History of the Inductive Sciences* which first appeared in 1837, and represented a landmark in that, although explicitly a discussion of the progress of science rather than of its errors, it was less Whiggish and hagiographical than most histories had been, and indeed continued to be. It is a classic in the historiography of science; and Whewell's remarks particularly upon the science of his own day, and of the generation before his, are of the greatest interest. Both the *History* and the *Philosophy* were revised considerably in their second editions, which were and are generally felt to be improvements on the first editions. Material from these volumes was subsequently reorganized in works bearing different titles: *Novum Organum Renovatum*, and *The Philosophy of Discovery*.

Whewell also published, anonymously, an essay *Of the Plurality of Worlds* in which he sought to overturn the arguments which such authors as Huygens, Fontenelle and Derham had made generally accepted. He demonstrated that the Sun was not a typical star; that the Moon and the other planets in the solar system would be unfavourable to life; and that upon the whole there was no good reason to suppose that the Earth was not unique in possessing material living beings. This provoked David Brewster, a Newtonian who wrote a standard life of Newton, and who edited the *Edinburgh New Philosophical Journal* and the *Edinburgh Encyclopedia*. Brewster's scientific work had been experimental studies on optics; he was one of the last defenders of the Newtonian emission theory of light against the undulatory theory of Young and Fresnel. His defence of the doctrine of a plurality of worlds is quite entertaining reading, though he lacked the style of Whewell; the debate gives one an insight into theoretical astronomy that one does not always find in more serious works. Another critic of Whewell – and Whewell seems to have enjoyed provoking critics – was Baden Powell, one of whose *Essays* of 1855 was directed to establishing the plurality of worlds, because the doctrine went well with his fervent belief in the uniformity of nature. Another essay argued, against Whewell, for the unity of sciences: whereas Whewell believed that the basic conceptions of each science were different, so that the frontiers between sciences were natural, and the reduction of, for example, biology to physics impossible, Baden Powell held that the axioms of physics must be in the last resort those of all sciences.

The anti-mechanistic ideas of Whewell can also be found in Coleridge, snippets of whose science can be found in *The Friend*, in the introductory essay to the *Encyclopedia Metropolitana*, in the very widely read *Aids to Reflection*, and in the *Hints towards the formation of a more comprehensive theory of life*, which appeared in 1848. In the first three we find the doctrine of polarities, of the union of opposites to produce a synthesis; and particularly in the fourth, the idea that the whole world is full of life, that it is an illusion to suppose that there is an inanimate realm. Coleridge and Whewell, and later

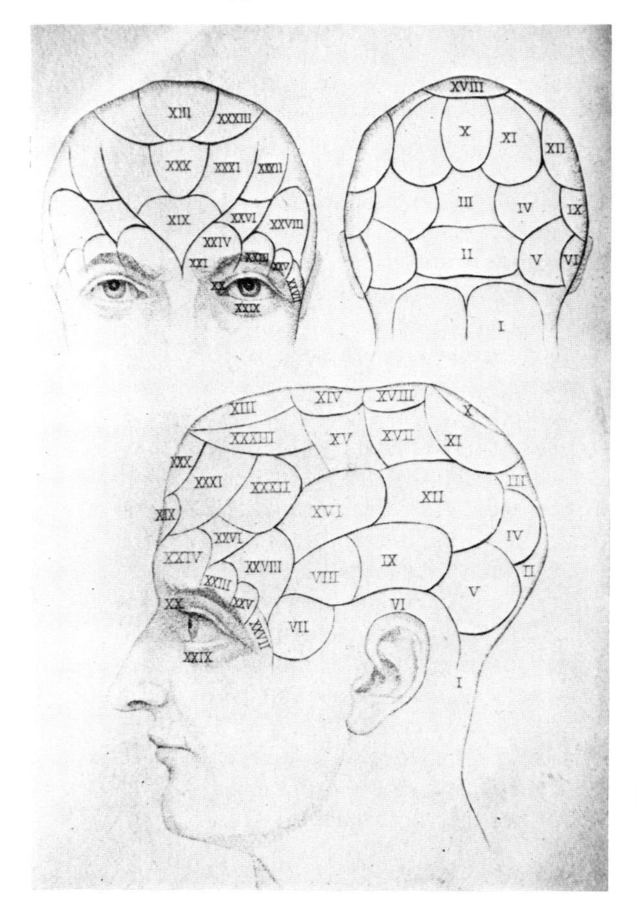

87 Craniological diagram, for phrenologists (Spurzheim, J. C., *Physiognomical System*, 1815, 2nd ed, frontispiece)

88 Craniological examination (Anon, *Craniology Burlesqued*, 1818, pl facing p 1)

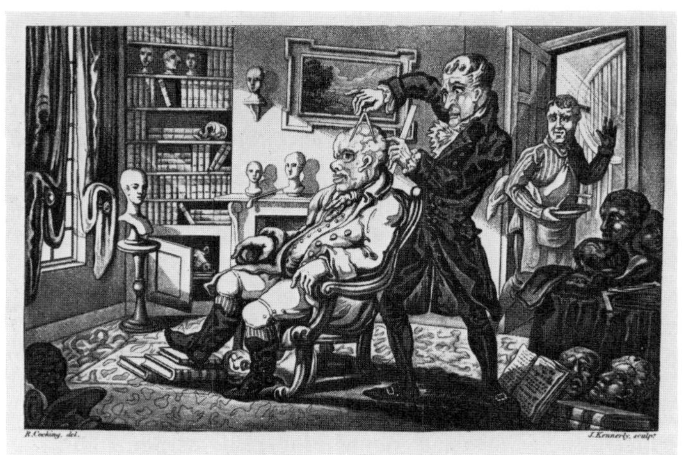

T. H. Huxley, had nothing but contempt for the French version of empiricism which Comte elaborated into his system of Positivism. Comte was a mathematician; he is chiefly remembered as a philosopher of science for his assertion that we could never know the chemistry of the stars, a few years before Bunsen and Kirchhoff applied the spectroscope and in fact inaugurated this very subject. It was as a pioneer of the social sciences that Comte exerted an influence in England; his writings were translated by Harriet Martineau, and admired by John Stuart Mill.

Many of the early Positivists were also admirers of that rather strange attempt at a mechanistic psychology, phrenology; the works of Gall, its founder, were translated in 1835; his pupil Spurzheim's in 1815. One of the most important devotees of phrenology in Britain was the educationalist George Combe, whose *Constitution of Man* first appeared in 1828 and was extraordinarily widely read in the early Victorian period; in America, the major advocate of the science was Nathan Capen. Phrenology cries out for investigation by an un-Whiggish historian; it grew out of physiognomy, exemplified in J. C. Lavater's extraordinary *Essays*, profusely illustrated with portraits and caricatures. Phrenological doctrines seem to have given adherents the belief that man's circumstances could be indefinitely improved; and did contribute to ideas on cerebral localization. An anonymous satire upon Gall and Spurzheim, *Craniology Burlesqued, in three Serio-comic lectures*, is entertaining reading.

The famous three stages of the Comtean scheme – the theological, the metaphysical, and the positive – gave prestige to the sciences, which had reached the last stage. In G. H. Lewes' *Biographical History of Philosophy* we find the view that philosophy, with its endless discussion of insoluble problems, should be dropped in favour of science, with its clear answers to distinct questions. Lewes in fact saw himself as writing an obituary of philosophy, to which Comte had applied euthanasia. Later authors took the struggle back a further stage, and framed the myth of the battle between theology, always in retreat, and science. The best-known examples of this genre are J. W. Draper's *History of the Conflict between Religion and Science* of 1875, and A. D. White's *Warfare of Science and Theology* of 1896; this schema provided useful insights in its day, but although adherents to it may still be uncovered occasionally, it could not be said to produce a very valuable key to the history of science, especially in English-speaking countries. But as sources for the intellectual history of the later nineteenth century, Draper and White are invaluable.

As well as translating and abridging Comte's *Positive Philosophy*, published in 1853, Harriet Martineau wrote with H. G. Atkinson the Comtean book *Letters on the laws of man's nature and development*, which appeared in 1851. G. H. Lewes had in 1847 published *Comte's Philosophy of the Sciences*; he himself wrote biological works, including *The Physiology of Common Life*, 1859–60, and *Problems of Life and Mind*, 1874–9, but it is as a popularizer of philosophy and as the man with whom George Eliot lived that he is remembered. J. S. Mill published his *Auguste Comte and Positivism* in 1865. Mill's *System of Logic* was intended as a riposte to Whewell; its attempts to provide a justification for induction has continued to interest philosophers, but the elaborate schemes of rules for inductive inference can never have been very close to what scientists do. More interesting are his famous *Autobiography*, and his *Three Essays on Religion*, published posthumously in 1873 and 1874 by Helen Taylor; the latter form a particularly interesting document of nineteenth-century agnosticism. While Mill was able to allow the possibility of a benevolent creator, but not to accept any doctrine of providence, Henry Maudsley in his *Body and Will* of 1883, and in rationalist Sunday lectures, advocated a materialistic theory of the mind. In somewhat similar vein, Leslie Stephen's *Essays on Freethinking and Plainspeaking*, published in 1873, should be noticed, although their relationship to science is vague; the same goes for the histories of rationalism by Lecky and by A. W. Benn. Indeed, the direct effect of actual scientific discoveries on religious faith in England in the last century seems to have been less than is sometimes supposed; moral and political objections were more potent, particularly when allied with Positivism.

Scientists had their own Newtonian empirical traditions, and Positivism did not come to play an important rôle until the end of the century, when the writings of Ernst Mach and Pierre Duhem, for whom theories were little more than mnemonic devices, proved congenial to some physicists in their, then current, perplexities. In the nineteenth century, the most important *Logic* before Mill's had been Richard Whately's; he went on to become the Anglican archbishop of Dublin. Again, scientists would have found little philosophy of science in his book. Alexander Bain, the biographer of James Mill, in his *Senses and Intellect* of 1855 and his *Mind and Body* of 1873 explored the border territory between philosophy and psychology; and also wrote a *Logic*, in two parts. Augustus de Morgan, the mathematician, wrote a logic; and George Boole was one of the most important logicians of the century, whose work led to a rebirth of formal logic. In his *Laws of Thought* he developed the Boolean algebra according to which $x+y=xy$; an algebra not of numbers but of classes, and a departure comparable to the discovery of the non-Euclidean geometries. Boolean algebra was used by Benjamin Brodie from 1866 in his attempt to replace chemical equations by an operational calculus, described in his posthumous *Ideal Chemistry*. The economist Stanley Jevons also wrote a work of logic and philosophy of science, *The Principles of Science*, which appeared in 1874; he believed, in opposition to Mill, that scientific discovery was a hypothetico-deductive process, and he urged the provisional nature of all scientific generalizations. His discussions of probability, of measurement, and of the limits of scientific method are particularly noteworthy; and the book is full of examples from the work of great scientists.

The old-fashioned popularization of science in the form of natural theology had three strong Scottish exponents in the first half of the nineteenth century: Thomas Chalmers, Thomas Dick, and Hugh Miller. Chalmers' *Series of Discourses on the Christian Revelation, Viewed in Connexion with the Modern Astronomy*, appeared in 1817 and were based upon addresses given at Glasgow. Chalmers was a minister, and he and Miller played an important part in the schism when the Free Church broke from the Established Church of Scotland. The *Discourses* are lucid and rhetorical; science delivered in the best pulpit style. Chalmers also wrote a Bridgewater Treatise trying to show how nature and society display the goodness of God. Dick was a schoolmaster alarmed at the small part played by religion in Mechanics' Institutes; he published in 1823 *The Christian Philosopher*, in 1837 *Celestial Scenery*, and about 1841, *On the Improvement of Society by the Diffusion of Knowledge*. Hugh Miller is important as a geologist as well as a popularizer. His autobiography, *My Schools and Schoolmasters* is splendid reading, showing how his interest in geology was fed by his first profession as a stonemason. His *Old Red Sandstone*, published in 1841, contained a chapter refuting Lamarck; and when *Vestiges* appeared he issued a counterblast, *Footprints of the Creator*, which appeared in 1849 and was enormously successful. The American edition had a preface by Agassiz. Miller followed this by *Testimony of the Rocks*; he also wrote *Popular Geology*; and *Rambles of a Geologist*, which enjoyed wide sales. Indeed he was the best of the critics of evolutionary theories from what we would call a fundamentalist point of view. His book is refreshing in that he neither argues simply for a First Cause, nor does he distort the Biblical account in order to make it fit. He rejected evolutionary theories because he believed in man's moral freedom and immortality.

After Darwin the old-fashioned natural theology was no longer tenable, as we have already seen. Coleridge, in his *Confessions of an Inquiring Spirit*, pointed to the dangers of bibliolatry; and his disciple F. D. Maurice similarly attacked the systems, notions, and theories by which he believed theologians were becoming hidebound. His *Kingdom of Christ* of 1838 and *Theological Essays* of 1853 show his sympathy with a dynamical science, remote from fact-grubbing or system-building; and his style is not nearly as daunting as it is, and was, often made out to be. Newman's *Essay on the Development of Christian*

Doctrine, of 1845, shows the place that the idea of development was occupying in men's minds in the decades before Darwin; this idea was taken up by J. R. Illingworth in the post-Darwinian series of essays, *Lux Mundi*, edited by Charles Gore. *Essays and Reviews*, of 1860, included Baden Powell on evidences and miracles and C. W. Goodwin on the Mosaic cosmogony; the latter was not particularly powerful or original, but it did constitute an epitaph for fundamentalist geology appearing as it did in an, admittedly explosive, theological work. In the second half of the nineteenth century, theologians taking account of science were operating in an unsympathetic climate of agnosticism and Positivism, and were bound to be on the defensive as they had not been in the early part of the century. Dean Stanley's funeral sermon on Lyell is conciliatory in tone; as are Frederick Temple's famous Bampton Lectures of 1885 on *The Relations between Religion and Science*. An excursion into theology by two eminent scientists, P. G. Tait and Balfour Stewart's *Unseen Universe*, reminds us that not all eminent Victorian scientists were agnostics: indeed the agnostics of the X club were probably no more than a powerful minority. Contemporary accounts of the relations between science and religion may be found in John Hunt's *Religious Thought in England in the Nineteenth Century*, published in 1896; and in R. H. Hutton's *Aspects of Religious and Scientific Thought*, a collection of essays which appeared in 1899. Hunt was a scholar and theologian; Hutton a critic and journalist, an editor of Bagehot and a keen admirer of F. D. Maurice.

All these authors had been concerned more with disseminating what they thought to be a truly scientific attitude rather than with making known the latest discoveries of empirical science. Harris' *Lexicon Technicum* had been concerned with science and technology; and Ephraim Chambers' *Cyclopedia* of 1728 emphasized current science and technology. The *Encyclopédie* of Diderot and d'Alembert, from which *Select Essays* were published in English in 1773, was noteworthy for its descriptions of technical processes; and the *Encyclopædia Britannica* also covered science and technology in some detail. The first edition of 1771 is curious and entertaining; the supplement of 1801 is of interest to the historian of science chiefly for the long article on Boscovich, in which atomism is discussed ably and at length, and for Thomas Thomson's chemical articles. As the century went on, successive editions grew larger and the articles attained a high level of expertise; particularly famous are the eighth and ninth editions. The first volume of the eighth edition, of 1860, was taken up, as its predecessors had been since the supplement to the fourth edition, with 'Preliminary dissertations on the history of the Sciences', in which the history of philosophy was described by Dugald Stewart and James Mackintosh, of theology by Archbishop Whately, and of physical science by John Playfair, John Leslie, and J. D. Forbes. The whole volume is permeated with the idea of progress; and is a very valuable source for intellectual history. The ninth edition contained articles by contributors such as Clerk Maxwell, whose article 'Atom' is a masterpiece of popularization without condescension. This edition was very different from its predecessor in that emphasis in the

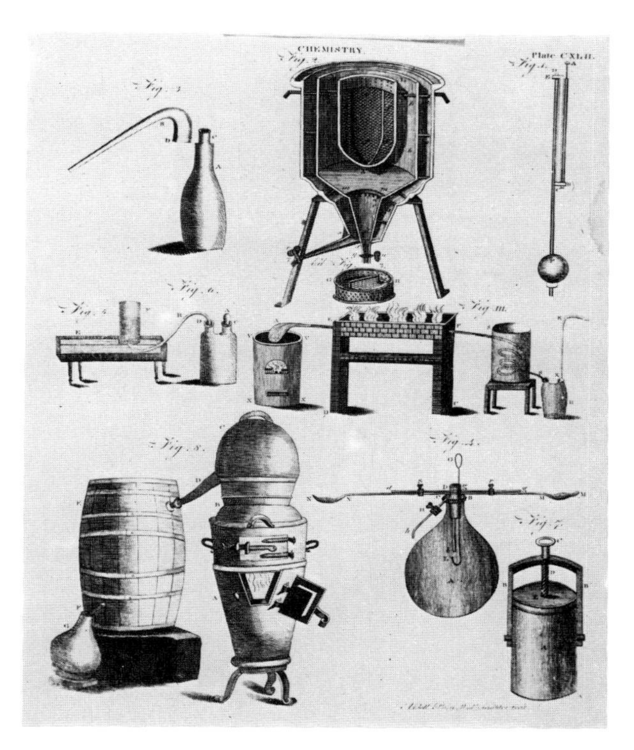

89 Chemical apparatus (*Encyclopædia Britannica*, 4th ed, 1810, vol IV, pl 142)

sciences had changed radically; and the preliminary dissertations had disappeared. The article on encyclopedias in this edition is a useful bibliographical source. In general, the nineteenth-century editions of the *Britannica* give useful accounts of contemporary science; and the plates, maps, and articles indicate the progress of technology, exploration, and science, as the century went on.

Although in the *Britannica* distinguished contributors were given reasonable opportunities to discuss the topics as they saw fit, the effect is bound to be fragmentary. One cannot learn science from even a very good encyclopedia of the ordinary type. The *Encyclopedia Metropolitana* represents an attempt to get around this difficulty. It, or at least the first thirteen volumes, were meant to be read from cover to cover, not consulted for particular points; and anybody who survived that far would indeed have acquired a formidable self-education. The plan was conceived by Coleridge, who wrote a characteristic introduction 'On Method', not to be confused with system; it was heavily edited, and the original version may be found in the 1818 edition of *The Friend*. This introduction was followed by articles of book-length, some of which are monographs of outstanding merit making no concessions whatever to the reader. Seventeenth-century works of natural history, like Topsell's, discussed the animals in alphabetical order; but by the end of that century, this arrangement had been abandoned in favour of a scientific classification in genera and species. The *Metropolitana* was to extend this to the whole of knowledge; the reader would not be at a loss when faced with the mathematical

90 Seals and Walrus (*Encyclopedia Metropolitana*, 1817–45, vol 27, Mammalia, pl 9)

physics of Sir John Herschel, because he would already have learnt mathematics, and similarly everything would be arranged in the proper methodical order. First came the pure sciences; the authors included Whately on logic, George Peacock on arithmetic, George Biddell Airy, the Astronomer Royal, on trigonometry, de Morgan on probability, and F. D. Maurice on morals and metaphysics. Then came treatises on law and theology; and then in volumes three to eight the mixed sciences. Here again the authors were very distinguished; they include, for example, Herschel on astronomy, light and sound; Peter Barlow, on various topics in physics; P. M. Roget on galvanism; Captain Kater on nautical astronomy; Airy on the figure of the Earth, and the tides; Nassau Senior on political economy; Daubeny and John Phillips on geology; and Babbage on manufactures. Then came five volumes of history and biography in a continuous series; and finally a great dictionary in twelve volumes, with useful quotations but some fanciful etymologies, and then three volumes of plates. The work appeared complete in 1845; but parts had appeared slowly over the preceding decades. The whole work is a great monument to the period in which it appeared, and of great value to the historian interested in intellectual history or in the history of science. It is perhaps not surprising that there were no more editions of a work so uncompromising, so relatively unrewarding to the desultory reader, and so relatively difficult to bring up to date.

Rees' *Cyclopedia* is not on the same level as the *Metropolitana* but it is well known, particularly for its articles and plates on technological subjects, and a lot of ground is covered in its thirty-nine volumes. These include five volumes of plates, and an atlas. The plates on agriculture include plans for model farmhouses, barns, and cottages; and illustrate all kinds of implements, hedges, fences, and gates. 'Architecture' has plates illustrating roof-trusses and bridges; and there is a splendid series of astronomical instruments by makers such as Troughton. The processes of cotton and iron manufacture are also depicted in detail; and so are electrical machines, ships, and balloons. The fifth volume of plates is devoted to natural history; the first plate of horses is by George Stubbs, and the other quadrupeds include an early depiction of the 'Kanguroo', with a young one in the pouch. Some of the other exotic mammals, such as the hippopotamus and the giraffe, seem to have been copied from earlier works rather than from nature; but the general effect is very splendid indeed. There is no doubt that encyclopedias lost in attractiveness, though they gained in convenience, when in the middle of the century woodcuts sprinkled about on the appropriate pages replaced the copperplate illustrations.

91 Balloons (Rees' *Cyclopedia*, 1819–20, plates vol IV, pl 2)

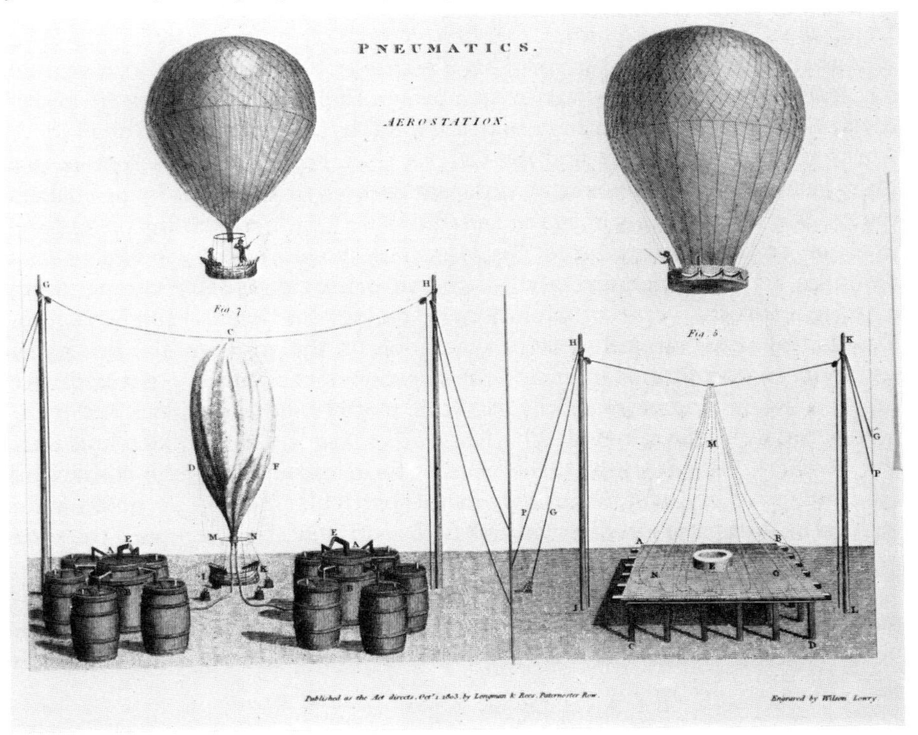

The *Britannica* was, throughout the nineteenth century, published in Edinburgh, and so were two other major encyclopedias: the *Edinburgh*, edited by David Brewster, and *Chambers's*, published by William and Robert Chambers, the latter being the author of *Vestiges*. The *Edinburgh* is valuable to the historian of science; the most notable article is perhaps that by Oersted on electromagnetism, describing his discovery in the third person. *Chambers's* was 'A Dictionary of Universal knowledge for the people'; it appeared in ten volumes, the contributors being mostly Scotsmen, and represented the culmination of the Chambers' publishing of popular but sound works. A shorter work was W. T. Brande's *Dictionary of Science, Literature, and Art*; the articles were very short, but useful information can be gained from them. Getting further from systematic science, we find Sir Richard Phillips' *A Million of Facts . . .* ; a collection of data on science, geography, chronology, and other topics that must have delighted schoolboys of the period; Phillips was an advocate of common-sense philosophy. And towards the end of the century, there is G. W. Ogilvie's *Encyclopaedia of useful information and atlas of the world*; this is full of praise for facts after the manner of Mr Gradgrind but carries such non-factual information as advice on etiquette, and even on how to use a knife and fork. Perhaps the working men to whom it was addressed were interested in social climbing as well as facts. The book also contains athletic records.

From the encyclopedias and other compilations, it would be possible to learn science at almost any level. There were other series of books on sciences, of which the most famous are perhaps Jardine's *Naturalists' Library* and Lardner's *Cabinet Cyclopedia*. The first volume of the latter series was John Herschel's *Preliminary Discourse*: his *Treatise on Astronomy* was also written for it; a work on chemistry was written by Michael Donovan; and a *History of Natural Philosophy*, of a rather uncritical kind, by Baden Powell. But all these works stand on their own, and the *Cabinet Cyclopedia* is not much more of a unity than are series of books by different authors brought out by publishers today. We have already met the indefatigable Lardner writing on railway economy; in 1859 he produced *The Museum of Science and Art* in twelve little volumes, which is an amazingly full and surprisingly readable compendium of science on the verge of evolutionary biology and statistical mechanics. Popular fallacies, the influence of the Moon on the weather, the history of transport and communication, and the question of the plurality of worlds, are topics aired, among more solidly scientific matter, and the science and technology are well intermingled. The *Naturalists' Library* is more of a unit than the *Cabinet Cyclopedia*; the little volumes form a useful set, each containing biographies of eminent naturalists in the particular field. The books were written by authorities in their various fields and they show very well the state of natural history before Darwin.

Popular science was, in the nineteenth century, written partly by professional scientists and partly by popularizers; the former class tend to survive best, especially when the author is a scientist of importance. Thus Faraday's

Chemical History of a Candle, published in 1861, is a masterpiece; it is the text of his Christmas Lectures for children delivered at the Royal Institution, and taken down in shorthand. Another series of Christmas Lectures, *On the Various forces of Matter*, was published in 1860; later editions have the title *On the Various forces of Nature.* The candle served as a means of discussing in simple terms combustion and other chemical processes; and the forces that formed the subject of the other series were those of gravity, cohesion, chemical affinity, heat, magnetism, and elasticity. The sixth and last lecture was on what Faraday still called 'The Correlation of the Physical Forces', or in our terms, conservation of energy. The lectures were accompanied by apt and memorable experiments. Aimed at rather different readership was Clerk Maxwell's *Matter and Motion*, an excellent introduction to Newtonian mechanics and, indeed, to physics in general.

The *Popular Scientific Lectures* of Helmholtz, which appeared in 1881, contains his enunciation of the principle of conservation of energy, and other essays, on vision, on music, and on the axioms of geometry. William Thomson's *Popular Lectures and Addresses* appeared in three volumes, reprinting his British Association addresses and other lectures on topics ranging from the size of atoms to geology. The geology papers include his attack, based upon thermodynamics, on the theories of Lyell and Darwin, on the grounds that the Sun could not have been blazing, nor the Earth habitable, for anything like the periods required by them. T. H. Huxley took up again his rôle of Darwin's bulldog; and his *Lay Sermons* includes 'Geological Reform', his attempt to reply. It also contains his attack on Positivism, a delightful lecture 'On a piece of chalk', and other papers. His *Six Lectures to Working Men*, on scientific method and the idea of 'cause' in biology, were published in 1863; and are reprinted with other essays in his *Darwiniana*, a selection of papers relating to Darwin's theory. The *Collected Essays*, of which *Darwiniana* is the second volume, appeared in nine volumes in 1893–4. They include essays, always lively and controversial, on philosophy of science, science and education, and science and religion.

One of Huxley's most important friends and allies in promoting evolutionary theory, agnosticism, and scientific education was John Tyndall; both he and Huxley have been identified with characters in W. H. Mallock's entertaining *New Republic.* Tyndall's popular writings have a certain windy rhetoric about them – one of his audience of working men remarked after a lecture that they had had plenty of icing and not much cake – but the drift of such productions as his notorious Belfast Address to the British Association, with its statement that 'Science demands the radical extirpation of caprice, and the absolute reliance upon law in nature', was unmistakable. This address, and other popular writings, are to be found in Tyndall's *Fragments of Science*, 1871, which contains papers on a whole range of subjects, including geology, conservation of energy, electricity, spontaneous generation, miracles, and prayer. Among the most interesting are the lecture on scientific materialism of 1868; and on the scientific use of the imagination, of 1870. Tyndall's

Essays on the Floating Matter of the Air, in relation to Putrefaction and Infection has already been mentioned in connection with Pasteur, whose discoveries Tyndall disseminated and extended. His most famous book was probably *Heat a mode of motion*, which was designed to set up the mechanical, or kinetic, theory of heat beside the theory of gravity and the undulatory theory of light as a pillar of modern physics. It is indeed an excellent book on the subject, having the virtues of clear arrangement and interesting presentation. Tyndall was an enthusiastic Alpinist, and his interest in snow and glaciers shows itself in his *Heat* and in his *Forms of Water*, a series of Christmas Lectures delivered in 1871. In describing his climbing, or in discussing circumscribed questions, Tyndall is admirable; in his more general writings, he comes closer perhaps than anybody to the image of the Victorian physicist complacently supposing that he holds the key to the universe.

Sir John Lubbock was another pundit; an agnostic biologist and correspondent of Darwin, who propagated the Darwinian theory. He became an M.P., and was largely responsible for the August Bank Holiday, which used to be called St Lubbock's Day in consequence; later he was created Lord Avebury. His well-known two-volume *Pleasures of Life* of 1887–9, which went through numerous editions, included essays on happiness; on books, with his famous list of one hundred good ones; on science; and on the beauties of nature. His writings on insects, particularly his *Ants, Bees, and Wasps*, which appeared in many editions, and his *Senses, Instincts, and Intelligence of Animals* were well-known and excellent examples of natural history; he also wrote on botany, his *British Wild Flowers, considered in relation to insects* and his *On Seedlings* being noteworthy. In the realm of anthropology and prehistory, he wrote *Prehistoric Times*, and *The Origin of Civilisation and the primitive condition of man*, which were important, though readable and popular, pioneering works.

While many Victorian scientists thus wrote popular works, and in many cases their popular or general writings have lived longer than their more-specialized studies, there were other professional popularizers who made no original contributions to knowledge or understanding of the world, but made known to laymen, and to scientists in other fields, the discoveries and opinions which men of science put forward in an obscure form. The most distinguished of them was Mary Somerville, who knew many eminent scientists, and received numerous honours from scientific societies, including election with Caroline Herschel as an Honorary Member of the Royal Astronomical Society. It was in a review of one of her books, *On the Connexion of the Physical Sciences*, in the *Quarterly Review* that the word 'scientist' first appeared. Her other writings included *The Mechanism of the Heavens, Physical Geography*, and *Molecular and Microscopic Science*. The books are very sound and informative; not easy or particularly attractive in style, but clear. She, like other observers of and participants in nineteenth-century science, was struck by the way disciplines converged; and she tried to take a broad synoptic view of the whole field. Her books provide, now as then, a very good introduction to the state of the sciences in the first half of the nineteenth century, presented in non-mathematical form; she can perhaps be compared to some of the best of the eighteenth-century Newtonians.

In the field of chemistry in the early nineteenth century there were various textbooks, of which Thomas Thomson's were outstanding. They were used by medical students, who had at that time to learn a lot of chemistry, and presumably by those training to become pharmacists, analysts, and industrial chemists. Among books which remind us that revolutions in the sciences are not as rapid and complete as they sometimes seem is Priestley's *Doctrine of Phlogiston Established*, published at Northumberland, Pennsylvania, in 1800. Priestley praised Davy's Essays on heat and light, concluding that his researches, with Rumford's, proved 'that *heat* is not produced by any proper *substance*, such as is now called *calorique*, and which is so essential to the new theory'. Davy presented a copy to the Royal Institution, and toyed himself with 'phlogistic' explanations in his work on chlorine; but Lavoisier's theory proved able to stand up when its seeming props were removed. This was not, however, obvious to one W. P. Stevenson whose *Non-Decomposition of Water* was published in 1848. It was a part of the phlogistic doctrine that hydrogen was a compound containing water, as metals were compounds containing their calces; unless, as some held, hydrogen was phlogiston. To declare that water was not decomposed into oxygen and hydrogen was an article of the phlogistic faith in the first years of the nineteenth century; but it must have seemed odd indeed in the mid-century. Indeed, as happened with Gothic architecture, phlogistic survival and revival must almost have overlapped; for in 1871 the distinguished chemist William Odling lectured 'On the Revived Theory of Phlogiston' which he compared to the concept of chemical potential energy. This may be found in his *Abstract of Lectures delivered at the Royal Institution* of 1874.

Popular works sometimes lagged behind the pioneers in other fields; as we find in the *Description of Three Hundred Animals* originally published in 1730 the information in which derived from Topsell's *History of Four-footed Beasts* of a century earlier. The animals, which include fish and insects – with some nice dragons and a unicorn – are illustrated; and the mammals look at the reader with such sad and human faces as to tempt to vegetarianism. In natural philosophy and chemistry, however, even books written for children attempted to keep up with the times. Two famous works on chemistry published in the early nineteenth century were in the form of dialogues; S. Parkes' *Chemical Catechism*, and Mrs Jane Marcet's *Conversations on Chemistry*, the first editions of both of which came out in 1806. Many of the greatest scientific books, including Galileo's *Two Great Systems of the World*, Boyle's *Sceptical Chymist*, and Réaumur's *Bees*, had been written as dialogues: but Parkes' and Mrs Marcet's books were dialogues between teacher and pupil. Both books appeared in a great number of editions, and must have met a crying need; they are very sound little works. It was 'Mrs Marcet' that introduced Faraday to chemistry when he was a bookbinder's apprentice. The success of her chemical conversations was such that she wrote similar

92 Gas works (Accum, F., *Description of the Process of manufacturing coal gas*, 1820, frontispiece)

works on natural philosophy, vegetable physiology, and the evidences of Christianity, which are also excellent introductions and delightful period-pieces without perhaps quite the charm of the chemical one.

William Henry's little *Epitome of Chemistry* of 1801 is less entertaining than Mrs Marcet; it was based upon lectures, and is interesting for its affinity diagrams, after Bergman. Henry was Dalton's greatest friend and most important scientific associate. His *Elements of Experimental Chemistry* went through numerous editions; a French translation of the sixth edition was published in 1812, and helped to make Dalton's theory known in France. In 1801 J. Murray, who was to be a spokesman for those who disagreed with Davy's views on chlorine, published at Edinburgh his *Elements of Chemistry*, containing strong arguments for heat being a caloric fluid; and describing, like most textbooks of the period, a Newtonian world of corpuscles and attractive forces of cohesion and affinity. Frederick Accum, the pioneer of coal-gas lighting, published in 1803 his *System of Theoretical and Practical Chemistry*, with quite a wide discussion of theories of heat and combustion. Rather earlier, J. F. Jaquin's *Elements of Chemistry* had been translated from the German; it contains an exposition of Lavoisier's doctrine of elements, which the author contrasts favourably with the imaginary indivisible atoms of the philosophers. In 1817 there appeared Joseph Luckock's *Essays on the theory of the tides, the figure of the earth, the atomical philosophy, and the moon's orbit*. He believed the tides to be produced by the alternate accelerated and retarded movements of the heavenly bodies, and attacked Daltonians on grounds both good and bad. The work is quite amusingly mad, and contrasts with the rather sober and sensible mainstream of popular books and text-books.

W. Weldon's *Popular Explanation of the Elements and General Laws of*

Chemistry is undated; but the Royal Institution's copy bears the date 1824. It is noteworthy for its discussion of the 'etherial chemical elements', heat, light, electricity, and magnetism, which Weldon thought it best, in the current state of knowledge, to treat as distinct elements. He believed that it was not necessary for every element to possess mass; that there might be a gradation of properties from ponderable matter, through the ethereal elements, to substances devoid of gravity and impenetrability. Of later popular works, those of Thomas Griffiths, Professor of Chemistry at St Bartholomew's Hospital, London, deserve mention. In 1842 he published *The Chemistry of the Four Ancient Elements*; and in 1846 *The Chemistry of the Four Seasons*. He was interested not in theory but in discoveries; and whereas in the latter work Davy and Faraday receive enthusiastic praise, Dalton is mentioned only as an investigator of the evaporation of water. The *Four Ancient Elements* contrasts the old and new ideas of 'element', which now meant only 'a confession of the limits of . . . analytical skill'. An earlier work of an elementary kind, hard to classify, was Jeremiah Joyce's *Scientific Dialogues* of 1807, covering most of physics. The author acknowledged his indebtedness to the ideas of Richard Lovell Edgeworth, the educationalist and inventor who belonged to the Lunar Society; but his book is not of the same standard as those of Parkes or Mrs Marcet. Nevertheless, John Stuart Mill wrote that he was never 'so wrapt up in any book', and was 'rather recalcitrant to [his] father's criticisms of the bad reasoning respecting the first principles of physics, which abound in the early part of that work'. Joyce produced other dialogues, and versions of the doctrines of Paley and Adam Smith. Another work intended as a fairly popular exposition of physics was the Jesuit Joseph Bayma's *Molecular Mechanics* of 1866, which was praised by A. R. Wallace; it is interesting chiefly for its ideas on atomism and on the relation of atomic theory to crystallography.

Historians of science in the nineteenth century include, beside Whewell, the extraordinary J. O. Halliwell, whose Historical Society of Science published in 1841 his *Collection of Letters* concerning science of the Elizabethan and early seventeenth-century period. The more important and scholarly works of Rigaud and of Edelston have already been mentioned. The Historical Society of Science also published in 1841 a volume edited by Thomas Wright, printing an Anglo-Saxon manual of astronomy, and a versified bestiary by Philip de Thaun. Of greater interest to the student of science in pre-Norman England is Thomas Cockayne's *Leechdoms, Wortcunning, and Starcraft*, of 1864–6. This prints the Early English texts, with a translation on facing pages. Sylvanus Thompson was chiefly responsible for the publication in 1900 of a translation of William Gilbert's *de Magnete*, handsomely produced by the Chiswick Press to match the original; a volume of *Notes* appeared in 1901. For later science, Bernard Becker's *Scientific London*, of 1874, is an indispensable guide to the scientific institutions and societies in nineteenth-century London; and Bence Jones' *The Royal Institution* remains a standard work.

Robert Watson's *History of the Literary and Philosophical Society of Newcastle upon Tyne* is an excellent guide to science in a provincial city. It was a resident of Newcastle, J. T. Merz, who published at the turn of the century what is still the best general account of the science of this period; his *History of European Thought in the Nineteenth Century*, the first two volumes of which are devoted to science, is a masterpiece. The third and fourth volumes, on the history of philosophy, can also be studied with advantage by the historian of science, who can only marvel at the industry and judgement of the author.

Both Darwin and Wallace had been influenced by Malthus, and Darwinian theory was soon taken up by political thinkers. Walker Bagehot's *Physics and Politics* of 1872 used Darwinian ideas to explore the reasons for progress; Herbert Spencer, who had already written the evolutionary *Principles of Biology*, advocated in his *The Man versus the State* of 1884 a *laissez-faire* and competitive world; Edward Clodd, who hoped that might would in the end give way to right, nevertheless saw man's state, in his *Story of Creation*, as sharp struggle and conflict; and Sir Henry Maine applied evolutionary and comparative ideas to law, and thereby exerted an influence back upon science by modifying the idea that laws were immutable or absolute. D. G. Ritchie, in his *Darwinism and Politics*, pointed out the dangers of social Darwinism, and indeed of applying biological theories uncritically in the fields of ethics and politics. He was alarmed particularly at the writings of D. F. Strauss, author of the notorious *Life of Jesus*, whose *The Old Faith and The New* seemed to give the authority of Darwinian theory to reactionary Prussian militarism. Benjamin Kidd, in his *Social Evolution* of 1895, showed the way things had changed when he remarked that fanatical opposition to religion seemed to have disappeared, and suggested that the question that the Darwinian must ask is whether religion has a function in the evolution of society. Darwinian science helped to bring about a state of mind akin to that of the Jesuits derided by F. D. Maurice for teaching the Copernican theory which they formally abjured; truth and falsehood had disappeared, and flux and convention were left behind.

10 THE DIFFUSION OF SCIENCE IN THE NINETEENTH CENTURY

Accum, F., *A System of Theoretical and Practical Chemistry*, 2 vols, 1803.
Description of the Process of manufacturing coal-gas . . ., 1820.
Anon., *Craniology Burlesqued, in three serio-comic lectures . . .*, 2nd ed., 1818; 1st ed. not seen.
Anon., *A Description of Three Hundred Animals*, 1730.
Bagehot, W., *Physics and Politics, or thoughts on the application of the principles of 'natural selection' and 'inheritance' to political society*, 1872.
Bain, A., *The Senses and the Intellect*, 1855.
Mind and Body: the theories of their relation, 1873.
Logic, 2 vols., 1873–9.

Barlow, P., *A New Mathematical and Philosophical Dictionary . . .*, 1813.
Bayma, J., S.J., *The Elements of Molecular Mechanics*, 1866.
Becker, B. H., *Scientific London*, 1874.
Bence Jones, H., *The Royal Institution, its Founder and its First Professors*, 1871.
Benn, A. W., *The History of English Rationalism in the nineteenth century*, 2 vols., 1906.
Boole, G., *An Investigation of the Laws of Thought . . .*, 1854.
Brande, W. T., *A Dictionary of Science, Literature and Art*, 1839.
Brewster, D., *A Treatise on New Philosophical Instruments*, Edinburgh, 1813.
More Worlds than one : The creed of the philosopher, and the hope of the Christian, 1854.
Capen, N., *Reminiscences of Dr Spurzheim and George Combe, and a review of the science of phrenology . . .*, New York, 1881.
Chalmers, T., *A Series of Discourses on the Christian Revelation, viewed in connection with Modern Astronomy*, Glasgow, 1817.
Chambers, E., *Cyclopedia, or an Universal Dictionary of Art and Science*, 2 vols., 1728.
Chambers's Encyclopedia, 10 vols., London and Edinburgh, 1860–8.
Clodd, E., *The Story of Creation*, 1888.
Cockayne, T. O. (ed.), *Leechdoms, wortcunning and starcraft of early England . . .*, 3 vols., 1864–6
Coleridge, S. T., *The Friend*, 3 vols, 1818; ed. B. E. Rooke, 2 vols., 1969.
Aids to Reflection, 1825.
Confessions of an Inquiring Spirit, 1840.
Hints towards the formation of a more comprehensive theory of life, ed. S. B. Watson, 1848.
Combe, G., *Essays on Phrenology*, 1819.
The Constitution of Man, Edinburgh, 1828.
Comte, A., *The Positive Philosophy*, tr. H. Martineau, 2 vols., 1853.
De Morgan, A., *Arithmetical Books, from the invention of printing to the present time*, 1847.
A Budget of Paradoxes . . ., 1872.
Newton : his friend and his niece, 1885.
Dick, T., *The Christian Philosopher . . .*, 1823.
Celestial Scenery; or, The Wonders of the Heavens Displayed, 1837.
On the Improvement of Society by the Diffusion of Knowledge, Glasgow, n.d.
Diderot, D., *et al.*, *Select Essays from the Encyclopėdy*, 1772.
Draper, J. W., *History of the Conflict between Religion and Science*, New York, 1875.
Edelston, J., *Correspondence of Sir Isaac Newton and Professor Cotes . . .*, 1850.
Edinburgh Encyclopedia, 18 vols., Edinburgh, 1808–30.
Encyclopædia Britannica, 3 vols., Edinburgh, 1771; 4th ed., 20 vols., 1810; 8th ed., 21 vols., 1860; 9th ed., 25 vols., 1875; 10th ed., 9th + 10 further vols. + index, 1902–3. *Supplement* to 2nd ed., 2 vols., 1801; and to 4th, 5th, and 6th eds., 6 vols., 1824; contained 'Dissertations'.
Encyclopedia Metropolitana, 29 vols., 1817–45.
Faraday, M., *A Course of six lectures on the non-metallic elements*, ed. J. Scoffern, 1853.
On the various Forces of Matter, 1860.
The Chemical History of a Candle, 1861.
Flint, R., *A History of Classifications of the Sciences*, 1904.
Gall, F. J., *On the functions of the brain and of each of its parts . . .*, ed. N. Capen, tr. W. Lewis, 6 vols., Boston, Mass., 1835.
Gilbert, W., *On the Magnet . . .* (tr. S. P. Thompson), 1900.
Notes on the de Magnete . . . (by S. P. Thompson), 1901.
Goldsmith, O., *An History of the Earth and Animated Nature*, 8 vols., 1774.
Survey of Experimental Philosophy, 2 vols., 1776.
Gore, C. (ed.), *Lux Mundi*, 1889.
Griffiths, T., *The Chemistry of the Four Ancient Elements*, 1842.

The Chemistry of the Four Seasons, 1846.
Halliwell, J. O., *A Collection of Letters illustrative of the progress of science in England . . .*, 1841.
Hamilton, W., *Lectures on Metaphysics and Logic*, ed. H. L. Mansel and J. Veitch, 4 vols., Edinburgh, 1859–60.
Helmholtz, H. von, *Popular Lectures on Scientific Subjects*, tr. E. Atkinson, *et al.*, 2 series, 1873–81.
On the Sensations of Tone . . ., tr. A. J. Ellis, 1875.
Treatise on Physiological Optics, ed. J. P. C. Southall, 3 vols., Rochester, N.Y., 1924–5.
Henry, W., *An Epitome of Chemistry*, 1801.
Elements of Experimental Chemistry, 2 vols., 6th ed., 1810; early eds. not seen.
Hunt, J., *Religious Thought in England from the Reformation to the end of the last century*, 3 vols., 1870–3.
Religious Thought in England in the Nineteenth Century, 1896.
Hutton, R. H., *Aspects of Religious and Scientific Thought*, 1899.
Huxley, T. H., *Collected Essays*, 9 vols., 1893–4.
Jacquin, J. F. von, *Elements of Chemistry*, 3rd ed., 1803; earlier eds. not seen.
Jevons, W. S., *The Principles of Science*, 1874.
Joyce, J., *Scientific Dialogues: intended for the instruction and entertainment of young people . . .*, n.d.; bibliography involved.
Kidd, B., *Social Evolution*, 1895.
Knight, C. (ed.), *The Penny Cyclopedia*, 29 vols., 1833–46.
The English Cyclopedia, 23 vols., 1854–62.
Lange, F. A. *History of Materialism*, tr. E. C. Thomas, 3 vols., 1877.
Lardner, D. (ed.), *The Cabinet Cyclopedia*, 133 vols., 1830–49.
The Museum of Science and Art, 1859.
Lavater, J. C., *Essays on Physiognomy; designed to promote the knowledge and the love of mankind*, tr. H. Hunter, 3 vols. in 5, 1789–98; tr. T. Holcroft, 2nd ed., 3 vols., 1804; 10th ed., 8°, 1858.
Lecky, W. E. H., *History of the rise and influence of the spirit of rationalism in Europe*, 2 vols., 1865.
Lewes, G. H., *The biographical history of philosophy*, 4 vols., 1845–6; 2nd ed., revised, 1857.
Comte's Philosophy of the Sciences, 1847.
Physiology of Common Life, 2 vols., 1859–60.
Problems of Life and Mind, 5 vols., 1874–9.
London Encyclopedia, The, 29 vols., 1829.
Lubbock, J., *Prehistoric Times*, 1865.
The Origin of Civilisation . . ., 1870.
On British Wild Flowers, considered in relation to insects, 1873.
Ants, Bees, and Wasps, 1882.
Pleasures of Life, 2 vols., 1887–9.
Senses, Instincts, and Intelligence of Animals, 1888.
A Contribution to our knowledge of seedlings, 2 vols., 1892.
Luckock, J., *Essays on the theory of the tides, the figure of the earth, the atomical philosophy, and the moon's orbit*, 1817.
Maine, H. S., *Ancient Law: Its connection with the early history of society, and its relation to modern ideas*, 1861.
(Mallock, W. H.), *The New Republic*, 2 vols., 1877.
Marcet, Mrs, *Conversations on Chemistry . . .*, 2 vols., 1806.
Conversations on Natural Philosophy . . ., 1819.
Conversations on the Evidences of Christianity . . ., 1826.
Conversations on Vegetable Physiology . . ., 2 vols., 1829.

Martineau, H., and Atkinson, H. G., *Letters on the laws of man's nature and development*, 1851.
Maudsley, H., *Body and Will . . .*, 1883.
Maurice, F. D., *The Kingdom of Christ*, 3 vols., 1838.
Theological Essays, 1853.
Maxwell, J. C., *Matter and Motion*, 1873.
Merz, J. T., *A History of European Thought in the Nineteenth Century*, 4 vols., 1896–1914.
Mill, J. S., *A System of Logic*, 1843.
Auguste Comte and Positivism, 1865.
Autobiography, ed. H. Taylor, 1873.
Nature, the Utility of Religion, and Theism. Three Essays on Religion, ed. H. Taylor, 1874.
Miller, H., *The Old Red Sandstone . . .*, Edinburgh, 1841.
Footprints of the Creator . . ., Edinburgh, 1849.
The Testimony of the Rocks . . ., Edinburgh, 1857.
The Cruise of the Betsy; or a Summer Ramble among . . . the Hebrides . . ., Edinburgh, 1858.
Sketches of Popular Geology . . ., Edinburgh, 1859.
Murray, J., *Elements of Chemistry*, 2 vols., Edinburgh, 1801.
Muspratt, S. (ed.), *Chemistry, Theoretical, Practical, and Analytical . . .*, 2 vols., 1860.
Newman, J. H., *Essay on the Development of Christian Doctrine*, 1845.
Odling, W., *Abstract of Lectures delivered at the Royal Institution*, 1874.
Ogilvie, G. W., *Encyclopedia of useful information . . .*, n.d.
Parkes, S., *Chemical Catechism*, 1806.
Phillips, R., *Protest against the prevailing principles of Natural Philosophy, with the development of a common-sense system*, 1830.
A Million of Facts . . . (1832).
Powell, Baden, *History of Natural Philosophy*, 1837.
Priestley, J., *The Doctrine of Phlogiston Established*, Northumberland, Penna., 1800.
Rees, A., *The Cyclopedia; or, Universal Dictionary of Arts, Sciences, and Literature*, 39 vols., 1819–20.
Reid, T., *Essays on the Intellectual Powers of Man*, Edinburgh, 1785.
Essays on the Active Powers of Man, Edinburgh, 1788.
Rigaud, S., *Correspondence of Scientific Men of the Seventeenth Century*, 2 vols., Oxford, 1841.
Ritchie, D. G., *Darwinism and Politics*, 1889.
Robison, J., *Proofs of a Conspiracy against all the Religions and Governments of Europe*, Edinburgh, 1797.
Elements of Mechanical Philosophy, Edinburgh, 1804.
A System of Mechanical Philosophy, 4 vols., Edinburgh, 1822.
Somerville, M., *The Mechanism of the Heavens*, 1831.
On the Connexion of the Physical Sciences, 1834.
Physical Geography, 2 vols., 1848.
Molecular and Microscopic Science, 2 vols., 1869.
Spencer, H., *Essays: Scientific, Political, and Speculative*, 3 vols., 1858–74.
The Principles of Biology, 2 vols., 1864–7.
The Man versus the State, 1884.
Spurzheim, J. C., *Outlines of the physiognomical system of Drs Gall and Spurzheim . . .*, 2nd ed. 1815; 1st ed. not seen.
Stephen, L., *Essays on Freethinking and Plainspeaking*, 1873.
Stevenson, W. P., *The non-decomposition of water distinctly proved . . .*, 1848.
Stewart, D., *Elements of the Philosophy of the Human Mind*, 1792–1827.
Philosophical Essays, Edinburgh, 1810.

Strauss, D. F., *The Old Faith and the New . . .*, tr. M. Blind, 2nd ed., 1873; 1st ed. not seen.
(Tait, P. G., and Stewart, B.), *The Unseen Universe, or physical speculations on a future state*, 1875.
Paradoxical Philosophy . . ., 1878.
Temple, F., *The relations between religion and science . . .*, 1884.
Thomson, W. (Lord Kelvin), *Popular Lectures and Addresses*, 3 vols., 1889–94.
Tyndall, J., *The Glaciers of the Alps*, 1860.
Heat considered as a Mode of Motion, 1863.
Fragments of Science, 1871.
The Forms of Water, 1872.
Watson, R. S., *The History of the Literary and Philosophical Society of Newcastle upon Tyne*, 1897.
Weldon, W., *A Popular Explanation of the Elements and General Laws of Chemistry* (1824).
Wells, W. C., *An Essay upon Single Vision with Two Eyes . . .*, 1792.
An Essay on Dew . . ., 1814. A new ed. of these, *Two Essays : one upon Single Vision . . . the other on Dew . . .*, 1818, contained an appendix 'anticipating' Darwin.
Whately, R., *Essays on some of the Peculiarities of the Christian Religion*, 1825.
Elements of Logic, 1826.
Whewell, W., *The Mechanical Euclid*, Cambridge, 1837.
History of the Inductive Sciences, 3 vols., 1837; 3rd ed., 3 vols., 1857.
The Philosophy of the Inductive Sciences, 2 vols., 1840; 2nd ed., 1847; this was then reformed into: *History of Scientific Ideas . . .*, 2 vols., 1858; *Novum Organum Renovatum*, 1858; and *On the Philosophy of Discovery*, 1860.
(Whewell, W.), *A Dialogue on the Plurality of Worlds*, 1854.
White, A. D., *A History of the Warfare of Science with Theology in Christendom*, 2 vols., 1896.
Wright, T., *Popular Treatises on Science written during the middle ages . . .*, 1841.
Wylde, J. (ed.), *The Circle of the Sciences*, 2 vols. (1862–7).

Recent Publications

Brock, W. H. (ed.), *The Atomic Debates*, Leicester, 1966.
Clark, G. K., *The Making of Victorian England*, 1962.
Coleridge, S. T., *The Notebooks*, ed. K. Coburn, 1957–.
Collison, R. L. E., *Encyclopedias : their history throughout the ages, a Bibliographical Guide*, New York, 1964.
Haight, G. S., *George Eliot*, Oxford, 1968.
Houghton, W. E., *The Victorian Frame of Mind*, New Haven, 1957.
Muir, P., *English Children's Books, 1600–1900*, 1954.
Munby, A. N. L., *The History and Bibliography of Science in England : the first phase, 1833–1845*, Berkeley and Los Angeles, 1968.
Reingold, N., *Science in Nineteenth-Century America*, 1966.
Russell, B. and P. (ed.), *The Amberley Papers*, 2 vols., 1937.
Tyndall, J.: A. S. Eve and C. H. Creasey, *The Life and Work of John Tyndall*, 1945.
Vidler, A. R., *F. D. Maurice and Company*, 1966.
Walsh, S. P., *Anglo-American General Encyclopedias, 1703–1967*, New York, 1968.
Whittaker, E., *A History of the Theories of Aether and Electricity*, 2 vols., 1951.
Young, G. M., *Victorian England. Portrait of an Age*, 1936.
Young, R. M., *Mind, Brain, and Adaptation in the Nineteenth Century*, Oxford, 1970.

11 *Scientific Publications in the Nineteenth Century*

The nineteenth century was a period in which the journal came to play an increasingly important rôle; the number of journals increased apace, and so did their degree of specialization. The *opus*, or the most significant part of it, of such scientists as Joule and Faraday appeared in the form of papers in journals and was never reorganized into monographs. It was, however, published as volumes of collected papers: and for a great number of nineteenth-century scientists we have such volumes, which save the historian much trouble in seeking out the journals, though one suspects that they originally appeared as a monument when current interest had long passed on to other topics. Apart from the journals, there were official or semi-official publications under the auspices of governments: under this heading would come material relating to the Great Exhibition of 1851, such as the catalogues and reports of juries, and to other international exhibitions; and the reports drawn up after expeditions, such as the *Beagle* material edited by Darwin, the immense and informative reports of the *Challenger* oceanographical expedition, and the Pacific Railroad Reports, with splendid illustrations, from America. Under this heading also come reports on scientific education, and on such topics as factory conditions, the state of industry, and the reform of universities. And finally there are biographies; it has been remarked that Victorian biographies of scientists did not make much contribution to biography as an art form. This is true, particularly as the century wore on and the tradition that it was a part of the duties of filial piety to compile *Life and Letters* grew invincible. Nevertheless, such biographies, with their ill-digested and often emasculated chunks of material from diaries and letters, can be very valuable guides and sources, in which the human figure of the scientist does emerge despite the heroic and inductivist mould into which the attempt is usually made to force him. There were also popular biographies in which devotion to fact came far behind the desire to edify or to instruct; some of these are a good joke.

In the second half of the eighteenth century, the egregious Benjamin Martin edited the useful *General Magazine of Arts and Sciences*, a vehicle of popular science, but closer to an encyclopedia than a journal. We have elsewhere mentioned the more important *Nicholson's Journal*, *The Philosophical Magazine*, and *Annals of Philosophy*, treating them as chemical journals; and indeed papers on chemistry and galvanism filled many of their pages in the first decades of the century. But all these journals also carried articles on physics and geology; and reviews and reports of meetings of scientific societies both in England and abroad. Translations or abstracts of papers in foreign languages often appeared in these journals; and so did reprints of papers from the more august pages of the *Philosophical Transactions*. In these journals, and later in such publications as *Chemical News*, a long paper would appear in somewhat arbitrary sections: our ancestors read their science as they read their novels, in parts. While the perusal of these three journals gives the reader an unequalled view of early nineteenth-century science, and reveals papers of an enormous variety of standards, the prestigious vehicle for publication at the time was the *Philosophical Transactions*. In the early years of the century, it contained Sir Everard Home's plagiarisms or continuations of the physiological researches of John Hunter; Young's papers on vision and on the wave theory of light; papers by Wollaston on electricity, energy, and crystallography; Davy's electrochemical researches; and other works, including the geophysical measurements of Edward Sabine and Henry Kater.

In 1809 there was published a valuable abridgement, in nineteen volumes, the last of which contains plates and index, of the *Philosophical Transactions* down to 1800. This work also contains brief biographical notices of eminent Fellows; book reviews have been amended to fit the perspective of the early nineteenth century; papers originally in Latin appear in translation; and important papers are reprinted in full. The title, and a terse editorial comment, is all that is left of some papers; and in the abridged articles it is not always as easy as one might wish to distinguish original and editorial matter, but this does not usually present serious problems. Of the Royal Society's other publications, the *Proceedings* began in the nineteenth century as a collection of abstracts of papers published in the *Philosophical Transactions*, and of those read to the Society but not published. At first each volume covered a number of years; but gradually it evolved into a journal in its own right, in which by the end of the century useful obituary notices of Fellows were *inter alia* printed. Ultimately, it and the *Philosophical Transactions* split into two parts, (*A*) concerned with physical sciences and (*B*) with biology, as the sciences became increasingly specialized.

Other general scientific journals included the *Edinburgh Philosophical Journal*, later the *Edinburgh New Philosophical Journal*, edited by David Brewster and often referred to as *Brewster's Journal*. There were also publications of the Royal Institution, its *Proceedings* and *Quarterly Journal*, which contained articles by those associated with the Institution, and also included book reviews. In America, this rôle was played by Benjamin Silliman's *American Journal of Science and Arts*, which gradually came to be devoted chiefly to biology and geology, but for the bulk of the nineteenth century was the only American scientific journal of any repute. It included papers by Europeans such as Berzelius, and reports and reviews, as well as articles by Americans. After 1858 the Smithsonian Institution in Washington D.C. included in its *Annual Reports* obituaries, often reprinted or translated from foreign journals, of eminent men of science from all over the world; the *Annual Reports* themselves are fascinating reading for anyone interested in the rise of science in a country which though technologically extremely advanced was scientifically still rather backward. In Britain, a very important early Victorian general journal was Richard Taylor's *Scientific Memoirs*, which is given over entirely to translations of papers which had appeared in foreign publications. There were seven volumes of the journal in all, the last two being a New Series in which the volume on natural history was edited by A. Henfrey and T. H. Huxley, and that on natural philosophy by J. Tyndall and W. Francis. The original series had not been divided; and had included numerous papers on electricity and magnetism, and on chemistry, by authors including Ohm, Gauss, Bunsen, and Berzelius; and a paper on spontaneous generation by C. G. Ehrenberg.

The *Philosophical Magazine* came to be increasingly a vehicle for important and original papers, chiefly on physics; and the need for a general journal in English was met by Norman Lockyer's journal *Nature*, which began in 1869. Its small type and double columns permitted the inclusion of a great deal of material and its editorials, articles, and reviews give an unequalled impression of the science of the period. In *Nature* were printed various lectures and addresses by eminent scientists, including the Presidential Addresses at the British Association meetings. These also appeared, in time, in the British Association *Reports*; the numbering of these is tiresome, because the first two meetings were reported in volume one, while subsequent meetings had a volume each, so the meetings and volume numbers are out of step. Victorians tended to refer to the meetings by place rather than date or numbers. There are also in each volume two parts, in both of which the numbering starts at page 1 and goes on; so references to this journal are not always unambiguous. The first part contains the Presidential Address and the reports of committees; and the second the abstracts of the papers read, some of which are so condensed as to be of little value. Later the pagination became continuous; and the addresses of the presidents of the various sections were also printed. They provided eminent men of science, or engineers, with an opportunity for taking stock and are, for the most part, well worth reading. Valuable papers read at the meetings were as a rule published in full in another journal, such as the *Philosophical Magazine*.

Other general journals which should be noticed are the *Popular Science Monthly* with both descriptive and historical articles; and more important, the *Transactions of the Royal Society of Arts*, covering all branches of science from a technological point of view. The *Transactions of the Royal Society of Edinburgh* contained many original papers of importance, as did the *Transactions of the Royal Irish Academy*. The *Memoirs of the Literary and*

Philosophical Society of Manchester, which after three series became in 1888 the *Memoirs and Proceedings*, uniting with the *Proceedings* which had begun in 1857, is very important for its articles by and about Dalton; and papers by J. P. Joule and other eminent men of science appeared in its pages. General journals intended for artisans were *The Mechanics Magazine*, *Chambers's Edinburgh Journal*, and the *Penny Magazine of the Society for the Diffusion of Useful Knowledge*. Charles Knight's *Penny Cyclopedia* appeared in parts but was not a journal; all these are valuable sources of popular science. *Retrospect of Discoveries* was a useful abstracting journal, with editorial comments.

But as the century wore on, an increasing number of important papers began to appear in the specialized journals often associated with the new kind of scientific society which took only one science for its province. The Linnean Society and the Geological Society, and in Edinburgh the Wernerian Natural History Society, published journals covering biology and geology, and so did the Zoological Society. The Royal Astronomical Society's *Memoirs* began in 1822; and its *Monthly Notices* in 1827. These latter were at first off-prints from the *Philosophical Magazine* distributed to members, containing abstracts of papers read to the society, of which the most important appeared in full in the *Memoirs*. The Chemical Society was founded later than these, and in reaction to speculation it did not publish in its journal papers of a purely theoretical nature. The bulk of the papers in the nineteenth century are therefore purely experimental; but the Presidential Addresses, and at intervals from 1867 the Faraday Lectures, are concerned with various aspects of chemical theory, education and explanation. In 1867 another chemical journal, *The Laboratory*, appeared: but no doubt because it paid its contributors, the first volume concluded the run, although valuable papers and reviews had been published in it. In the 1850s there had been journals called *The Chemist* and *The Chemical Gazette*; and in 1860 William Crookes began *Chemical News*. This appeared weekly, in the quarto format *Nature* was to adopt, with small type and double columns. It printed addresses and papers read to the British Association, important theoretical papers, including those of J. A. R. Newlands and D. I. Mendeleev on the Periodic Table of the chemical elements, and papers by Crookes himself; and its correspondence and editorials show in a fascinating way the emergence of chemistry as a profession during the second half of the century. Other more strictly professional journals include those published by the Physical Society and the Institute of Electrical Engineers. The various scientific journals had their indexes, but the Royal Society's *Catalogue of Scientific Papers* is invaluable to the historian, or to anybody who wishes to follow up a topic or an author, including as it does both British and foreign journals and scientists.

While the historian of science must consider these journals, he would do well also to look at the general publications of the era before it could be claimed that there were two cultures. The *Spectator* of Addison and Steele had contained articles on science and natural theology; the *Gentleman's Magazine* gives sometimes the only obituaries we have of men of science, and its pages were occasionally occupied by scientific questions; and the mathematical problems set in the *Ladies' Diary* were famous, and set Dalton on the road towards a scientific career. The *Monthly Review* discussed scientific books and papers; and so did the formidable *Edinburgh Review*, which took particular notice of Davy's researches; and the *Quarterly Review*, which even reviewed *inter alia* chemistry textbooks. In 1859 it carried an excellent review of geological science, on the eve of the Darwinian explosion, and an almost hysterical attack on Baden Powell for his views on miracles; and in the next year it had Bishop Wilberforce's assault upon *The Origin of Species*.

The *British Critic* in the first quarter of the century, and the *Westminster Review* in the second, also had discussions of science. The *British Critic* then became increasingly theological; the *Westminster* had been set up as a Radical and Utilitarian organ with James and John Stuart Mill playing an important part in it; it was then taken over by John Chapman, the free-thinking publisher, who installed George Eliot as editor of the new series that began in January 1852. Under her editorship, articles on atomism, on geology and on biology appeared as before. Another review that should be mentioned is the *North British*, which for example in 1868 used the occasion of a new edition of Lucretius to discuss the atomic theory in modern chemistry. In the *Reviews* the articles were anonymous; but some were later published under their author's name in collections of essays, and the authors of most became known. W. A. Copinger published in 1895 his *Authorship of the first hundred numbers of the Edinburgh Review*; and W. A. Houghton and his colleagues are, in the *Wellesley Index*, continuing the work. In a different class from the *Reviews*, the weekly magazine *The Athenaeum* is sometimes a valuable source of information. In the years before *Nature* and *Chemical News* began, it carried reports of meetings of the British Association, and for some discussions and impromptu addresses, such as that on Prout's hypothesis by J. B. Dumas in 1851, it is almost the only source we have. In the 1860s it carried reports and correspondence on chemical atomic theory.

As the journal and the review increased in importance, so, particularly in the physical sciences, do we begin to find an increase in the number of *Collected Papers* which appeared. Sometimes, perhaps usually, these have an aspect of filial piety about them, like three-decker biographies; a son, a brother, or pupil of the deceased great man lovingly assembled papers by then of little interest into unwieldy tomes. It was not of course a new idea to bind up together all the works of a scientist; Birch's edition of Boyle's *Works* was an invaluable compilation of the eighteenth century. In 1779–85 Samuel Horsley prepared an edition of Newton's *Works*, although not until the twentieth century was there an edition of his correspondence or his mathematical papers, both of which are currently in progress. Hooke's *Works* were never published together until in this century R. T. Gunther assembled them in his *Early Science in Oxford*; Henry Cavendish's published and unpublished papers were edited in the nineteenth century, the edition of the electrical researches by Clerk Maxwell being particularly noteworthy.

Also in the nineteenth century there appeared Robert Willis' edition of *The Works of William Harvey*, in which the Latin was translated; although it should be noted that for scholarly purposes Willis' translations have been superseded. Dalton's papers, although they include an interesting discussion of the term 'particle' in *Nicholson's Journal*, and an exchange on atomic theory with Berzelius in *Annals of Philosophy*, have not yet been collected together and published; nor have W. H. Wollaston's important papers on crystallography, atomism, energy, and other topics. Perhaps for this reason he enjoys less than his contemporary reputation. Thomas Young's *Miscellaneous Works* were published; and so were Davy's *Works* in nine volumes, which include passages from his notebooks and his lecture-notes as well as all his major published works, certain reviews being omitted. Faraday's *Experimental Researches*, those on electricity in three volumes, and those on chemistry and physics in one, reprint his papers; mostly from the *Philosophical Transactions* and *Philosophical Magazine*. Maxwell's *Scientific Papers* were collected into an edition of two princely volumes, which includes papers of a technical and of a general kind; this has recently appeared in one volume in a greatly reduced facsimile which is much easier to handle but harder on the eyes. The *Collected Papers* of poor J. J. Waterston were not published until the twentieth century; those of the great physicists Stokes, William Thomson and Lord Rayleigh are of great interest to anybody studying the development of classical physics.

Among chemists, Thomas Graham's *Papers* were published in a limited edition for private distribution, edited by his colleague R. A. Smith; his important papers on diffusion, on the occlusion of hydrogen in metals, and on polybasic acids are included. The *Scientific Papers* of Thomas Andrews appeared in 1889, and contained his fundamental researches on the continuity of the liquid and vapour states, and his investigations of ozone and of heat of chemical combination. J. P. Joule's *Scientific Papers* has already been mentioned; the slimmer second volume contains papers written in collaboration. The papers of Adams, the co-discoverer of the planet Neptune, appeared in two volumes at the end of the century; as did those of Osborne Reynolds, on viscous fluids, lubrication, and the propulsion of ships.

If the rise of specialized societies and journals is one aspect of the development of science in the nineteenth century, another is the increasing participation by government in science. The voyages of Captain Cook, with the scientific objective of investigating the natural history of the Pacific regions, and observing the transit of Venus; of Bligh in the *Bounty*, which was to have conveyed breadfruit to the West Indies in the hopes that it would grow there; and earlier of Halley in the *Peregrine* pink, come to mind. The material collected by Banks and Solander was not published; on Cook's second voyage, the naturalists were J. R. and G. Forster, who published their *Observations* in 1778. The *Fishes* and the *Flora* from the *Endeavour* are now at last being published. In the nineteenth century, Captain Beechey sailed into the northern Pacific; on his return the zoology and botany of the voyage were published, Edward Lear doing some of the drawings for the former, and W. J. Hooker

93 Christmas Harbour and Observatory, Kerguelen Island (Ross, J. C., *Voyage of Discovery and Research*, 1847, vol I, frontispiece)

for the latter. Captain Flinders, on his voyage to Australia, had been accompanied by the great botanist Robert Brown; and Captain Fitzroy in the *Beagle* had Darwin as his naturalist. Darwin's account of the voyage is justly well-known; he also supervised the formal writing-up and publication of the scientific results of the voyage.

Joseph Hooker sailed with James Clark Ross, on what was principally a voyage investigating magnetism, to the Antarctic from 1839 to 1843; Ross' account of the *Voyage* includes passages by Hooker, and some of the illustrations are from Hooker's drawings. It is also a very important and exciting account of the long and hazardous voyage. Henry Foster, a surveyor and Fellow of the Royal Society, accompanied Parry on his expeditions in the North Polar regions in the 1820s, and was awarded the Copley Medal for his observations. He then commanded the *Chanticleer* on a voyage to the South Seas to make magnetic and gravitational observations, but he died in an accident and the account of the voyage was written by Webster, the ship's surgeon. But the greatest of these voyages was the circumnavigation of the globe by the *Challenger* between 1872 and 1876, under the scientific direction of Charles Wyville Thomson, who had in 1872 published the results of other less ambitious expeditions in his book *The Depths of the Sea.* The full reports of the expedition were eventually published in fifty large volumes, which are a classic of marine biology and oceanography; Thomson also wrote an account of the Atlantic part in *The Voyage*, published in 1877. After his death, the publication of the results was supervised by his assistant, John Murray, whose own books *The Depths of the Ocean* and *The Ocean* are valuable accounts of what oceanography is about.

94 Civet cat, pineapple, and scorpion (Smith, W., *A New Voyage to Guinea*, 1744, pl facing p 148)

In America, the first expedition of this kind to be organized was under Charles Wilkes; it encountered all sorts of administrative difficulties, and on his return Wilkes was court martialled; but the reports which eventually appeared were valuable, though in extremely small editions and by the time of publication rather stale. Wilkes unfortunately described land where Ross penetrated the pack-ice to discover the Ross Sea. Naturally, exploration of the West preoccupied American scientists, with such highlights as the expedition of Lewis and Clark; publication of Western expeditions culminated in the splendid *Pacific Railroad Reports*, in twelve volumes, the printing of which cost the government a million dollars. The bibliography of the *Reports* is confusing, and they sometimes appear as thirteen volumes; seven of the volumes are illustrated with coloured lithographs, and the whole work is fully discussed in Robert Taft's *Artists and Illustrators of the Old West*. American governmental interest in science is also shown in the publications of the Smithsonian Institution.

There was nothing new in government in nineteenth-century Britain supporting scientific voyages; but African explorations also began to receive government support. Works such as William Smith's *New Voyage to Guinea* of 1744 had contained 'An Account of their Animals, Minerals etc.' with illustrations; but with the suppression of the slave trade, and the beginnings of missionary activity on a large scale, came government support for the specifically scientific aspects of expeditions. Thus Henry Salt, in the handsome quarto account of his mission to Abyssinia, added an appendix describing the animals, plants, and birds indigenous to the country; as indeed had James Bruce in his famous, voluminous, and not-always-accurate *Travels to Discover the Source of the Nile* of 1790. The Ethiopian fauna had been previously illustrated in the armchair-explorer Hiob Ludolf's *History of Ethiopia*, which appeared in English in 1682. Thomas Bowditch's well-known account of his *Mission to Ashantee*, a quarto volume of 1819 with coloured plates, similarly contained a chapter, written by the expedition's surgeon, on *materia medica* and diseases.

The most famous explorer of West Africa at this period was Mungo Park, who travelled on behalf of the Association for the Discovery of the Interior Parts of Africa, usually known as the African Association, of which Banks was the first secretary. Park's *Travels* are one of the great classics of African exploration; he died in Africa, and subsequent travellers tried to find out the circumstances of his death. The expedition to Nigeria, orginally led by James Richardson, but after his death by Heinrich Barth, was under the auspices of the British government. Barth's diaries, his *Travels and Discoveries in North and Central Africa*, are a mine of botanical, anthropological, and geographical knowledge. In 1854, Auguste Petermann had written up Barth's earlier dispatches into his *Account of the Progress of the Expedition to Central Africa*, a rare and splendid folio volume. The expedition of Livingstone and Kirk on the Zambezi from 1858 to 1863 was also government sponsored, and instructions were drawn up by J. D. Hooker for Kirk, the botanist of the party. Livingstone's *Narrative* was published in 1865; his writings need no recommendation here. The exploration of the White Nile, narrated in the writings of Burton, Speke, Grant, and Baker, while it makes splendid reading is perhaps of less interest to the historian of science, narrowly considered. In the Orient, Stamford Raffles, founder of modern Singapore and first President of the Zoological Society, supported the researches of Horsfield, and included natural history and anthropology in his *History of Java*, which island he administered during the British Occupation in the Napoleonic Wars. William Marsden's *Sumatra*, especially the handsome third edition, is similarly valuable; and Commodore Perry, later in the century, brought to light new material on Japan.

While government supported scientific expeditions, it also intervened actively to improve the quality of scientific instruction, although nobody could claim that the British government was particularly forward in this. In the nineteenth century, the problem in universities was perhaps the excess of academic freedom. Thus at Glasgow University certain professors were more equal than others, and ran the university as a kind of closed corporation; to the wrath of Thomas Thomson, who did not hold one of the powerful chairs. Government investigation at length led to reforms. The universities of Oxford and Cambridge were also investigated, and *Reports* were published in 1852 which are fascinating documents. Later the Devonshire Commission published its three volumes of *Reports* on scientific instruction and the advancement of science, between 1872 and 1875. Other reports were concerned *inter alia* with the teaching of science at various levels; the Fords' *Guide to Parliamentary Papers* is very valuable for anybody wishing to pursue such topics.

The relations of government and universities gradually became more involved as the century passed. The profits of the Great Exhibition of 1851, which was the first and last such event to make a profit, were used partly to move the British Museum's Natural History collections to South Kensington; and the lectures and teacher-training there, in which T. H. Huxley played a notable rôle, were of a high standard. Similarly in the Royal School of Mines, and the Royal College of Chemistry, the government found itself supporting bodies concerned with higher education in science; indeed Imperial College grew out of these various institutions. Meanwhile, new universities were springing up; the impulse had been given by the founding of University College, London, quickly followed by King's College and by the University of Durham, which secured its charter before London. Individual donors and civic munificence led to the founding of further colleges; of which the most important, at least from the point of view of the history of science,

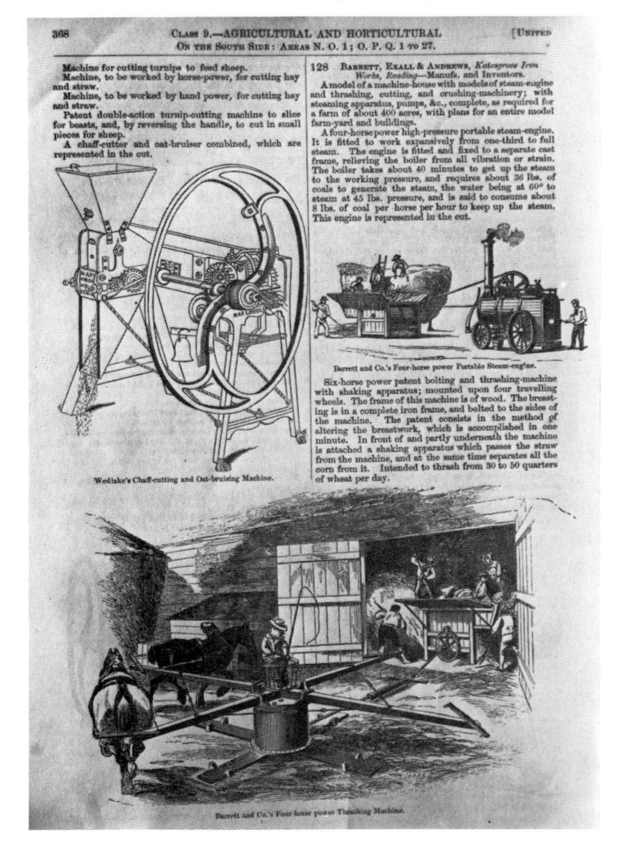

368 CLASS 9.—AGRICULTURAL AND HORTICULTURAL [UNITED

ON THE SOUTH SIDE: AREAS N. O. 1; O. P. Q. 1 TO 27.

Machine for cutting turnips to feed sheep.

Machine, to be worked by horse-power, for cutting hay and straw.

Machine, to be worked by hand power, for cutting hay and straw.

Patent double-action turnip-cutting machine to slice for beasts, and, by reversing the handle, to cut in small pieces for sheep.

A chaff-cutter and oat-bruiser combined, which are represented in the cut.

Wedlake's Chaff-cutting and Oat-bruising Machine.

128 BARRETT, EXALL & ANDREWS, *Katesgrove Iron Works, Reading*—Manufs. and Inventors.

A model of a machine-house with models of steam-engine and thrashing, cutting, and crushing-machinery; with steaming apparatus, pumps, &c., complete, as required for a farm of about 400 acres, with plans for an entire model farm-yard and buildings.

A four-horse power high-pressure portable steam-engine. It is fitted to work expansively from one-third to full steam. The engine is fitted and fixed to a separate cast frame, relieving the boiler from all vibration or strain. The boiler takes about 40 minutes to get up the steam to the working pressure, and requires about 36 lbs. of coals to generate the steam, the water being at 60° to steam at 45 lbs. pressure, and is said to consume about 8 lbs. of coal per horse per hour to keep up the steam. This engine is represented in the cut.

Barrett and Co.'s Four-horse power Portable Steam-engine.

Six-horse power patent bolting and thrashing-machine with shaking apparatus; mounted upon four travelling wheels. The frame of this machine is of wood. The breasting is in a complete iron frame, and bolted to the sides of the machine. The patent consists in the method of altering the breastwork, which is accomplished in one minute. In front of and partly underneath the machine is attached a shaking apparatus which passes the straw from the machine, and at the same time separates all the corn from it. Intended to thrash from 30 to 50 quarters of wheat per day.

Barrett and Co.'s Four-horse power Thrashing Machine.

95 Agricultural machinery (*Great Exhibition Catalogue*, 1851, vol 1, p 368)

was that at Manchester. At first as Owens College, it gave what might now be described as 'further education'; but it then became, with institutions in Leeds and Liverpool, part of the Victoria University and finally a university in its own right. The teaching of chemistry at Manchester under such men as Edward Frankland and Henry Roscoe, was particularly important; but under Balfour Stewart physics was not neglected either, and J. J. Thomson was among the most notable alumni. The history of the college by J. Thompson, and the *Essays and Addresses* by the faculty in 1874, give an idea of what it was like. By the end of the century the government had begun to give grants to the newer universities, chiefly to support the teaching of science.

The biographies of such men as Lyon Playfair, Henry Roscoe, and William Ramsay, which cast light upon these developments, will be mentioned below. We can end our brief survey of government and science with the various exhibitions, beginning with the Great Exhibition of 1851. The *Official Catalogue*, in three volumes and a fourth supplementary volume, and its one-volume abridged version, provides an unequalled vista of industry in the mid-century. The *Prospectus of Exhibitors* was also published; and so were the advertisements they put out. The *Reports by the Juries* in the thirty classes in which exhibits were judged make fascinating reading; and so do the *Lectures* on the exhibition published under the auspices of the Royal Society of Arts. An entertaining off-shoot is B. F. Powell's ferociously rationalistic and materialistic *Testament of 1851*. The exhibits from Britain were notably successful; but in some departments the Americans attracted attention, and a team including Joseph Whitworth was sent to America to study their methods of manufacture. Lyon Playfair among the organizers recognized that the exhibition represented the end of an era; and indeed as other countries became industrialized, Britain never did so well again. In 1862 there was another international exhibition in London; but it had not the Crystal Palace to house it, it had not the charm of novelty, and the proceedings were clouded by the death of the Prince Consort the previous winter. Nevertheless, the *Official Illustrated Catalogue* is a useful source, and so again are the *Reports by the Juries*; that by A. W. Hofmann on chemistry being particularly noteworthy. Finally, following the publication of the Devonshire Report, there was an exhibition at South Kensington in 1876 of a Loan Collection of Scientific Apparatus, some of which became the nucleus of the holdings of the Science Museum. In connection with this, conferences were held, with lectures by eminent men on the state, history and method of science and technology, and in particular on how the apparatus displayed should be used. The lectures were published; and with the reports of juries from the exhibitions they provide an excellent inside view of Victorian science and technology.

A view of science to complement this can be got from biographies; and it is to them that we should now turn. The various journals as a rule published obituaries of important scientists, as did newspapers; and similar brief accounts of the life of an author are not infrequently included in editions of his books published after his death. But the ample scope of the Victorian *Life and Letters* saved the biographer the trouble of deciding what to leave out. This makes for a less-than-gripping narrative, all too often; but it can help to show the cross-connections between different branches of science, and between science and arts. The biographies tend to expatiate upon the man's family, his early years, and the honours he received, so that his main achievement is frequently blurred. We shall refer first to biographies of men of science who flourished before the nineteenth century; then to a few biographies of engineers and technologists; and then to those of nineteenth-century scientists, taking first the disseminators of science, then the naturalists, and finally the physical scientists.

The distinguished astronomer and historian of astronomy, J. L. E. Dreyer, published in 1890 his life of Tycho Brahe, which still remains a standard work on the last great pre-telescopic observer. He later published a *History of the Planetary Systems* which also remains very valuable. The memoirs of the inventor of logarithms and of the decimal point, John Napier of Merchiston, were published in 1834; but the processes of thought which guided him remain obscure. For Seth Ward, a pillar of the early Royal Society, we have a readable and interesting biography by Walter Pope, published in 1697. For Newton, whom Ward proposed as a fellow of the Royal Society, there is the lengthy and somewhat uncritical, but still useful, *Memoirs* published by David Brewster in 1855; an earlier, shorter version had appeared in 1831. Baily published in 1835 his life of John Flamsteed, towards whom Newton did not behave very well, and whose reputation had perhaps suffered accordingly. William Whiston, a cantankerous Unitarian, and successor of Newton at Cambridge, published his autobiographical *Memoirs* in 1753; which provides a unique and gossipy record of Newtonian science. Whiston also wrote a biography of Samuel Clarke, the Newtonian divine, whose orthodoxy was also open to doubt. For Joseph Black there is a *Life and Letters* compiled by the great chemist William Ramsay and published in 1918; but while it is useful, this work does not really come to life. Ramsay's own experiments which led to the discovery of argon, and then of the other inert gases of the atmosphere, were set off when he read of Cavendish's experiments on nitrogen, in George Wilson's *Life of Cavendish* of 1851. For the life of Joseph Priestley, we have an autobiography, down to 1795, which was continued down to Priestley's death by his son, and published at Northumberland, Pennsylvania, in 1806; from this it is evident that Priestley considered himself first and foremost a theologian, and there is little about his scientific work. *The Life and Letters of Gilbert White of Selborne* was published in 1901, and gives useful background information; and White's *Journals*, a magnificent piece of unselfconscious reporting of nature, in 1931. Ernst Krause's biography of Erasmus Darwin includes an introduction by Charles Darwin.

We may count as a technologist the cleric, agriculturalist, and inventor Edmund Cartwright, whose life by Strickland appeared in 1843; his mechanical loom formed the basis of modern power looms. A memoir of the tough

Wilkinson dynasty of iron-founders was written by John Randall. Without John Wilkinson's skills, particularly in cannon-boring, James Watt's engine might have remained impracticable; and Wilkinson was the first to use a Boulton and Watt engine for purposes other than pumping when he installed one for the blast at his furnaces. J. P. Muirhead's account of *James Watt's Mechanical Inventions* appeared in 1854, and his *Life of Watt* in 1858; both of these are based upon primary materials and are standard sources for Watt's early life. George Williamson's *Memorials . . . of James Watt*, which was published in quarto in 1856, is also valuable. Of Watt's colleagues in the Lunar Society, the potter Josiah Wedgwood, the industrial chemist James Keir, and the educationalist-inventor R. L. Edgeworth, found biographers. Keir's *Life*, by J. K. Moillet, was privately printed; Wedgwood's, by Eliza Meteyard, is a standard two-volume affair; Edgeworth's is an autobiography, written in collaboration with his famous daughter Maria.

We have already referred, in Chapter 6, to Samuel Smiles' biographies of engineers; and in 1883 was published the *Autobiography* of James Nasmyth, the inventor of the steam hammer, which Smiles edited. Of the great bridge builders and civil engineers, we have the *Memoirs* of Thomas Telford, edited by J. Richman, which appeared in 1838; the *Autobiography* of John Rennie Jr, of 1875; and Beamish's *Memoir of Sir Marc Isambard Brunel*, the builder of the first Thames tunnel, and father of Isambard Kingdom Brunel. Of Sir William Fairbairn, the builder of about a thousand bridges, there is a *Life* by W. Pole, published in 1877; the telegrapher Samuel Morse's *Letters and Journals* appeared in 1914; and the *Life and Labours of John Mercer*, the textile chemist, were described by E. A. Parnell in 1886. In the electrical industry in the nineteenth century, two giants who were at once technologists, entrepreneurs and scientists were the Siemens brothers. The biography of William, who came to England and had a most distinguished career, was written by Pole; his elder brother Werner remained in Germany, and wrote his own vivid *Personal Recollections*, which has recently appeared in a new and handsome edition and is extremely readable as well as valuable to the historian. He laid telegraph cables all over the world, both overland and under water; built, in 1881, the first electric tramway, near Berlin; and ruled his firm with a benevolent paternalism.

Of those whom we may call disseminators of science, there are also biographies to notice. *William Paley's Life and Writings* was written by Edmund Paley, in 1825; Paley's importance as the transmitter of natural theology to countless undergraduates is enormous. The more interesting figure of Thomas Chalmers, the preacher of astronomical sermons and first moderator of the Free Church after the schism in Scotland, was described in Hanna's ample biography. Charles Gibbon's *Life of George Combe* is a useful source for information about the famous educationalist and phrenologist; Katherine Lyell's *Memoir* of the first warden of London University, Leonard Horner, provides useful background; as do the *Recollections* of Sir Henry Holland, who wrote on geology for the *Quarterly Review*; and the three-volume

Passages of a Working Life of Charles Knight, of the *Penny Cyclopedia*. More important in the internal history of science were Hugh Miller, the geologist, and William Whewell. Miller's autobiographical *My Schools and Schoolmasters* is deservedly a classic; his *Life and Letters* were edited by Bayne. Todhunter's *William Whewell* is sound, but dry as dust; the biography by Janet Douglas is less monumental, and more personal. The *Life and Letters of Herbert Spencer*, another very important figure in Victorian intellectual life, thought highly of in his day as a philosopher of biology, appeared in 1908.

Passing on to naturalists, we may notice M. R. Audubon's *Audubon and his Journals* of 1887, which includes J. J. Audubon's autobiography; but this agreeable mixture of fact and fiction was put straight in 1917 by F. H. Herrick in his *Audubon the Naturalist*, after years of research among primary materials. The life of the American oceanographer M. F. Maury was written by his daughter, Mrs Corbin; he was responsible for the coordination of data from sailors of all nations, with which he revised wind and current charts; and he wrote the first textbook of modern oceanography, *Physical Geography of the Sea*. A *Memoir* of Robert Jameson, the Wernerian geologist, was published by Lawrence Jameson in 1854; and *Memoirs of William Smith*, who originated the science of stratigraphy, by Phillips in 1844. W. D. Conybeare wrote a brief autobiography, which was published in 1905 by F. C. Conybeare, appended to his *Letters and exercises of the Elizabethan Schoolmaster John Conybeare*, which also includes notes by W. D. Conybeare. Archibald Geikie's *Life of Sir Roderick I. Murchison* is one of the best of the Victorian *Lives*; Murchison's career from country squire to a position of great distinction and influence in science is fascinating. The *Life* of Adam Sedgwick casts light on science in Cambridge in the nineteenth century; and that of Lyell is a valuable source for the history of geology as seen in the activities of one of the greatest of all geologists. The biography of Andrew Ramsay, Professor of Geology at University College, London, was also written by Geikie; who in 1924 published his own *Long Life's Work*, and also wrote *Founders of Geology*. His writings all have a vigour and charm; and his autobiography, written shortly before his death at eighty-eight, helps us to recapture a past era.

Darwin's *Life and Letters* provides useful background; it contains an autobiography, which was edited and has only recently been published in full; and an extremely interesting chapter by T. H. Huxley on the reception of Darwin's theory. Two volumes of *Further Letters* appeared subsequently. The *Life and Letters of Thomas Huxley* was written by Leonard Huxley, and appeared in 1900; and Leonard Huxley also wrote the *Life and Letters of Sir J. D. Hooker*, the eminent Darwinian botanist, which was published in 1918. Something of T. H. Huxley's enormous energy and range of interests can be gathered from his biography; and the Hooker volumes contain interesting material from his voyage with Ross, and his travels in India, which were fully described in his *Himalayan Journals* of 1854. Much more idiosyncratic than these was A. R. Wallace, the co-discoverer of the principle of natural selection, whose autobiography is full of splendid passages like that describing his attempts to convert a flat-earth fanatic. The *Life of Richard Owen*, the palaeontologist who opposed Darwin, was written by his grandson; there are two two-volume *Lives of Louis Agassiz*, the chief opponent of Darwin in America, which print numerous letters illustrating his strange career. The *Letters* of Asa Gray, the chief American Darwinian, were published by Jane Loring Gray, in 1893.

In the next generation, G. J. Romanes, whose *Life and Letters*, edited by his wife, appeared in 1896, is an interesting example of a Darwinian who lost his religious faith and then found it again. The *Reminiscences* of W. C. Williamson, a founder of the science of palaeobotany, came out in 1896; he made extensive studies on fossil plants in coal. Francis Galton, the geneticist and cousin of Charles Darwin, published his autobiography in 1908; he is chiefly remembered for his book *Hereditary Genius* of 1869, in which he showed how 'eminence' runs in families. Of foreign biologists, a translation of Vallery-Radot's standard biography of Pasteur appeared in 1885. A *Life* of Pasteur by Percy Frankland and his wife appeared in 1898 as part of the Century Science Series of short biographies; and an English translation of Iltis' *Life of Mendel* was published in 1932.

If Thomas Beddoes' most important discovery was Humphry Davy, he can count among the physicists and chemists; his biography was written by J. E. Stock, and contains a bibliography. Davy has always attracted biographers; Paris' book is lively, unreliable, and seems to have been written partly to debunk its subject; whereas John Davy's reply to it was intended as a monument. A compressed version of this forms the first volume of Davy's *Works*; and after Lady Davy's death, John Davy published further material as *Fragmentary Remains*. A fictionalized and edifying version of Davy's career is set out in Mayhew's period-piece *Wonders of Science*, which compares oddly with his writings on the London poor.

The standard lives of Dalton are those by W. C. Henry, who undertook without much enthusiasm a task inherited from his father; by R. A. Smith, who included a general discussion of the atomic theory; and by Roscoe and Harden, who based their study upon Dalton's notebooks, which have since for the most part been destroyed in an air-raid. The *Life* of Dalton's pupil Joule was written by Osborne Reynolds; of Davy's protégé Faraday, the standard life is that by Bence-Jones, but Tyndall's *Faraday as a discoverer* is a vivid portrait, and J. H. Gladstone's *Michael Faraday* is useful. *The Home Life of Sir David Brewster* was produced by M. M. Gordon in 1869; other eminent Scots scientists included J. D. Forbes, whose biography appeared in 1873; George Wilson, who was the subject of a charming *Memoir* by J. A. Wilson; and Thomas Graham, whose *Life and Works* were written up by R. A. Smith, the editor of his *Works*.

A brief biography of Justus von Liebig by W. A. Shenstone appeared in the Century Science Series in 1895. Edward Frankland was an important chemist, a pioneer of valency theory, and a member of the X Club; his autobiography, edited and concluded by his two daughters, was published in 1902. Henry Roscoe was Professor of Chemistry at Manchester, and became a Member of Parliament; his autobiography is an interesting, if slow-moving, document. A biography of him by T. E. Thorpe appeared in 1916. Roscoe was a great advocate of scientific and technical education, as was Lyon Playfair, whose rôle in the Great Exhibition has already been mentioned; a biography of Playfair by T. W. Reid appeared in 1899. Sir William Crookes was famous for his work on cathode rays, and his investigations of spiritualism; his *Life* was written by E. E. Fourier d'Albe in 1923. Crookes edited *Chemical News*; the founder of *Nature* was the astrophysicist Norman Lockyer whose standard biography appeared in 1928. Lockyer identified helium in the spectrum of the Sun; it, with the other inert gases, was found on earth by William Ramsay, whose *Life* was written by William Tilden and, more recently, by his collaborator, Morris Travers; Travers' *Life* particularly is a most useful source for the study of chemistry at the turn of the century. The *Life of William Thomson*, Lord Kelvin, was written by Sylvanus Thompson; it brings out the tremendous range of his interests, and the immense influence he had over all fields of science in the later nineteenth century. A more intimate study is *Lord Kelvin's Early Home*, based upon the recollections of his sister. The *Life* of Thomson's friend Helmholtz was written by L. Koenigsberger; an English translation, somewhat abridged, was published in 1906. The standard *Life* of Clerk Maxwell is that by Campbell and Garnett; of Lord Rayleigh, that by R. J. Strutt; while for J. J. Thomson we have his own *Recollections and Reflections*.

With few exceptions, these biographies are uncritical and loosely organized, and the editorial standards are often not very high; passages are excised from letters without this being made very clear, and so on. But they do provide a picture of the great men they commemorate, and of the times in which they lived, and they are an invaluable source of information and pleasure. Shorter historical works, such as Thorpe's *Essays in Historical Chemistry* and Tilden's *Famous Chemists*, have worn a good deal less well, though they can still be consulted with advantage. We have now covered most of our field; and in the Epilogue we shall glance at certain late Victorian and early twentieth-century books which are of interest both to the historian of science and to the collector.

11 SCIENTIFIC PUBLICATIONS IN THE NINETEENTH CENTURY

Adams, J. C., *The Scientific Papers*, ed. W. A. Adams and R. A. Simpson, 2 vols., Cambridge, 1896–1900.

Agassiz, L.: Marcou, J., *Life, Letters, and Works of Louis Agassiz*, 2 vols., New York, 1896; Agassiz, E. C., *Louis Agassiz, his life and correspondence*, 2 vols., Boston, Mass., 1885.

Ampère, A. M., *The Story of his Love: being the journal and early correspondence . . .*, ed. H. C(hevreux), 1873.
Annals of Natural History, *1* (1838)–*14* (1844); 2nd series, *1* (1845)–*20* (1857); 3rd series, *1* (1858)–*20* (1867); 4th series, *1* (1868)–*20* (1877); etc.
Annals of Philosophy, *1* (1813)–*16* (1820); new series, *1* (1821)–*12* (1826).
Athenaeum, The, 1828–1921.
Audubon, J. J.: Audubon, M. R., *Audubon and his Journals*, 2 vols., 1898; Herrick, F. H., *Audubon the Naturalist*, 2 vols., New York, 1917.
Bacon, F., *Works*, ed. J. Spedding, R. L. Ellis, and D. D. Heath, 14 vols., 1857–74.
Barth, H., *Travels and Discoveries in North and Central Africa*, 5 vols., 1857–8.
Beddoes, T.: Stock, J. E., *Memoirs of the Life of Thomas Beddoes*, 1811.
Beechey, Capt., F. W., *Narrative of a Voyage to the Pacific and Bering's Strait . . .*, 2 pts, 1831.
The Botany of Captain Beechey's Voyage, 10 pts, 1830–41.
The Zoology of Captain Beechey's Voyage, 1839.
Black, J.: Ramsay, W., *Life and Letters of Joseph Black*, 1918.
Bowditch, T. E., *Mission from Cape Coast Castle to Ashantee . . .*, 1819.
Boyle, R., *Works*, ed. T. Birch, 5 vols., 1744; 6 vols, 1772.
Brahe, T.: Dreyer, J. L. E., *Tycho Brahe*, 1890.
Brewster, D.: Gordon, M. M., *Home Life of Sir David Brewster*, Edinburgh, 1869.
British Association, Report, *1* (1831–2)–.
British Critic, The, *1* (1793)–*42* (1813); 2nd series, *1* (1814)–23 (1825); 3rd series, *1* (1825)–*3* (1826); 4th series, *1* (1827)–*36* (1843).
Brougham, H. S., *A Discourse of Natural Theology*, 1835.
Lives of Men of Letters and Science who flourished in the time of King George III, 2 vols., 1845–6.
Bruce, J., *Travels to discover the Source of the Nile*, 5 vols, 1790.
Brunel, M. I.: Beamish, R., *Memoir of the Life of Sir Marc Isambard Brunel*, 1862.
Cartwright, E.: S(trickland), M., *A Memoir of the Life, Writings, and Mechanical Inventions of Edmund Cartwright*, 1843.
Cavendish, H.: Wilson, G., *The Life of the Honourable Henry Cavendish*, 1851.
Chalmers, T.: Hanna, W., *Memoirs of the life and writings of Thomas Chalmers*, 4 vols., Edinburgh, 1849–52.
Chambers's Edinburgh Journal, *1* (1832)–*12* (1843); new series, *1* (1844)–*20* (1855).
Chemical Gazette, The, *1* (1842–3)–*17* (1859); it was absorbed into
Chemical News, *1* (1860)–*145* (1932).
Chemical Society, Journal, *1* (1848)–.
Proceedings, *1* (1889)–.
The Jubilee . . . 1841–1891, 1896.
Memorial Lectures, I, 1901; II, 1914; III, 1933.
Faraday Lectures, 1869–1928, 1928.
Chemist, The, *1* (1824)–*2* (1825); another journal, *1* (1840)–*8* (1848); new series, *1* (1849)–*4* (1852); new series, *1* (1854)–*5* (1857).
Clarke, S.: Whiston, W., *Historical Memoirs of the life of Dr Samuel Clarke*, 1730.
Combe, G.: Gibbon, C., *The life of George Combe*, 2 vols, 1878.
Conybeare, W. D.: Conybeare, F. C., *Letters and Exercises of the Elizabethan Schoolmaster John Conybeare . . .*, 1905.
Copinger, W. A., *The Authorship of the first hundred numbers of the Edinburgh Review*, Manchester, 1895.
Crookes, W.: d'Albe, E. F. F., *The Life of Sir William Crookes*, 1923.
Dalton, J.: Henry, W. C., *Memoirs of . . . John Dalton*, 1854; Smith, R. A., *Memoir of John Dalton*, 1856; Roscoe, H. E., and Harden, A., *A New View of the Origin of Dalton's Atomic Theory . . . , 1896*.
Darwin, C. R.: *The Life and Letters of Charles Darwin*, ed. F. Darwin, 3 vols., 1887;

More Letters . . ., ed. F. Darwin and A. C. Seward, 2 vols., 1903.
Darwin, E.: Krause, E., *Erasmus Darwin*, tr. W. S. Dallas, 1879.
Davy, H.: *Collected Works*, ed. J. Davy, 9 vols., 1839–40; Paris, J. A., *The Life of Sir Humphry Davy*, 2 vols., 1831; Davy, J., *Memoirs of the Life of Sir Humphry Davy*, 2 vols., 1836, and *Fragmentary Remains, Literary and Scientific, of Sir Humphry Davy*, 1858; Mayhew, H., *The Wonders of Science, or Young Humphry Davy*, 1855; Hartley, H., *Humphry Davy*, 1966.
Dreyer, J. L. E., *History of the Planetary Systems from Thales to Kepler*, Cambridge, 1905.
Edgeworth, R. L. and M., *Memoirs of Richard Lovell Edgeworth*, 2 vols., 1820.
Edinburgh Philosophical Journal, *1* (1819)–*14* (1826); *Ed. New Phil. J.*, *1* (1826)–*57* (1854); new series, *1* (1855)–*19* (1864).
Edinburgh Review, *1* (1802)–.
Fairbairn, W.: Pole, W., *Life of Sir W. Fairbairn*, 1877.
Faraday, M.: Bence Jones, H., *The Life and Letters of Faraday*, 2 vols., 1870; Gladstone, J. H., *Michael Faraday*, 1872.
Tyndall, J., *Faraday as a Discoverer*, 1868.
Flamsteed, J.: Baily, F., *An Account of the Rev. John Flamsteed*, 1835; supplement, 1837.
Flinders, M., *A Voyage to Terra Australis*, 2 vols., 1814.
Forbes, J. D.: Shairp, J. C., Tait, P. G., and Adams-Reilly, A., *Life and Letters of James David Forbes*, 1873.
Frankland, E.: *Sketches from the Life of Sir Edward Frankland*, ed. by his two daughters, 1902.
Galton, F., *Hereditary Genius*, 1869.
Memories of my Life, 1908.
Geikie, A., *The Founders of Geology*, 1897; 2nd ed., 1905.
A Long Life's Work, 1924.
Gentleman's Magazine, *1* (1731)–*303* (1907).
Geological Society, Transactions, *1–5*, 2nd series, *1–7* (1811–1856); *Proceedings*, *1* (1834)–4 (1846); *Quarterly Journal*, *1* (1845)–.
Graham, T.: Smith, R. A., *The Life and Works of Thomas Graham*, Glasgow, 1884.
Gray A.: Gray, J. L., *Letters of Asa Gray*, 2 vols., Boston, Mass., 1893.
Great Exhibition, *Official Descriptive and Illustrative Catalogue*, 3 vols. + supplement, 1851; abridged ed., 1851.
Prospectus of Exhibitors, 16 vols., 1851.
Reports by the Juries . . ., 1852.
Harvey, W., *The Works*, ed. and tr. R. Willis, 1847; *Anatomical Lectures*, ed. G. Whitteridge, Edinburgh, 1964; *De Motu Locali Animalium*, tr. G. Whitteridge, Cambridge, 1959; *De Circulatione Sanguinis*, tr. K. J. Franklin, Oxford, 1968; *De Motu Cordis*, tr. K. J. Franklin, Oxford, 1957.
Helmholtz, H. von: Koenigsberger, L., *Hermann von Helmholtz*, tr. F. A. Welby, Oxford, 1906.
Henry, J., *The Scientific Writings*, 2 vols., Washington, D.C., 1886.
Memorial of Joseph Henry, ed. W. B. Taylor, Washington, D.C., 1881.
Holland, H., *Recollections of Past Life*, 1872.
Fragmentary Papers on Science and other subjects, ed. F. J. Holland, 1875.
Hooker, J. D., *Notes of a tour in the plains of India . . .*, 2 pts, 1848.
Himalayan Journals . . ., 2 vols., 1854.
A Century of Indian Orchids . . ., Calcutta, 1887.
and Gray, A., *The Vegetation of the Rocky Mountain Region*, Washington, 1881.
and Ball, J., *Journal of a Tour in Morocco and the Great Atlas*, 1878.
Huxley, L., *Life and Letters of Sir Joseph Dalton Hooker*, 2 vols., 1918.
Hooker, W. J., *Niger Flora*, 1849. (Plants found by Vogel.)

Huish, R., *A Narrative of the Voyage . . . of Captain Beechey*, 1836.
The Travels of Richard and John Lander . . . for the discovery of the course of the Niger, 1836.
Huxley, T. H.: Huxley, L., *The Life and Letters of Thomas Huxley*, 2 vols., 1900.
Institution of Electrical Engineers, *Journal*, *1* (1872–3)–.
International Exhibition, 1862, *Official Illustrated Catalogue*, 12 pts [1862].
Reports by the Juries, ed. J. F. Iselin and P. le N. Foster, 1863.
Jameson, R.: Jameson, L., *Biographical Memoir of the late Professor Jameson*, Edinburgh, 1854.
Joule, J.: Reynolds, O., *Memoir of James Prescott Joule*, 1892.
Keir, J.: Moillet, Mrs A., *Sketch of the Life of James Keir* (1868).
Knight, C., *Passages of a working life*, 3 vols., 1864–5.
Laboratory, The, *1* (1867); no more published.
Ladies Diary, The (1704)–(1840).
Leeuwenhoek, A. van, *The Select Works . . .*, tr. S. Hoole, 1800; 3 pts of vol. 1; no more published.
Lewis, M., and Clark, W., *History of the expedition . . . to the sources of the Missouri, thence . . . to the Pacific Ocean*, by N. Biddle, ed. P. Allen, 2 vols., New York and Philadelphia, 1814; ed. E. Coues, 4 vols., New York, 1893.
Original Journals . . ., ed. R. G. Thwaites, 8 vols., New York, 1904–5.
Liebig, J. von: Shenstone, W. A., *Justus von Liebig . . .*, 1895.
Linnean Society, *Transactions*, *1* (1791)–*30* (1875); 2nd series, zoology *1* (1879–; botany, *1* (1880)–.
Livingstone, D. and C., *Narrative of an expedition to the Zambesi . . .*, 1865.
Lockyer, N.: Lockyer, T. M. and M. W., *The Life and Work of Sir Norman Lockyer*, 1928.
Ludolph, H., *A new history of Ethiopia*, tr. J. P., 1682.
Lyell, C.: *The Life, Letters, and Journals of Sir Charles Lyell*, ed. by his sister-in-law, Mrs Lyell, 2 vols., 1881.
Lyell, K. M., *Memoir of Leonard Horner*, 2 vols, 1890.
Manchester Literary and Philosophical Society, *Memoirs*, *1* (1785)–*5* (1802); 2nd series, *1* (1805)–*15* (1860); 3rd series, *1* (1862)–*10* (1887); *Memoirs and Proceedings*, *1* (1888)–*10* (1896); 2nd series. *1* (1897)–.
Marsden, W., *History of Sumatra*, 1783; 3rd ed., 1811.
Martin, B., *General Magazine of Arts and Sciences*, 14 vols., 1755–65.
Maury, M. F., *The Physical Geography of the Sea*, 1855.
Corbin, D. F. M., *A Life of Matthew Fontaine Maury*, 1888.
Maxwell, J. C.: *The Scientific Papers*, ed. W. D. Niven, 2 vols., Cambridge, 1890; Campbell, L., and Garnett, W., *Life of James Clerk Maxwell*, 1882.
Mechanics' Magazine, The, *1* (1823)–97 (1873).
Mendel, J. G.: Iltis, H., *Life of Mendel*, tr. E. and C. Paul, 1932.
Mercer, J.: Parnell, E. A., *The life and labours of John Mercer*, 1886.
Miller, H., *My Schools and Schoolmasters*, Edinburgh, 1854.
Bayne, P., *The Life and Letters of Hugh Miller*, 2 vols., 1871.
Monthly Review, *1* (1749)–*81* (1789); 2nd series, *1* (1790)–*108* (1825); 3rd series, *1* (1826)–*15* (1830); 4th series, *1* (1831)–*45* (1845).
Morse, S. F. B., *Letters and Journals*, ed. E. L. Morse, Boston and New York, 1914.
Moseley, H. N., *Notes of a Naturalist on the 'Challenger' . . .*, 1879.
Murchison, R. I.: Geikie, A., *Life of Sir Roderick I. Murchison*, 2 vols., 1875.
Murray, J., *The Ocean*, 1913.
and Hjort, J., *The Depths of the Ocean*, 1912.
Napier, J.: Napier, M., *Memoirs of John Napier of Merchiston*, Edinburgh, 1834.
Knott, C. G. (ed.), *Napier Tercentenary Memorial*, Edinburgh, 1915.
Nasmyth, J., *An Autobiography*, ed. S. Smiles, 1883.

Nature, 1 (1869–70)–.
Newton I., *Opera quae exstant omnia*, ed. S. Horsley, 5 vols., 1779–85. Brewster, D., *Memoirs of the life, writings, and discoveries of Sir Isaac Newton*, 2 vols., Edinburgh, 1855; an expanded and revised version of his *Life of Sir Isaac Newton*, 1831.
(*W. Nicholson's*) *Journal of Natural Philosophy, Chemistry, and the Arts, 1* (1797)–*5* (1802), 4°; *1* (1802)–*36* (1813). 8°.
North British Review, 1 (1844)–*53* (1871).
Owen, R.: *The Life of Richard Owen*, by his grandson, the Rev. Richard Owen, 2 vols., 1894.
Owens College: *Essays and Addresses by Professors and Lecturers . . .*, 1874.
Paley, W.: Paley, E., *An account of the life and writings of William Paley, D.D.*, 1825.
Park, M., *Travels in the Interior Districts of Africa*, 1799.
Parliamentary Papers: Scottish Universities Commission, 1831; Glasgow University Commission, 1839; Oxford University Commission, 1852; Cambridge University Commission, 1852; Devonshire Commission, 1872–5.
Parry, W. E., *Nautical Astronomy by night . . .*, Bath, 1816.
Journal of a Voyage for the Discovery of a North-West Passage . . ., 1821–4.
Journal of a Second Voyage . . ., 2 pts, 1824–5.
Journal of a Third Voyage . . ., 2 pts, 1826.
Thoughts on the parental character of God, 1841.
Pasteur, L.: Vallery-Radot, R., *Louis Pasteur, his life and labours*, tr. C. Hamilton, 1885.
Perry, M. C., *Narrative of the expedition of an American Squadron to the China Seas and Japan . . .*, 3 vols., Washington D.C., 1856(–58).
Petermann, A., *An Account of the Progress of the Expedition to Central Africa . . . under Messrs. Richardson, Barth, Overweg, and Vogel . . .*, 1854.
Philosophical Magazine, 1 (1798)–*68* (1827); 2nd series, *1* (1828)–*11* (1832); 3rd series, *1* (1832)–*37* (1850); 4th series, *1* (1851)–*50* (1875); 5th series, *1* (1876)–*50* (1900); etc., 8th series now in progress.
Physical Society, *Proceedings, 1* (1874)–.
Playfair, L: Reid, T. W., *Memoirs and Correspondence of Sir Lyon Playfair*, 1899.
Popular Science Monthly, 1 (1872)–.
Powell, B. F., *The Bible of Reason, Supplement; or, Exhibition-testament of 1851* (1851).
Prichard, J. C., *Researches into the Physical History of Man*, 1813.
A Review of the Doctrine of a Vital Principle . . ., 1829.
The Natural History of Man, 1843.
Priestley, J.: *Memoirs of Dr. Joseph Priestley . . .*, 2 vols., Northumberland, Penna., 1806; London, 1806.
Quarterly Journal of Science, The, 1 (1864)–7 (1870); 2nd series, *1* (1871)–*8* (1878); 3rd series, *1* (1879)–7 (1885).
Quarterly Review, The, 1 (1809)–.
Raffles, S., *History of Java*, 2 vols., 1817.
Ramsay, A. C.: Geikie, A., *Memoir of Sir A. C. Ramsay*, 1895.
Ramsay, W.: Tilden, W. A., *Memoir of Sir William Ramsay*, 1918; Travers, M. W., *A Life of Sir William Ramsay*, 1956.
Randall, J., *Our Coal and Iron Industries . . . the Wilkinsons*, Madeley, Salop, n.d. (1879?).
Rayleigh, Lord, *Scientific Papers*, 6 vols., Cambridge, 1899–1920.
The Theory of Sound, 2 vols., 1877–8.
Strutt, R. J., *J. W. Strutt; Third Baron Rayleigh*, 1924.
Rennie, J. (jr), *Autobiography*, 1875.
Reynolds, O., *Papers on Mechanical and Physical Subjects*, 3 vols., Cambridge, 1900–3.
Retrospect of Philosophical, Mechanical, Chemical, and Agricultural Discoveries, 1 (1805)–*8* (1815).
Romanes, G. J.: *The Life and Letters of George John Romanes*, by his wife, 1896.

Roscoe, H. E., *The Life and Experiences of Sir Henry Enfield Roscoe*, 1906.
 Thorpe, T. E., *The Right Hon. Sir Henry Enfield Roscoe . . .*, 1916.
Ross, J. C., *A Voyage of Discovery and Research in the Southern and Antarctic Regions . . .*, 2 vols., 1847.
Royal Astronomical Society, *Memoirs*, *1* (1825)–.
 Monthly Notices, *1* (1827–31)–.
Royal Institution, *Journals*, *1* (1802)–*2* (1803).
 Journal of Science and the Arts, *1* (1816)–*20* (1826); 2nd series, *1* (1827)–*7* (1830); *Journal*, *1* (1830–1)–*2* (1831).
 Proceedings, *1* (1851–4)–.
Royal Society of Arts, *Transactions*, *1* (1783)–*55* (1843–4); a 2nd series began in 1846–7, but was discontinued after 1848.
Lectures on the results of the Great Exhibition of 1851 . . ., 2 series, 1852–3.
Royal Society of Edinburgh, *Transactions*, *1* (1788)–.
Royal Society of London, *Catalogue of Scientific Papers, 1800–1900*, 19 vols., 1867–1925.
 The Philosophical Transactions . . . from . . . 1665 to . . . 1800; abridged by C. Hutton, G. Shaw, and R. Pearson, 19 vols., 1809.
Scientific Memoirs, *1* (1837)–*5* (1852); new series, 2 vols., 1852–3.
Scoresby, W., *An Account of the Arctic Regions with a history and description of the Northern Whale Fishery*, 2 vols., Edinburgh, 1820.
Sedgwick, A.: Clark, J. W., and Hughes, T. M., *The life and letters of the Rev. Adam Sedgwick*, 2 vols., Cambridge, 1890.
Siemens, C. W.: Pole, W., *Life of Sir William Siemens*, 1888.
Siemens, E. W., *Personal Recollections of Werner von Siemens*, 1893.
(B. Silliman's) American Journal of Science and Arts, *1* (1818)–*50* (1845); 2nd series, *1* (1846)–*50* (1870); 3rd series, *1* (1871)–*50* (1895); 4th series, *1* (1896)–.
Smith, W., *A New Voyage to Guinea . . .*, 1744.
Smith, W.: Phillips, J., *Memoirs of William Smith, LL.D.*, 1844.
Smithsonian Institution, *Annual Report*, *1* (1847)–.
 Contributions to Knowledge, *1* (1848)–.
Society for the Diffusion of Useful Knowledge: *Natural Philosophy*, 4 vols., 1829–38; *Mathematics*, 2 vols., 1830–47; *Penny Magazine*, *1* (1832)–*16* (1846); *Lives of Eminent Persons*, 1837; *A Manual for Mechanics' Institutions*, 1839.
South Kensington Museum, *Conferences held in connection with the special loan collection of Scientific Apparatus, 1876*, 2 vols., (1877).
Spectator, The, 1711–14.
Spencer, H.: Duncan, D., *The Life and Letters of Herbert Spencer*, 1908.
Spix, J. B., and Martius, C. F. P., *Travels in Brazil in the years 1817–20*, 2 vols., 1824.
Stokes, G. G., *Mathematical and Physical Papers*, 5 vols., Cambridge, 1880–1905.
 Memoir and Scientific Correspondence, ed. J. Larmor, 2 vols., Cambridge, 1907.
Thompson, J., *The Owens College . . .*, Manchester, 1886.
Thomson, C. W., and Murray, J. (ed.), *Report on the Scientific Results of the Voyage of H.M.S. Challenger . . .*, 40 vols. in 44, + 6 vols. atlas, 1880–95.
Thomson, C. W., *The Depths of the Sea*, 1873.
 The Voyage of the Challenger – the Atlantic, 2 vols., 1877.
Thomson, J. J., *Recollections and Reflections*, 1936.
Thomson, W., *Mathematical and Physical Papers*, 6 vols., Cambridge, 1882–1911.
 Lord Kelvin, Professor of Natural Philosophy . . . 1846–99, Glasgow, 1899.
 King, E. and E. T., *Lord Kelvin's Early Home*, 1909.
 Thompson, S. P., *The Life of William Thomson*, 2 vols., 1910.
Thorpe, T. E., *Essays in Historical Chemistry*, 1894.
Tilden, W. A., *Famous Chemists*, 1921.
U.S. Congress, *Official Explorations for Pacific Railroads, 1853–1855*, 12 vols., Washington, 1855–61.

Wallace, A. R., *My Life, a record of events and opinions*, 2 vols., 1905.
Ward, S.: Pope, W., *The Life of . . . Seth, Lord Bishop of Salisbury*, 1697.
Watt, J.: Muirhead, J. P., *The Origins and Progress of the Mechanical Inventions of James Watt*, 3 vols., 1854; *Life of James Watt, with selections from his correspondence*, 1858.
W(illiamson), G., *Memorials of the Lineage, Early Life, Education and Development of the Genius of James Watt*, Edinburgh, 1856.
Watts, H., *A Dictionary of Chemistry*, 5 vols., 1870.
Webster, W. H. B., *Narrative of the Voyage to the Southern Atlantic Ocean . . . in H. M. Sloop Chanticleer*, 2 vols., 1834.
Wedgwood, J.: Meteyard, E., *The life of Josiah Wedgwood*, 2 vols., 1865–6.
Wernerian Society, *Memoirs*, *1* (1811)–*8* (1839).
Westminster Review, The, *1* (1824)–*180* (1914).
Whewell, W.: Todhunter, I., *William Whewell . . .*, 2 vols., 1876; Douglas, J. M., *The life, and selections from the correspondence of William Whewell*, 1881.
Whiston, W., *Memoirs of the Life and Writings of W. W.*, 3 pts, 1749–50.
White, G., *Journals*, ed. W. Johnson, 1931.
White, R. H., *The Life and Letters of Gilbert White of Selborne*, 1901.
Wilkes, C., *Narrative of the United States Exploring Expedition*, 5 vols. + atlas, Philadelphia, 1844.
Voyage around the World, Philadelphia, 1849.
Williamson, W. C., *Reminiscences of a Yorkshire Naturalist*, 1896.
Wilson, G.: Wilson, J. A., *Memoir of George Wilson*, Edinburgh, 1860.
Woodward, H. B., *The History of the Geological Society of London*, 1907.
Young, T., *Miscellaneous Works*, 1855.
Zoological Society, *A record of the progress of the Zoological Society during the nineteenth century*, 1901.

Recent Publications

Cambridge History of American Literature, The, 4 vols., New York, 1917–21.
Ford, P. and G., *Select List of British Parliamentary Papers, 1833–99*, 1953.
(ed.) *Hansard's Catalogue . . . , 1696–1834*, 1953.
A Guide to Parliamentary Papers, 1956.
Hayden, J. O., *The Romantic Reviewers, 1802–1824*, 1969.
Houghton, W. E. (ed.), *The Wellesley Index to Victorian Periodicals, 1824–1900*, Toronto, 1966–.
Huxley, T. H., *Diary of the voyage of H.M.S. Rattlesnake*, ed. J. Huxley, 1935.
Jones, H. M., *O Strange New World*, 1965.
Perry, M. C., *Personal Journal*, ed. R. Pineau, Washington D.C., 1968.
Rosenberg, N. (ed.), *The American System of Manufactures*, Edinburgh, 1969.
Taft, R., *Artists and Illustrators of the Old West, 1850–1900*, New York, 1953.
Ward, W. R., *Victorian Oxford*, 1965.

Additional Reading

Brock, W. H., & Meadows, A. J., *The Lamp of Learning*, 1984.
Forgan, S. (ed.), *Science and the Sons of Genius*, 1980.
Gooding, D., & James, F. J. L. (eds.), *Faraday Rediscovered*, 1985.
Knight, D. M., *Sources for the History of Science*, Cambridge, 1975.
MacLeod, R. & Collins, P., *The Parliament of Science*, 1981.
Meadows, A. J. (ed.), *The Development of Science Publishing in Europe*, Amsterdam, 1980.
Morrell, J. & Thackray, A., *Gentlemen of Science*, Oxford, 1981.

Epilogue

The last quarter of the nineteenth century and the first decades of the twentieth saw the fruition of many of the seeds which we had seen being planted in the earlier chapters. In physics the principle of conservation of energy was combined by Einstein with that of conservation of mass, with alarming consequences; the emission theory of light reclaimed some territory from the undulatory, and wave–particle dualism was postulated first for light and then for matter as well. In chemistry, the atomic theory made ever-increasing headway, as the theory of valency, that each atom has a definite combining power, was reinforced by theories of the arrangement of atoms in space which proved extraordinarily powerful in organic, and then in inorganic, chemistry. The arrangement of the chemical elements in families, the so-called periodic table, by Mendeleev, gave support to the view that all the elements could not be irreducibly different. The spectroscope was used first to identify hitherto unknown elements, then to investigate the chemical constitution of the Sun and the stars, and finally, with first optical and then X-ray spectra, to reveal the structure of the atom. Thus the physicist's atoms and the chemist's elements were at last fitted into a hierarchy. And in biology, the work of Mendel made its sudden and explosive impact at the turn of the century, diverting biologists from the construction of family trees towards experiments in genetics, and giving rise to widespread doubt about the rôle of natural selection. This doubt persisted for a quarter of a century, until the mathematicians Ronald Fisher and J. B. S. Haldane showed how excellently Mendelian genetics and Darwinian evolution could be fitted together. The experimental method was also applied in new directions, notably to psychical research; but although eminent philosophers and scientists took part in these inquiries, nothing generally satisfactory came out of them.

A. W. Williamson was one of the most distinguished of British theoretical chemists, but once established in his professorial chair in London he published little except a basic textbook, *Chemistry for Students.* He had been a pupil of Comte in Paris; and his inaugural lecture, *Development of difference the basis of unity*, published 1849, was an attempt to disseminate Positivist doctrines. A Comtean flavour is said to be discernible in some of his later addresses, usually of a rambling kind; but he was an advocate of the atomic theory, and opposed the abandonment of theoretical entities. His contemporary, William Odling, the translator of Laurent, also enjoyed a high reputation and published various useful textbooks; his lectures on the chemical changes of carbon, published in 1869, are worthy of note. He was for many years Professor of Chemistry at Oxford, and crowned his career with a work on *The Technic of Versification* published in 1916. Edward Frankland, a pioneer in the theory of chemical bonding, published his *Experimental Researches* in 1877; and a book on water analysis in 1880. The rôle of the chemical analyst grew rapidly during the century with legislation covering the purity of water, food, and drugs. The great standard textbook of the late nineteenth century was Roscoe and Schorlemmer's *Treatise*, in three volumes. Roscoe alone had written a chemistry textbook in 1866, and a work on *Spectrum analysis*, his own particular field, both of which became standard works.

As we have already noted, chemistry in Britain after about 1830 was not very strong; but A. W. Hofmann was invited to England to the Royal College of Chemistry, partly at least through the efforts of the Prince Consort. His pupils included W. H. Perkin, who prepared the first synthetic dye. Hofmann's textbook, an *Introduction to Modern Chemistry*, appeared in 1865. In 1868 came a translation of *Modern Chemistry*, by the Frenchman A. Naquet, with interesting discussion of the phenomenon of dissociation; Naquet took an interest in Brodie's non-atomic calculus of chemical operations. More important French chemists were C. A. Wurtz and Marcellin Berthelot; Wurtz upheld the atomic theory against Berthelot, and his book on the subject became a standard work. His *Elements* was a useful textbook; and his *History of Chemical Theory* was famous for its remark that chemistry is a French science; which becomes less absurd when we remember that Wurtz was from Alsace and that this was the period of the Franco-Prussian war. Of Berthelot's writings, which include studies of alchemy, of thermochemistry, and of total syntheses of organic materials from their elements, only those on explosives were translated into English. A useful account of the various controversies and theories in nineteenth-century chemistry is Ernst von Meyer's *History of Chemistry*, which appeared in English in 1891. And invaluable for anybody interested in the period are the *Alembic Club Reprints*, which began to appear in 1898, and include translations of papers, or parts of papers, by eminent Continental chemists.

One of the most important developments in chemistry was the theory that the atom of a given element can only unite to a certain number of other atoms. The number is called the 'valency' of the element, and is fixed, except that some elements do exhibit two or more different valencies in different sorts of compounds. The most important pioneer in this theory, after Frankland, was August Kekulé, none of whose books found their way into English. The theory was then extended by van 't Hoff and le Bel, who were able to explain many phenomena of organic chemistry on the assumption that the four other atoms to which a carbon atom can be united are disposed at the corner of a regular tetrahedron, of which the carbon atom occupies the centre. Chemical structures became three-dimensional instead of plane. Their papers, with memoirs by Pasteur on asymmetrical crystals, and by Wislicenus developing their ideas further, were translated by G. M. Richardson in 1901; van 't Hoff's *Chemistry in Space* had appeared in English in 1891. These ideas were extended into the inorganic realm by Werner, whose *New Ideas on Inorganic Chemistry* was published in 1911, and paved the way for understanding the arrangement of atoms in crystals.

Van 't Hoff was also responsible for inaugurating a new era in chemical dynamics with his studies on osmosis, the passage of solvent through a membrane into a solution. He showed that the gas laws could be applied to this phenomenon; and Svante Arrhenius developed the theory to account for electrolytes, which he supposed to be dissociated into ions in solutions even when no electric current was applied. Arrhenius' *Theories of Chemistry* was published in English translation in 1907. The German chemist Wilhelm Ostwald systematized their ideas; he went so far as to argue that solution is the type of chemical combination, and that thermodynamic concepts applied to solutions and alloys should be employed in chemistry in place of the atomic theory. His book *Solutions*, and his textbooks, became standard and very important works; and his incorporation of thermodynamics into chemistry was an extremely important step, which survived his belated conversion to atomism. He edited a series of *Klassiker*, translations and new editions of important works of science, the equivalent of which the English-speaking world lacks; and after a distinguished career in chemistry took up general philosophical questions. His *Natural Philosophy*, an outcome of this, appeared in English in 1911.

Of equal importance in chemistry was the periodic table of the chemical elements, in which they are arranged in families; this was proposed by Mendeleev in 1869. There had been a number of precursors; an inferior table by J. A. R. Newlands had appeared before Mendeleev's, and by strident assertions of priority in his *Periodic Law* Newlands secured some recognition and a medal from the Royal Society. Mendeleev's *Principles of Chemistry* appeared in 1897; and his slim *Chemical Conception of the Ether* in 1904. The ether had been invoked before as the prime matter out of which by a process of inorganic evolution the various groups of elements had emerged.

Lord Rayleigh, in an attempt to determine the truth of Prout's hypothesis, investigated the atomic weight of nitrogen; and William Ramsay joined in the work. Together they were responsible for the discovery of argon, an inert gas which it proved very difficult to fit into the periodic table since it resembled no other elements. But within a very few years, Ramsay and his collaborators, of whom Morris Travers was in the later stages the most important, succeeded in identifying in the atmosphere, or occluded in rocks, the whole family of rare gases, which have since become of considerable economic importance. Ramsay described his work as far as it had got, in *The Gases of the Atmosphere*, of 1896; and Travers wrote up the whole investigation in *The Discovery of the Rare Gases*, in 1928. In 1912 Ramsay published *Elements and Electrons*, a groping towards the view that electrons are somehow responsible for linking atoms together.

Moving to the field of physics, we should notice Lord Rayleigh's *Theory of Sound*, of 1877–8, and his six-volume *Scientific Papers*; he was an extremely able experimental and mathematical physicist. Joseph Larmor was another great figure of this period; he edited the papers of Cavendish, Stokes, Maxwell, and James Thomson, and published his own Adams Prize Essay on 'Ether and Matter' in 1900; his collected papers appeared in 1929. Sir Oliver Lodge was an indefatigable writer, whose *Electrons* of 1906 and *Ether and Reality* of 1925 deserve mention. William Crookes was primarily a chemist, and his *Chemical Analysis* became a standard work and went through various editions. His speculative chemical and physical writings, particularly concerned with the radiometer and then with cathode rays, appeared in journals, particularly his own *Chemical News*. Norman Lockyer, the founder of *Nature*, published *Contributions to Solar Physics* in 1874, and *The Chemistry of the Sun* in 1887; these being based upon the application of new techniques in spectroscopy. He was a keen cosmologist, and in 1890 published *The Meteoric Hypothesis*, following it in 1900 with *Inorganic Evolution*. He shared with Crookes and others the belief that the numerous different chemical elements had somehow been produced from one kind of prime matter. He was also interested in history, and published his *Dawn of Astronomy*, on

96 The Observatory at Bekul; watching the solar eclipse (Lockyer, N., *Solar Physics*, 1874, pl 4)

ancient Egyptian worship and mythology, in 1894, and his *Stonehenge . . . astronomically considered* in 1906.

The researches of Crookes on cathode rays were followed by those of J. J. Thomson, whose *Conduction of Electricity through Gases* came out in 1903, and *Corpuscular Theory of Matter* in 1907. He had earlier published a work on the vortex atom, and important works on the applications of dynamics to physics and chemistry, and on the mathematical theory of elasticity. Thomson's researches on the cathode rays established that the corpuscles, later called electrons, were components of all matter; and Thomson suggested that neutral atoms consisted of a positive plum-pudding in which the negative electrons were embedded like currants. Meanwhile Planck had proposed, to account for the radiation of heat, that energy was not continuous but available in discrete 'quanta'; and Niels Bohr was able to apply this theory to the spectrum of hydrogen to give a mathematical account of the planetary atom which Rutherford had on experimental grounds already put forward. The Bohr atom, with its electrons rotating in fixed, quantized orbits about the heavy positive nucleus, is familiar; it was the last atomic model which was easy to visualize and make a model of. The alarming suggestion of de Broglie that electrons might, like light, conform to wave or particle equations in different circumstances, put an end to any hope of simple models. Bohr's atom was applied by G. N. Lewis to account for valency in terms of electron-sharing in his *Valence*, published in 1923 by the American Chemical Society.

On the other great development in physics about the turn of the century, relativity, one might mention the *Scientific Writings* of Fitzgerald, who called attention in England to the writings of Hertz and provided an explanation of Michelson and Morley's experiment, which seemed to prove that the Earth was not moving relative to the ether, in terms of a logically undetectable contraction of all bodies in the direction of their motion through the ether. This phenomenon was more elegantly accounted for by Einstein; whose popular account of his theory, *Relativity*, came out in 1920. A major influence upon Einstein was the critique of the metaphysics behind Newton's physics by Ernst Mach, whose *Science of Mechanics* appeared in English in 1902; criticisms of Newtonian physics from a somewhat different standpoint had also been made by Heinrich Hertz, in his *Principles of Mechanics*, translated in 1899.

An odd kind of physics, popular about 1900, as before and since, was speculation about the inhabitants of other worlds, within or beyond the solar system. The success of the theory of evolution in biology supported those who believed, in the face of a total absence of empirical evidence, that life must have developed in the course of things somewhere else in the universe. *Other Worlds than Ours* by R. A. Proctor, a popularizer of both astronomy and whist, appeared in 1870 and in numerous later editions. Among more serious scientists, A. R. Wallace, in *Man's Place in the Universe*, took up the question, coming, like Whewell, to the conclusion that there was probably not life elsewhere. Arrhenius, in his *Worlds in the Making*, came down on the other side; and so did Percival Lowell, in his *Mars and its Canals* of 1906, and *Mars as the Abode of Life* of 1909. The arguments are ingenious, and one feels sad that the canals are no longer believed in. In 1915 Lowell published *A Memoir on a trans-Neptunian Planet*, predicting Pluto, which was observed some years later.

This concludes our rapid survey of the physical sciences, and we turn to biology. Despite the resistance of such Paleyans as Edmund Beckett, author of *On the Origin of the Laws of Nature*, the Darwinian view made rapid progress. Ernst Haeckel became the exponent of an extremely materialistic version of Darwinism, stressing development due to natural causes, and urging the view that ontogeny recapitulates phylogeny; that is, that each individual in its embryonic development actually lives through a speeded-up history of its race. G. J. Romanes' major works were *Animal Intelligence*, *Mental Evolution in Animals*, containing unpublished material by Darwin, including an essay on instinct, and *Mental Evolution in Man*; and his *Essays*, which were published after his death. He sought to complete Darwin's edifice by establishing that the human mind was the product of evolution just like man's physical characteristics. He began as an opponent of religion, in his anonymous *Candid Examination of Theism* of 1877; but his posthumously published *Thoughts on Religion* show a complete change, due in large part to the influence of Charles Gore, the editor of *Lux Mundi*.

Indeed opposition to materialism began to grow towards the end of the century among men of science; for example, the animal magnetism of Anton Mesmer became respectable as hypnotism, as described in Blumenbach and in the *Animal Magnetism* of Binet, a pioneer of intelligence-tests, and Féré, which was published in 1887. The distinguished physiologist W. B. Carpenter inveighed equally against materialism and spiritualism, urging, in his *Mental Physiology* of 1874 that one must distinguish voluntary and automatic actions. The fourth edition, of 1876, carried an introduction assailing Tyndall and Huxley. The book is written in a discursive and readable way, and all kinds of human activities are covered. In his posthumously published *Essays* there are included a biography and a bibliography; and the essays and addresses there printed are also well worth reading.

Spiritualism had arrived from America about the middle of the century, and it seemed to at least some scientists that it could offer solid evidence of life beyond the grave, and hence disprove materialism, as the dogmatic and obsolete teachings of the Church could not. The prime movers in the establishment of the Society for Psychical Research were the Sidgwicks; Henry, a philosopher whose *Methods of Ethics* was very important, and his wife Eleanor who became principal of Newnham College, Cambridge, in 1892. Three leading members of the Sidgwicks' group published in 1886 *Phantasms of the Living*, which was mostly written by Edmund Gurney, who had embarked upon medical and legal training before taking up psychical research, and had written a book on aesthetics, *The Power of Sound*, which broke new ground. In 1887 he published a fascinating collection of essays, *Tertium Quid.* Of his collaborators on *Phantasms of the Living*, F. W. H. Myers became increasingly convinced of the truth behind at least some of the phenomena of spiritualism, and in 1903 wrote *Human Personality and its Survival of Bodily Death*; while Frank Podmore became increasingly sceptical as frauds were detected or suspected in so many cases. His *Modern Spiritualism, a History and a Criticism*, appeared in 1902. Two scientists not immediately associated with the Sidgwicks who investigated physical phenomena were William Crookes, whose *Researches in the Phenomena of Spiritualism* appeared in 1874; and Oliver Lodge, whose *Life and Matter* came out in 1906, and *Conviction of Survival* in 1931. Numerous other scientists took an interest in the phenomena; and their collected essays and biographies often include references to séances. William James' *Principles of Psychology* and *Varieties of Religious Experience* need no recommendation and were extremely influential in shifting attention towards what was actually experienced; and towards value rather than origin in matters of religion.

All this has diverted us from biology in a stricter sense; to which we should now return. The evolutionary histories of various species, most notably the horse, were worked out from fossils discovered in America, although modern horses were not found there. Two wealthy palaeontologists, O. C. Marsh and E. D. Cape, competing with the amorality of the most ruthless industrial tycoons, between them built up great collections of fossils, particularly from the West. Marsh's monograph on extinct toothed birds, and Cape's essays on evolution, *The Origin of the Fittest*, describe their researches; both published numerous short papers, which often led to squabbles over priority. One of the most important biological theorists of the period was August Weismann, whose *Studies in the theory of descent* and *Essays upon heredity and kindred biological problems* became classics. The critical idea in Weismann's writings was that the germ cells are distinct from the body cells, so that whatever characters the body may acquire through use or disuse cannot be passed on. In England, the most important student of inheritance was Francis Galton; his first books described African travels, and then he began his various inquiries on inheritance; *Hereditary Genius* was published in 1868; studies on *English men of science* in 1874; and further works in the 1880s. In 1893 he published his essay on finger-prints; which gradually thereafter came to be used for identifying people.

E. Ray Lankaster was a distinguished zoologist, the son of Edwin Lankester, the editor of a rather defective edition of John Ray's *Letters*, and a busy popularizer and educationalist. In 1886 Ray Lankester wrote an interesting little book *Degeneration*, showing that natural selection did not necessarily lead to what one thinks of as higher forms; among parasites, for example, it will pare away whatever is unnecessary until little more than a stomach and generative organs are left. Lankester's *Extinct Animals* of 1905 is quite interesting; and his newspaper articles, published under the titles of *From an Easy Chair* and *Diversions of a Naturalist*, are worth looking at as natural history and as illustrating the *Zeitgeist*. C. C. Coe, in his *Nature versus Natural Selection*, 1895, argued that while evolution had undoubtedly happened, the Darwinian mechanism was incompetent to account for it. This view received notable support from William Bateson, who in 1902 published a translation of Mendel's paper on inheritance. His *Materials for the Study of Variation* had been published before he knew of Mendel, in 1894; and in 1913 he published a series of lectures on *Problems of Genetics*. He encouraged the study of inheritance and cross-breeding rather than the construction of histories of species. Hugo De Vries, one of the 'discoverers' of Mendel, argued for the importance of mutations, sharp changes in the gene, rather than the almost imperceptibly slow process envisaged by Darwin. His lectures, *Species and Varieties*, were published in 1905; and his great treatise, *The Mutation Theory*, was translated shortly afterwards.

The year 1909 marked the centenary of Charles Darwin's birth, and the golden jubilee of *The Origin of Species*; and a symposium, *Darwin and Modern Science*, edited by A. C. Seward was published to mark the occasion, with contributions from eminent biologists. In the same year, Francis Darwin published, as *The Foundations of the Origin of Species*, the drafts that Darwin had drawn up in 1842 and 1844, laying down his theory. But in biology the turn of the century also saw a revulsion in some quarters against the mechanistic explanations of both Darwinians and Mendelians. D'Arcy Thompson, at once a biologist and a classical scholar, translated Aristotle's *Historia Animalium* and drew up a glossary of Greek birds; he also produced a bibliography of writings on invertebrates, and in 1917 published his famous work, *On Growth and Form*, which is one of the great classics of biology. Its emphasis is on morphology, and there is criticism of those who mistakenly supposed that analogy of form indicated community of descent; but the tone is not polemical and the book can be enjoyed by anybody.

Further support for vitalism came from the writings of Henri Fabre, whose observations of insects, and particularly on their instincts, showed how much more complicated the world was than some laboratory workers seemed to have supposed. Henri Bergson's *Creative Evolution*, translated in 1911, developed these ideas; as did Lloyd Morgan, who had studied animal intelligence and instinct and in 1923 published his *Emergent Evolution*, in which the blind chance which seemed to characterize the Darwinian scheme disappears. In 1916 Jacques Loeb had published his holistic *The Organism as a Whole*. This vitalism led to fresh attempts to reconcile science and religion; among which Charles Raven's *Creator Spirit* may be noted. But in *Science, Religion, and Reality*, a valuable symposium of which he was the editor, Joseph Needham declared that mechanistic biology must triumph; but that it was possible to admit mechanism in its own sphere, and religion in its. Frances Mason's symposium, *The Great Design*, is on the other hand vitalist; for this debate is one that cannot readily be won. But Ronald Fisher's *Genetical Theory of Natural Selection* of 1929, by showing that natural selection and Mendelian genetics were compatible, and indeed fitted together very well, put an end to the period of flux which had begun at the end of the last century.

In natural history in England, the last decades of the century had seen the appearance of Lord Lilford's great bird book, the illustrations of which were by Archibald Thorburn; whose own *British Birds* came out in 1915–16, and is a collector's piece. Slightly earlier, Alfred Newton, Professor of Zoology at Cambridge, compiled a *Dictionary of Birds*; and Henry Seebohm, an industrialist, wrote a *History of British Birds* including much field observation. Finally, although it falls well outside our period, one might mention Lord Grey of Fallodon's delightful book, *The Charm of Birds*, of 1927, with pleasant woodcuts by Robert Gibbings which, although it does not set out to be a scientific work, is an excellent and evocative piece of natural history.

With the exponential rise in the number of scientists and in scientific writing, it becomes hopeless to try to cover the whole of science beyond the Victorian era; and this epilogue has clearly been very selective. The trend which we have noticed, by which the paper became relatively more important than the book, continued; but the monograph has not disappeared, and even in the physical sciences there will be books from our century which succeeding generations will value for their insights, or venerate as monuments. But with the immense growth of American science in this century, and with the increase of scientific publication in English in countries where it is not the native language, science in English has become much more cosmopolitan, and no longer reflects what happens in Britain or America. Nevertheless, it still seems to be true that certain sciences and certain approaches to nature flourish in certain countries at certain times; and it is the business of the historian to describe and, if possible, to account for it.

EPILOGUE

Alembic Club Reprints, Edinburgh, 1898–; translations include: Gay-Lussac, and *4*, Avogadro; *8*, Scheele; *13*, Scheele, Berthollet, and Gay-Lussac and Thenard; *14*, Pasteur; *15*, Kolbe; *18*, Cannizzaro; *19*, van 't Hoff and Arrhenius; *20*, Stas and Marignac; *22*, Röntgen.

Arrhenius, S., *Theories of Chemistry . . .*, ed. T. S. Price, 1907.

Worlds in the Making, tr. H. Borns, 1908.

Bateson, W., *Materials for the study of variation . . .*, 1894.

Mendel's Principles of Heredity, Cambridge, 1902; incl. translation.

Problems of Genetics, 1913.

Essays and Addresses, ed. B. Bateson, Cambridge, 1928.

Beckett, E. (previously Denison, E. B.), *On the Origin of the Laws of Nature*, 1879.

Bergson, H., *Creative Evolution*, tr. A. Mitchell, 1911.

Berthelot, P. E. M., *Explosives and their Power*, ed. and tr. C. N. Hake and W. Macnab, 1892.

Binet, A., and Féré, C., *Animal Magnetism*, 1887.

Bohr, N., *The Theory of Spectra and Atomic Constitution*, tr. A. D. Udden, Cambridge, 1922.

de Broglie, L., *Matter and Light*, tr. W. H. Johnston, 1939.
Carpenter, W. B., *Principles of Mental Physiology*, 1874.
Nature and Man, 1888.
Coe, C. C., *Nature versus Natural Selection*, 1895.
Cope, E. D., *The Origin of the Fittest*, New York, 1887.
Crookes, W., *Researches in the Phenomena of Spiritualism*, 1874.
Select Methods in chemical analysis . . ., 1871.
Darwin, C., *The Foundations of the Origin of Species*, ed. F. Darwin, Cambridge, 1909.
De Vries, H., *Species and Varieties; their origin by mutation*, 1905.
The Mutation Theory, tr. J. B. Farmer and A. D. Darbishire, 2 vols., 1910–11.
Einstein, A., *Relativity . . .*, 1920.
Fabre, J. H. C., *Insect Life . . .* (tr. M. Roberts), ed. F. Merrifield, 1901.
Social Life in the Insect World, tr. B. Miall, 1912.
The Wonders of Instinct, tr. A. de Mattos and B. Miall, 1918.
Fisher, R. A., *The Genetical Theory of Natural Selection*, Edinburgh, 1929.
Fitzgerald, G. F., *The Scientific Writings . . .*, ed. J. Larmor, Dublin, 1902.
Frankland, E., *Experimental Researches in pure, applied, and physical chemistry*, 1877.
Water Analysis for sanitary purposes, 1880.
Franklin, B., *et. al.*, *Animal Magnetism*, 2nd ed., Philadelphia, 1837; 1st ed. not seen.
Galton, F., *Narrative of an Explorer in tropical South Africa*, 1853.
Art of Travel, 1855.
English Men of Science; their Nature and Nurture, 1874.
Inquiries into Human Faculty and its Development, 1883.
Natural Inheritance, 1889.
Finger Prints, 1893.
Grey of Fallodon, Viscount, *The Charm of Birds*, 1927.
Gurney, E., *The Power of Sound*, 1880.
Tertium Quid; Chapters on various disputed questions, 2 vols., 1887.
Myers, F. W. H., and Podmore, F., *Phantasms of the Living*, 2 vols., 1886.
Haeckel, E. H. P. A., *The History of Creation . . .*, tr. E. R. Lankester, 2 vols., 1876.
The Evolution of Man . . ., 2 vols., 1879.
The Riddle of the Universe . . ., tr. J. McCabe, 1900.
Haldane, J. B. S., *The Causes of Evolution*, 1932.
Hertz, H., *The Principles of Mechanics*, tr. D. E. Jones and J. T. Walley, 1899.
Hofmann, A. W., *Introduction to Modern Chemistry*, 1865.
James, W., *The Principles of Psychology*, 2 vols., 1890.
The Varieties of Religious Experience, 1902.
Lamarck, J. B., *Zoological Philosophy . . .*, tr. H. Elliot, 1914.
Lankester, E. R., *Degeneration; a chapter in Darwinism*, 1880.
The Advancement of Science, 1890.
Extinct Animals, 1905.
From an Easy Chair, 1908.
Diversions of a Naturalist, 1915.
Larmor, J., *Aether and Matter*, Cambridge, 1900.
Mathematical and Physical Papers, 2 vols., Cambridge, 1929.
Lewis, G. N., *Valence and the Structure of Atoms and Molecules*, New York, 1923.
Lilford, Lord (Powys, T. L.), *Coloured Figures of the Birds of the British Islands*, 7 vols., 1885–97.
Lockyer, J. N., *Contributions to Solar Physics*, 2 pts, 1874.
The Chemistry of the Sun, 1887.
The Meteoric Hypothesis . . . a spectroscopic inquiry into the origin of cosmical systems, 1890.
The Dawn of Astronomy, 1894.
Inorganic Evolution as studied by spectrum analysis, 1900.

Stonehenge . . . astronomically considered, 1906.
Lodge, O. J., *Electrons*, 1906.
Life and Matter, 1906.
Ether and Reality, 1925.
Conviction of Survival, 1930.
Past Years; an Autobiography, 1931.
Loeb, J., *The Organism as a Whole*, 1916.
Lowell, P., *Mars and its Canals*, New York, 1906.
Mars as the Abode of Life, New York, 1909.
A Memoir on a trans-Neptunian Planet, Flagstaff, Ariz., 1915.
Mach, E., *The Science of Mechanics*, tr. E. J. McCormack, Chicago, 1902.
Marsh, O. C., *Odontornithes; . . . extinct toothed birds . . .*, Washington and New Haven, 1880.
Mason, F. (ed.), *The Great Design*, 1934.
Mendeléeff, D. I., *The Principles of Chemistry*, tr. G. Kamensky, ed. T. A. Lawson, 2 vols., 1897.
An Attempt towards a Chemical Conception of the Ether, tr. G. Kamensky, 1904.
Meyer, E. von, *History of Chemistry*, tr. G. McGowan, 1891.
Morgan, C. Lloyd, *Animal Life and Intelligence*, 1890.
Habit and Instinct, 1896.
Emergent Evolution: The Gifford Lectures, 1923.
Myers, F. W. H., *Human Personality and its survival of Bodily Death*, 2 vols., 1903.
Naquet, A., *Modern Chemistry*, tr. W. Cortis, 1868.
Needham, J. (ed.), *Science, Religion, and Reality*, New York, 1925.
Newlands, J. A. R., *The Periodic Law*, 1884.
Newton, A., *A Dictionary of Birds*, 1893.
Odling, W., *A Course of Six Lectures on the Chemical Changes of Carbon*, ed. W. Crookes, 1869.
Ostwald, F. W., *Outlines of General Chemistry*, tr. J. Walker, 1890.
Solutions . . ., tr. M. M. P. Muir, 1891.
The Principles of Inorganic Chemistry, tr. A. Findlay, 1902.
Natural Philosophy, tr. T. Seltzer, 1911.
Planck, M., *Treatise on Thermodynamics*, tr. A. Ogg, 1903.
The Theory of Heat Radiation, tr. M. Masius, 1914.
Podmore, F., *Modern Spiritualism; a History and a Criticism*, 2 vols., 1902.
Proctor, R. A., *Other Worlds than Ours*, 1870.
Ramsay, W., *The Gases of the Atmosphere*, 1896.
Elements and Electrons, New York, 1912.
Raven, C. E., *The Creator Spirit*, 1927.
Richardson, G. M. (ed. and tr.), *The Foundations of Stereochemistry: Memoirs by Pasteur, van 't Hoff, le Bel, and Wislicenus*, New York, 1901.
Romanes, G. J., *A Candid Examination of Theism, by Physicus*, 1877.
Animal Intelligence, 1882.
Mental Evolution in Animals, 1883.
Mental Evolution in Man, 1888.
Thoughts on Religion, ed. C. Gore, 1895.
Essays, 1897.
Roscoe, H., *Lessons in Elementary Chemistry*, 1866.
Spectrum Analysis, 1869.
and Schorlemmer, C., *A Treatise on Chemistry*, 3 vols, 1877–84.
Rutherford, E., *Radio-activity*, Cambridge, 1904.
Radioactive Substances and their Radiations, Cambridge, 1913.
Seebohm, H., *A History of British Birds with coloured illustrations of their eggs*, 4 vols., 1883–5.

Classification of Birds, 1890.
The Birds of Siberia, 1901.
Seward, A. C. (ed.), *Darwin and Modern Science*, Cambridge, 1909.
S(idgwick), A. and E. M., *Henry Sidgwick: A Memoir*, 1906.
Thompson, D'Arcy W., *A Bibliography of Protozoa, Sponges, Coelenterata and Worms . . . for the years 1861–1883*, Cambridge, 1885.
A Glossary of Greek Birds, Oxford, 1895.
On Growth and Form, Cambridge, 1917.
Thomson, J. J., *A Treatise on the Motion of Vortex Rings*, 1883.
Applications of Dynamics to Physics and Chemistry, 1888.
Elements of the Mathematical Theory of Electricity and Magnetism, Cambridge, 1895.
The Conduction of Electricity through Gases, Cambridge, 1905.
The Corpuscular Theory of Matter, 1907.
Thorburn, A., *British Birds*, 4 vols., 1915–16.
Travers, M. W., *The Discovery of the Rare Gases*, 1928.
Van 't Hoff, J. H., *Chemistry in Space*, ed. and tr. J. E. Marsh, Oxford, 1891.
Studies in Chemical Dynamics, ed. E. Cohen, tr. T. Ewen, Amsterdam, 1896.
Lectures on Theoretical and Physical Chemistry, tr. R. A. Lehfeldt, 3 pts (1899–1900).
Wallace, A. R., *The Scientific Aspect of the Supernatural . . .*, 1866.
The Wonderful Century, its successes and its failures, 1898.
Man's Place in the Universe, 1903.
Weismann, A., *Studies in the Theory of Descent*, tr. R. Meldola, 3 pts, 1880–2.
Essays upon Heredity and kindred biological problems, 2 vols., Oxford, 1889.
The Germ-plasm, tr. W. N. Parker and H. Rönnefeldt, New York, 1893.
On Germinal Selection as a source of definite variation, 1896.
Werner, A., *New Ideas on Inorganic Chemistry*, tr. P. Hedley, 1911.
Williamson, A. W., *Development of difference the basis of unity*, 1849.
Chemistry for Students, Oxford, 1866.
Wurtz, C. A., *A History of Chemical Theory from the Age of Lavoisier to the Present Time*, tr. H. Watts, 1869.
The Atomic Theory, tr. E. Cleminshaw, 1880.
Elements of Modern Chemistry, tr. W. H. Greene, 1881.

Recent Publications

Bohr, N.: Rozental, S. (ed.), *Niels Bohr*, Amsterdam, 1967.
Gauld, A.: *The Founders of Psychical Research*, 1968.
Harré, R. (ed.), *Scientific Thought, 1900–1960*, Oxford, 1969.
Hohenberg, P. M., *Chemicals in Western Europe, 1850–1914*, Chicago, 1967.
Kekulé, A.: *Kekulé Centennial Celebrations*, Washington, 1966.
Lodge, O.: *A Bibliography*, by T. Besterman, 1935.
Sorby, H. C.: Higham, N., *A Very Scientific Gentleman*, Oxford, 1963.
Spronsen, J. W. van, *The Periodic System of Chemical Elements*, 1969.
Wright, S. (ed.), *Classical Scientific Papers, Physics*, 1964.

Additional Reading

Harman, P. M., *Energy, Force and Matter*, Cambridge, 1982.
Heath-Stubbs, J., & Salmon, P. (eds.), *Poems of Science*, 1984.
Hendry, J., *Cambridge Physics in the Thirties*, Bristol, 1984.
Meadows, A.J., *Science and Controversy*, 1972.
McCormach, R., *Night Thoughts of a Classical Physicist*, Cambridge, Mass., 1982.
Pais, A., *Subtle is the Lord*, Oxford, 1982.

Index

Note: Figures in italics denote illustrations